East Asian Low-Carbon Community

Weisheng Zhou • Xuepeng Qian •
Ken'ichi Nakagami
Editors

East Asian Low-Carbon Community

Realizing a Sustainable Decarbonized Society from Technology and Social Systems

Springer

Editors
Weisheng Zhou
College of Policy Science
Ritsumeikan University
Ibaraki, Osaka, Japan

Xuepeng Qian
College of Asia Pacific Studies
Ritsumeikan Asia Pacific University
Beppu, Oita, Japan

Ken'ichi Nakagami
College of Policy Science
Ritsumeikan University
Ibaraki, Osaka, Japan

ISBN 978-981-33-4338-2 ISBN 978-981-33-4339-9 (eBook)
https://doi.org/10.1007/978-981-33-4339-9

This Springer imprint is published by the registered company Springer Nature Singapore Pte Ltd.
The registered company address is: 152 Beach Road, #21-01/04 Gateway East, Singapore 189721, Singapore

Preface

According to the Mauna Loa Observatory in Hawaii, the monthly average atmospheric carbon dioxide concentrations measured in Hawaii in May 2019 exceeded 415 ppm. This is the highest concentration within the observable dataset over the last 800,000 years. Climate change is no longer a phenomenon yet to occur in the distant future; we are now confronting a global climate crisis. This crisis is no longer reversible, nor is it only in its infancy—it is now a very real and rapidly accelerating issue. The Fifth Assessment Report published by the Intergovernmental Panel on Climate Change (IPCC) in 2014 (AR5) asserts that it is extremely likely (95–100%) that most of the observed increase in global average temperatures since the mid-twentieth century has been a result of elevated greenhouse gas (GHG) concentrations from anthropogenic sources. It estimates that carbon dioxide exerts the greatest influence on global temperatures among the anthropogenic GHG emissions. The Paris Agreement that had come into effect in November 2016 requires all countries to participate in measures to mitigate global warming, regardless of whether they are a developed or developing nation. The goal is twofold: (1) limit the increase in average global temperatures to less than 2 °C above pre-industrial revolution levels (with a target of 1.5 °C); and (2) to reduce greenhouse gas emissions to near zero by the second half of the twenty-first century. According to the IPCC Special Report "Global Warming of 1.5 °C" issued on October 8, 2018, if global average temperature rise is to be controlled within 1.5 °C, carbon dioxide emissions must be reduced to 55% of 2010 levels by 2030, while actual carbon dioxide emissions need to be reduced to zero by 2050.

Although there are still many uncertainties associated with climate change, humankind no longer has the luxury of time to debate over climate change issues. We must take concrete action as soon as possible despite the uncertainty and fulfill the Paris Agreement commitments. There is an urgent need to accelerate the mobilization of resources and forces across all areas of society to collectively address climate change and promote sustainable development. The realization of a low-carbon society in order to mitigate global warming should become a common goal that all developed and developing nations are working toward. We need optimal countermeasures that are able to concurrently resolve domestic poverty, pollution, and global environmental problems. These are also essential conditions required to establish a sustainable human society.

The emission and emissions reduction of carbon dioxide (CO_2) (GHG) anywhere on the planet have almost the same effect on global warming. Carbon dioxide mitigation measures always lead to additional benefits, such as improved environmental quality. We believe that establishing a cross-border, wide-area, low-carbon society is possible and is able to aid in the realization of global sustainability and the sustainable development of developing nations. This realization requires the development and transfer of innovative, low-carbon technologies, the creation of the low-carbon economic and industrial systems, and the transformation to low-carbon social systems through changes to lifestyles and the eco-design of energy and materials cycles based on international cooperation.

If the energy consumption per unit of gross domestic product (GDP) and carbon dioxide emissions per unit of GDP in Southeast Asia, South Asia, Central Asia, the Middle East, and countries of the former Soviet Union, other than Central and Eastern Europe, reach the current annual levels in Japan, then a 60–70% reduction in annual energy consumption and carbon dioxide emissions is possible. Japan has the technological edge, while China boasts a market advantage. The two countries are complementary to each other in terms of capital, technology, labor and engineering, procurement, and construction. China is rapidly advancing on the technological front and surpassing Japan in some fields. Their cooperation will benefit them and third-party countries, representing a step forward toward the Sustainable Development Goals (SDGs) of the United Nations.

Therefore, the establishment of a wide-area, low-carbon society is expected to create a society full of vitality through the harmonious development of economy, environment, and society. A community is the union of people sharing common interests and goals. Global warming is a critical issue for a global community that shares the common interests and goals of humankind from climate mitigation and adaptation, and reductions to environmental impact and damage. The urgency of climate change, necessity of realizing a low-carbon society, and the add-on benefits of carbon dioxide countermeasures highlight the importance of building a regional, low-carbon community that transcends national borders. This may be achieved by any mechanisms such as financing, technology transfer, and capacity building. In this book, we introduce research on the East Asian Low-Carbon Community Concept and its Realization focusing on Japan, China, and Korea. This is discussed primarily from the nexus of three key systems: science and engineering, economic and industrial, and wide-area social systems. The research discussed is largely based on theoretical and empirical studies.

The Chinese philosopher Confucius said, "If those who are near get pleased, those who are far away will be attracted to come." The realization of the East Asian Low-Carbon Community concept is impossible without establishing relationships of mutual trust between countries and regions. Without learning and creating history, it is difficult to build a low-carbon society at the local scale. There are many issues that need to be addressed in the future, including economic development (overcoming poverty), local environmental problems (overcoming pollution), global environmental problems (reducing carbon emissions), joint solutions to air, water, and soil problems, human exchange and human resource development, and the consideration of financial mechanisms to realize this concept. This book pioneers the development of further research and practice in this field.

Ibaraki, Osaka, Japan — Weisheng Zhou
Beppu, Oita, Japan — Xuepeng Qian
Ibaraki, Osaka, Japan — Ken'ichi Nakagami

Contents

Contributors

Kyungah Cheon Research Center for Sustainability Science, Ritsumeikan University, Ibaraki, Osaka, Japan

Hirotaka Haga Faculty of Regional Design and Development, Department of Business Economics, University of Nagasaki, Nagasaki, Japan

Kenzo Ibano Graduate School of Engineering, Osaka University, Osaka, Japan

Baoju Jia Ritsumeikan Research Center for Sustainability Science, Ritsumeikan University, Ibaraki, Osaka, Japan

Yishu Ling Graduate School of Policy Science, Ritsumeikan University, Ibaraki, Osaka, Japan

Huan Liu Economics and Social Development Institute, Hainan Construction Project Planning and Design Research Institute Co., Ltd., Haikou, China

Ken'ichi Nakagami College of Policy Science, Ritsumeikan University, Ibaraki, Osaka, Japan

Yoshifumi Ogami College of Science and Engineering, Ritsumeikan University, Ibaraki, Osaka, Japan

Nabila Prastiya Puturi Graduate School of Science and Engineering, Ritsumeikan University, Ibaraki, Osaka, Japan

Xuepeng Qian College of Asia Pacific Studies, Ritsumeikan Asia Pacific University, Beppu, Oita, Japan

Hongbo Ren College of Energy and Mechanical Engineering, Shanghai University of Electric Power, Shanghai, China

Tatsuo Sakai Research Organization of Science and Technology, Ritsumeikan University, Ibaraki, Osaka, Japan

Faming Sun Research Institute of Global 3E, Kyoto, Japan

Xuanming Su Research Institute for Global Change/Research Center for Environmental Modeling and Application/Earth System Model Development and Application Group, Yokohama, Japan
Agency for Marine-Earth Science and Technology (JAMSTEC), Yokohama, Japan

Ryohei Yamada Graduate School of Science and Engineering, Ritsumeikan University, Ibaraki, Osaka, Japan

Yusuke Yamamoto Graduate School of Science and Engineering, Ritsumeikan University, Ibaraki, Osaka, Japan

Masato Yamazaki Disaster Mitigation Research Center, Nagoya University, Nagoya, Japan

Shuya Yoshioka College of Science and Engineering, Ritsumeikan University, Ibaraki, Osaka, Japan

Chong Zhang Zhejiang Gabriel Biotechnology Co., Ltd, Zhejiang, China

Weisheng Zhou College of Policy Science, Ritsumeikan University, Ibaraki, Osaka, Japan

Part I
Concept and Framework of the East Asian Low-Carbon Community

Chapter 1
Climate Change and Low-Carbon Society: Coping with Uncertainty

Weisheng Zhou

Abstract The rise and fall of ancient civilizations and the transition of world cities are inextricably linked to climate change. Atmospheric carbon dioxide (CO_2) concentrations had already exceeded 420 ppm in 2019, and in recent years, there has been an increased frequency in abnormal weather alongside rising temperatures. Developing countries have exposed their weaknesses in response to extreme weather events. The Intergovernmental Panel on Climate Change (IPCC) warns that the probability of a temperature rise is more than 95% likely. However, climate change issues are always faced with uncertainties in terms of causality, impact level, and the cost and effectiveness of countermeasures. Human beings implement sustainable measures and proceed with international cooperation when dealing with uncertainty. The realization of a wide-area, low-carbon society that transcends national borders is indispensable and offers benefits beyond the realization of the Sustainable Development Goals (SDGs) of the United Nations, for broader human society.

1.1 Introduction

Most ancient civilizations in the world have experienced the process of formation, development, and decline; oftentimes, the causes underpinning formation and decline remain a mystery. Researchers usually use political, economic, and social changes to explain as causes. In recent decades, the accuracy of chronological dating across many countries around the world has improved. With this technology, research into the Holocene (the current geological epoch, beginning approximately 11,700 ca. years before the present) has enabled progress in higher-resolution climate change research and the origin and decline of ancient civilizations. The relatively accurate textual research, the continued deeper understanding of the Holocene climate mutation phenomenon, and the role of climate mutation in the formation and decline of ancient civilizations have attracted greater attention from

W. Zhou (✉)
College of Policy Science, Ritsumeikan University, Ibaraki, Osaka, Japan
e-mail: zhou@sps.ritsumei.ac.jp

W. Zhou et al. (eds.), *East Asian Low-Carbon Community*,
https://doi.org/10.1007/978-981-33-4339-9_1

the academic community. An increasing number of studies have shown that while the human civilizations had developed due to several reasons and development laws, the impact of climate and environmental changes has also played a significant role. To this day, this impact is at time decisive to a certain extent. For example, abrupt climate change has played an essential role in the formation and decline of ancient civilizations such as the Tigris and Euphrates basin civilizations in the Mesopotamian Plain, the Egyptian Nile basin, the Indus basin civilization, and the Yellow River/Yangtze River basin civilization in China (Weiss et al. 1993; Weiss and Bradley 2001; Fagan 2004; Zhu 1973; Wang 2005).

These studies clarify the role of climate change in the formation and decline of ancient civilizations, improving the historical reference for human beings to adapt to possible climate mutations in the future. This research is of great significance to predict and cope with the impact of future climate mutations on human society.

1.2 Rise and Fall of Ancient Civilizations and Abrupt Changes in Climate

1.2.1 Changes in Human Society

As shown in Fig. 1.1, the development of human society has had an evolutionary history of approximately 6–7 million years. This development may be divided into five types of social forms, based on the characteristics at different stages of development:

Hunter-gatherer society: hunter-gatherer culture was the way of life for early humans from approximately 8000 B.C. to around 11–12,000 years ago; this is known as the "Neolithic Age." The lifestyle of hunter-gatherers was based on hunting animals and foraging for food.

Agricultural society: this stage lasted until approximately the "First Industrial Revolution" of 1750 A.D. This period followed the "Neolithic Age" where human beings continued their hunting and gathering life for more than 6 million

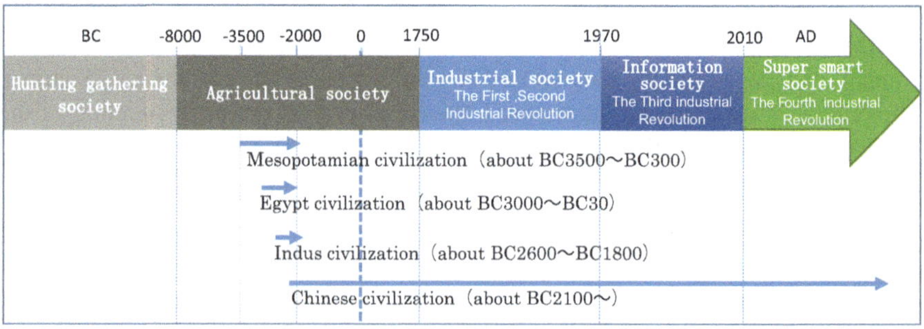

Fig. 1.1 Changes in human society and civilizations

years. Once they began developing and using irrigation techniques for farming and to raise livestock, people began to settle. In this farming society that lasted for nearly 10,000 years, humankind created a well-grounded ancient civilization (the agricultural civilization), including the four ancient civilizations.

Industrial society: from approximately 1750 A.D., the first industrial revolution was characterized using coal, ironmaking, steam engines, and railway technology. From ~1870, the use of petroleum, electricity, plastics, and other new materials and the proliferation of automobiles and home appliances characterized the second industrial revolution. As humankind entered an industrial society, it launched into an era of mass production, mass consumption, and mass pollution. In particular, the massive consumption of fossil fuels such as coal and petroleum greatly improved social productivity and the living standard of society. However, this led to a large amount of CO_2 emissions, causing a rapid increase in the atmospheric concentration of greenhouse gases (GHGs). The atmospheric CO_2 concentration has been increasing since this first industrial revolution. The CO_2 concentration of 280 ppm at that time has risen to approximately 420 ppm in the present day; this represents a more than 45% increase.

Information society: from the 1970s, the third industrial revolution has been characterized using computers, high informatization, electronics, and digitalization.

Super-intelligent society: the fourth industrial revolution, beginning from around 2010, has been able to positively integrate cyberspace (virtual space) and physical space (real space), the real economy, and digital economy through the use of artificial intelligence (AI) and the Internet of Things (IoT). In the fourth industrial revolution, information technology (IT), biotechnology (BT), environmental technology (EnvT), energy technology (EneT), nanotechnology (NT), space technology (ST), and other science and technological advancement will provide the social, technological, and economic systems to realize a sustainable society. This society will be characterized by low energy and resource consumption, low environmental load, and high social utility (Japan Cabinet Office 2015).

It has been recognized that the speed of the industrial revolution has been accelerating when compared to the length of each industrial revolution. Additionally, the energy consumption and CO_2 emissions per unit of gross domestic product (GDP) have also been decreasing; however, the total energy consumption and emissions are still rapidly increasing.

1.2.2 Rise and Fall of the Four Ancient Civilizations and Climate Change

It is difficult to completely deny that the formation, rise, and fall of civilizations were influenced by its inherent factors and from significant changes in environmental factors such as direct or indirect climate change.

How were ancient civilizations born while being at the mercy of an ever-changing climate? Did they perish again?

As shown in Fig. 1.1, the formation of the oldest four significant civilizations including Mesopotamian civilization (approximately 3500 B.C.), Egyptian civilization (approximately 3000 B.C.), the Indus River civilization (approximately 2300 B. C.), and Chinese civilization (Yellow River/Yangtze civilization) (approximately 2500 B.C.) between 3500 and 2000 B.C. have still not been completely elucidated. Research suggests that 5500 years ago, the living environment where humans had dispersed had become severe due to dryness and global cooling. As such, the population was concentrated by migration to lands surrounding rivers with better conditions. There are theories that have been attributed to the significant role in influencing the emergence of civilization, such as large-scale farming practices to support the population (Fagan 2004; Zhu 1973). In other words, climate change is considered a significant cause of the emergence of an ancient civilization.

For example, 6000–7000 years ago, the current Sahara Desert was an oasis replete with lakes at that time. It was an ideal home for our African ancestors. However, the population was relatively scattered, such that the population pressure was small, and it was not easy to form a civilized society. Around 5500 years ago, the climate in Africa changed significantly to colder and drier conditions. The desert oasis began to disappear, as lakes dried up and dunes activated, and these ancestors were forced to leave. Herdsmen living in the Sahara Desert migrated to the Nile River valley or delta plain. The population in Upper Egypt began to expand northward to Lower Egypt, and the migration of people from different regions to the Nile Valley or delta region increased the population pressure in this area. To resolve this pressure from population growth, humans began to artificially irrigate and develop adaptive technological innovations. Agricultural production began to increase, and there was surplus wealth. The increase in surplus wealth promoted the division of labor in the handicraft industry.

These developments had triggered the birth of a nation. Therefore, it may be inferred that climate change around 5500 years ago promoted the formation of ancient Egyptian civilization (Fagan 2004; Wang 2005).

Firstly, for civilization to occur, the production of food by farming and the production of surplus agricultural products are necessary. Therefore, the formation and development of human civilization is considered to be inextricably linked to climate change. The following conditions are key for the development and prosperity of ancient civilization:

1. The presence of a large river: common to all four civilizations is agricultural development, and as such, abundant water resources are required. All of the four largest ancient civilizations were located within a river basin; the Egyptian civilization was located around the Nile River, the Mesopotamian civilization was located around the Tigris and Euphrates rivers, the Indus civilization was located around the Indus River, and the Yellow River and Yangtze civilizations were located around the Yellow and Yangtze rivers (Fig. 1.2).

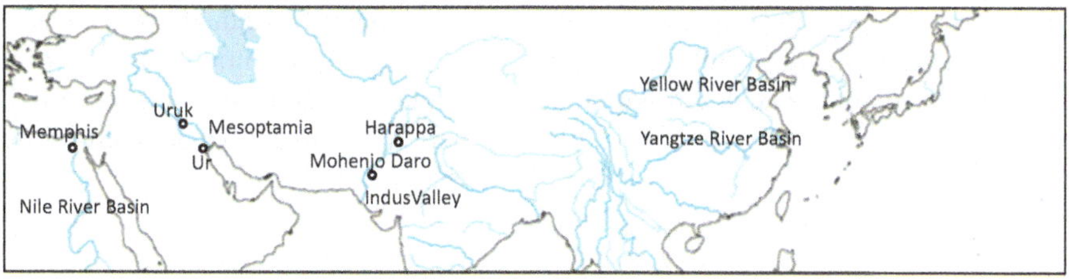

Fig. 1.2 Distribution of the four significant ancient civilizations

2. The presence of a broad plain: a large population provides advantages to the prosperity of a civilization. Therefore, a broad plain that may be easily populated is needed.
3. The settlement location must be in the mid-latitude area: the area where the season is relatively straightforward. This is because the production of food is dependent on understanding the appropriate time to sow crops. All four ancient civilizations are located between a latitude of 25°N and 55°N and occur in basins with a large river and a mild climate.

The relationship between the rise and fall of civilizations and the environment, such as climate change, is not always clearly understood; this has resulted in conflicting views. Regardless, environmental changes have influenced civilizational origin, while civilization activities in turn have affected the environment. In at least some civilizations, it is believed that changes in the environment due to the influence of the civilization itself had become a significant cause of its destruction (Ministry of the Environment 1995).

For example, Egyptian civilization is said to have begun around 3050 B.C. in the green areas of the Nile River. The culture of dynasty society, festivals, and beliefs developed amid repeated wet and dry climate change. While the dynasties collapsed multiple times, the collapse around 2000 B.C. was attributed to a dry climate and lower water levels on the Nile River. From the analysis of stripes discovered in a lake downstream of the Nile River over the past 10,000 years, it is thought that the extraordinary drought event around 2150 B.C. may be related to the end of the Old Dynasty (Fagan 2004; Fukui News 2019).

In terms of the decline and destruction of the Indus civilization, there are various theories such as the "desertification theory." This is the theory that the disappearance of the Indus civilization was due to desertification caused by climate change and deforestation around 2000 B.C. The "river flow theory" suggests that the crustal movement that occurred around 2000 B.C. led to the flood inundation that moved the flow path of the Indus River. Finally, the "climate change theory" states that the Indus civilization declined due to climate change.

Humans once moved in search of livable spaces with the deterioration of climate and returned when climate improved, leading to an increase in population. However, rapid warming after the Ice Age prompted hunting, gathering, and farming, and irrigation facilities and cities were built to increase productivity. As a result, short-term droughts and unprecedented heavy rains often occurred. Despite increasing resilience to disasters, the population was increased to the limit of the environmental capacity of the land. Furthermore, when significant climate change occurred, civilizations could no longer adapt to these changes and many people died; those who survived were scattered in various places (Fagan 2004).

A comprehensive analysis of historical, archaeological, and paleoclimatological evidence has demonstrated that a sudden change to drought occurred in the Nile, Mesopotamian, Indus, and Yellow basins from 2200 to 2000 B.C. This sudden change is standard for the mid-latitudes. This was a robust cold event since the Holocene entered the Great Warm Period. The decline of the Nile civilization occurred in the first intermediate period (2181–2040 B.C.), and the decline of Mesopotamian civilization began with the disintegration of the Akkad Kingdom, until the establishment of the Kingdom of Babylon, at around 2200–1900 B.C. The Indus civilization then suddenly declined in 1800 B.C. The cold events 4000 years ago proved to be a fatal blow to the civilizations of these three basins. China established the Xia Dynasty in 2070 B.C., which commenced the era of Chinese civilization. Although there were "partial fractures" in the middle, this civilization has continued to this day as the Chinese civilization (Wang 2005).

The four ancient civilizations were all agricultural civilizations, and the formation, development, and decline of agriculture were all affected by climate change. The following sequence of events may be considered the model of the formation, rise, and fall of ancient civilizations from the climate change perspective:

Climate change such as cold events and droughts $\Rightarrow$ migration to warmth/humidity/river areas $\rightarrow$ population increase $\rightarrow$ civilization occurrence/development $\rightarrow$ drought/abnormal weather $\rightarrow$ overcome by technological innovation in the short-term/civilization collapses in the long term

Of course, there may be other reasons that have caused the decline of civilizations, such as population growth, the destruction of the environment by human activities, and wars. However, a sudden change in climate may be one of the most influential factors.

In recent years, many studies have identified that the climate change during the Holocene may have been strongly related to the rise and fall of human civilization. As there is concern regarding rapid climate change due to global warming, it is essential to understand the Holocene climate change quantitatively and with high time resolution, to clarify its mechanism and its impact on humankind in predicting its impact on society.

1.2.3 Rise and Fall of Chinese Civilization and Climate Change

The Chinese civilization that originated in the Yellow and Yangtze basins is the only ancient civilization that has not been interrupted among the four ancient civilizations. Based on ancient records, the cuneiform writing of Mesopotamia, the hieroglyphs of ancient Egypt, and the writings of ancient Indian civilization have all entered the history museum. However, although the oracle bone inscriptions of the Shang Dynasty in China evolved into the regular script of today through the seal and official scripts, the structure of the characters has remained unchanged, and the "LIU SHU (Chinese character classification)" is still the root of Chinese characters. However, the phenomenon of the "partial fracture" had occurred in the origin of ancient Chinese civilization, and some ancient bronze civilizations also declined. The former includes the Liangzhu culture in the mid and lower reaches of the Yangtze River in China, the Pujiang Shangshan culture in the Qiantang basin, and the Shandong Longshan culture in the lower Yellow River area. The latter includes the developed Sanxingdui bronze civilization in China's Sichuan basin. Like other civilizations, the Chinese civilization, which repeats the rise and fall of the dynasty, has not been able to escape the threat of climate change and environmental destruction.

1.2.3.1 China's Recent 5000 Years of Climate Change

Chinese geographer and meteorologist, Kezhen Zhu, published an article in 1972 titled, "A Preliminary Study on China's Climate Change in the Past Five Thousand Years," combining the history, phenology, local records, and instrumental observations over the past nearly 5000 years (approximately from 2200 to 2200 B.C.). Climate change until 1900 may roughly be divided into four warm and four cold period, as shown in Fig. 1.3.

The first warm period was approximately from 2100 to 1100 B.C. (Xia and Shang period), approximately 900 years ago. The temperature was approximately 2–3 °C

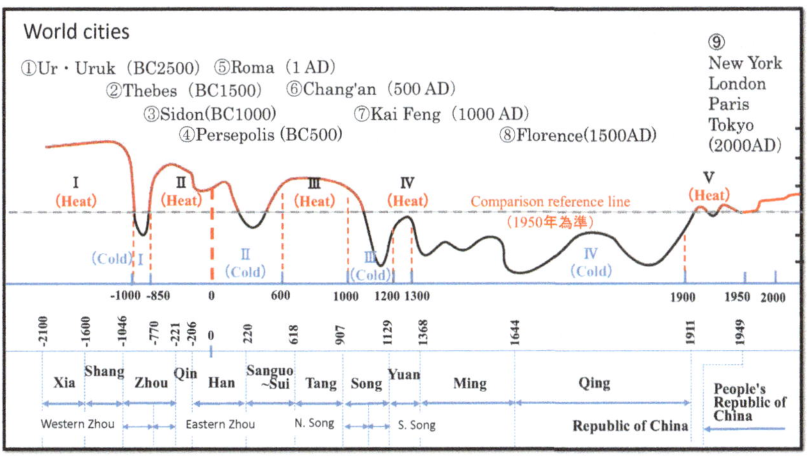

Fig. 1.3 Climate change curves in China in the last 5000 years and changes in world cities and the Chinese dynasty. (Source: based on Zhu 1973)

higher than the 1950s. However, before the Holocene entered the Great Warm Period (~2200 B.C.), there was a severe cold event, not mentioned in Zhu (1973). This cold event is the cause of the decline of the Mesopotamian, Egyptian, and Indus civilizations (China Meteorological Administration 2002).

The first cold period, from 1100 to 850 B.C. (early Western Zhou Dynasty), lasted approximately 250 years.

The second warm period, from 770 B.C. to 8 A.D. (spring and autumn, warring states, Qin, and Western Han periods), lasted approximately 230 years.

The second cold period from 25 A.D. to 600 A.D. (Eastern Han Dynasty, Three Kingdoms, Southern and Northern Dynasties period) lasted approximately 600 years.

The third warm period, from 600 to 1000 A.D. (Sui, Tang, and early Northern Song Dynasty), lasted approximately 400 years.

The third cold period, from 1000 to 1200 A.D. (Northern Song Dynasty and Southern Song Dynasty periods), lasted approximately 200 years.

The fourth warm period, from 1200 to 1400 A.D. (late Southern Song Dynasty, Yuan Dynasty), lasted approximately 200 years.

The fourth cold period, from 1400 to 1900 A.D. (Ming and Qing periods), lasted approximately 500 years.

As shown in Fig. 1.3, the 5000-year climate change curve was wave-like, and the reasons for climate change were different from the current temperature rise. It can be said that it was not initially of anthropogenic origin (although there are many opinions, this study will not demonstrate all of them here). The cause of abrupt changes in the past climate is still under continuous research. The extent to which the development of civilization was affected by sudden changes in climate needs further discussion. The decline of every civilization in every region cannot be attributed to the sudden impact of climate change.

In addition to the warm and cold periods, from 1900 to the present (since the first industrial revolution), a period of approximately 120 years, may be referred to as the fifth warm period.

In the 50 years from 1900 to 1950, the temperature rose by approximately 0.5 °C (Zhu's research was until 1950). In the 60 years from 1950 to 2010, there was a total rise of ~1 °C. Since 1951, China's climate has generally shown a warming trend, especially since the 1980s, where temperature has been relatively high. 2007, 2015, and 2016 were years with the highest annual average temperature records over the past 66 years.

The warming of the climate system is beyond doubt. Since 1950, many unprecedented changes have been observed that have not occurred over hundreds or even thousands of years. The atmosphere and oceans have warmed, the amount of snow and ice has decreased, the sea level has risen, and the concentration of GHGs have increased.

1.2.3.2 Changes in Chinese Civilization and Climate Change

There are many theories regarding the origin of Chinese civilization and the length of Chinese history. Chinese history is approximately 3300 years (from 1300 B.C.). This spans from the mid-Shang (Yin) dynasty, when the oracle bone inscriptions first appeared in the history of Chinese characters; the Erlitou culture site (the central plains of China, the core of the culture spanning the Neolithic and Bronze Ages, ~3700 years old). It was approximately 4100 years from the legendary Xia Dynasty in Western Zhou literature (from 2070 B.C.); according to Confucius, the legendary age of the Three Emperors and Five Sovereigns was approximately 4700 years (from 2698 B.C.). From ~5300 years ago from the site of the earliest discovered city in Liangzhu culture (one of the Neolithic cultures in China developed on the coast of Taihu Lake in the lower reaches of the Yangtze River). From the Neolithic Pujiang Shangshan culture (located in Pujiang County in the upper reaches of the Puyang River), discovered in 2001, it is the earliest origin of rice farming. From the Hemudu culture ~7000 years ago (the Neolithic in the area south of the lower reaches of the Yangtze River in China), and other prehistoric civilizations dating back 3000 years between ~11,400 and 8400 years ago. It also spans from the Paleolithic period of Peking Man and Lantian ape man approximately 68 to 1 million years ago (Fig. 1.4).

From the cycle of warm and cold periods shown in Fig. 1.3, the decline of dynasties in Chinese history almost corresponds to the low-temperature interval on the graph. The Qin, Tang, and Song dynasties (Northern Song and Southern Song) and the eras of the Yuan, Ming, and Qing dynasties all fell below the average temperature to become extremely cold in the past 2500 years. During the warm period, a strong central government was generally established, and at this time, the territory was vast. In contrast, during the cold period, the power of the Han people had been significantly reduced, and the frontier ethnic minorities become prosperous, with tremendous pressure from outsiders.

Ancient China was a continental country with oceans on the east and south and the tall Qinghai-Tibet plateau and vast deserts on the west. Due to the terrain and climate, foreigners could not execute a 500 km raid on the Han regime in the east. Therefore, the Han regime had been under pressure from the northern peoples in

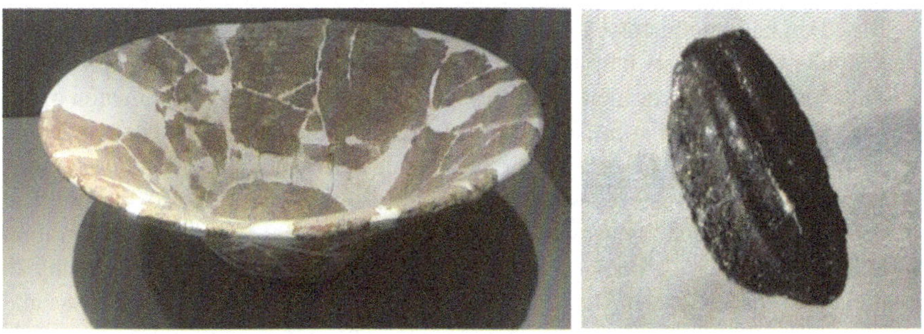

Fig. 1.4 Flaring-walled basins (left) and carbonized rice (right) found in Shangshan culture site

ancient times, such as the Huns, Turks, Jurchens, Mongolians, and Manchus. As the northern minority areas had a nomadic economy, they relied entirely on grasslands to feed themselves. When the climate became cold, the grasslands grew poorly, and these people could not obtain sufficient food supplies; this is when they attacked the Han regime in the south to obtain food. At this time, the south was also getting colder, and the output of grain and other materials was drastically reduced, resulting in a decline in overall national strength. The frequent invasion of foreigners in the north was a result of the pressure of survival in the context of a cold climate. Therefore, during the cold climate, the Central Plains Dynasty faced internal and external troubles. As shown in Fig. 1.3, the Tang Dynasty was the third warm and humid period. The warm and humid climate created conditions for the development of the agricultural economy in this dynasty, and the development of the agricultural economy laid the foundation for the overall development and prosperity of the social economy.

In the mid-eighth century (755–907 A.D.), the climate changed from warm to cold. The cold climate caused the deterioration of the living environment of the nomadic peoples in northern China, and they moved south to dominate northern China and establish regional dynasties. Following the Anshi Rebellion (755–763 A. D.), the central government of the Tang Dynasty gradually lost control over most of the country and slowly declined and perished. The Anshi Rebellion was a landmark event that marked the beginning of this southward migration. Following another cold mutation in the twelfth century in the Southern Song Dynasty, the northern nomadic Mongolians began to rule the central plains and established a unified dynasty, the Yuan Dynasty. There was a long cold zone between 1600 and 1700 A.D. During this period the Ming Dynasty fell, and the Qing Dynasty was established.

The Yangtze River civilization, particularly the rice cultivation culture in the Yangtze River civilization, had a far-reaching influence on East Asian civilizations and the global civilization. It began in 8000 BC, more than 10,000 years ago. It is listed as the two significant sources of the Chinese civilization alongside the Yellow River civilization. However, around 4200 years ago, the rice cultivation culture in the Yangtze River civilization had suddenly disappeared for approximately 300 years. It has been postulated that a rapid and sizeable cooling event may have damaged the rice cultivation culture and contributed to the collapse of the Yangtze delta society and the subsequent fall of the Yangtze civilization (Kajita et al. 2018).

Based on Chinese history, it appears that the warm period is better for economic development and social stability than a cold period. At the same time, in the past 2000 years, the decline of most dynasties and the corresponding phenomenon of the low-temperature range has not been coincidence. However, we cannot conclude that the rise and fall of dynasties are entirely a result of climate change; many reasons affect the course of history, with climate being one factor. Due to the political corruption of the feudal dynasty itself, coupled with the poor harvest and starvation from low temperatures, peasant uprisings and wars may be triggered, leading to changing dynasties. During cold periods, the southward migration of grassland pastures will also lead to the invasion and southward migration of northern nomads (Liu 2009).

1.2.4 Transition of World Cities and Climate Change

The result of the formation and development of civilizations is the emergence of cities. Vere Gordon Childe named the result the Urban Revolution. He pointed out that during this period, major revolutionary innovations occur such as language, currency, punishment, and bronzeware in cultural, social, judicial, and technical systems (Gordon 1950).

In each era, there is a world city (a global city) representative of the world. The world city is a city that has mainly economic, political, and cultural central functions and if of the most global importance and influence in that era. Since the Neolithic era, irrigated agriculture has been carried out, especially in the river basins, promoting large-scale settlement and the formation of cities since 3500 B.C. Figure 1.5 shows the world cities and their distribution from ancient to modern times.

Around 2500 B.C., the most important cities in the world were Ur and Uruk (the Sumerian cities and city-states of ancient Mesopotamia).

Around 1500 B.C., the world city was Thebes, in Egypt. Thebes was the capital of Egypt during the Middle and New Kingdoms and is in the mid-reaches (upper Egypt) about 1000 km south of Alexandria at the mouth of the Nile River. It was the center of faith and the wealthiest city during this period. Some have mentioned that in 1000 B.C., the city that could dominate the world was Sidon in Lebanon however. In 500 B.C., the world city was Persepolis in Persia; in 1 A.D., it was Rome; in 500 A.D., it was Chang'an in China; in 1000 A.D., it was Kaifeng around the Yellow River in China; and in 1500 A.D., it was Florence.

The modern world city is the center of the world economic system or the organizational node of the world city network system. Globalization of the economy, multi-polarization of politics, computerization of society, and cultural diversity are essential characteristics of the twenty-first century. Currently, the comprehensive power of the world's major cities has been evaluated and ranked based on six areas

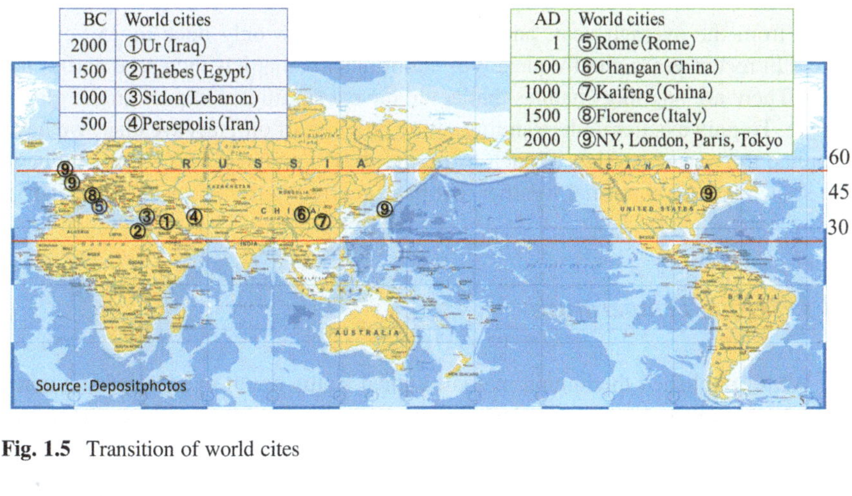

Fig. 1.5 Transition of world cites

of the economy: research/development, culture/interchange, residence, environment, and transportation/access. This is known as the Global Power City Index (GPCI). This GPCI analyzes the ranking itself and also the components of the ranking to understand the strengths, weaknesses, and challenges of each city has in a changing world. The present top four cities in the world according to the GPCI are London, New York, Tokyo, and Paris.

One of the characteristics of these world cities is that they are located in the mid-latitude civilization belt (25°N–55°N), similar to the origin of ancient civilizations. It is thought that the rise and fall were significantly influenced by climate change.

Among them, City Ur is modern-day Iraq; this great city was once one of the earliest civilizations in the world and is symbolic of Mesopotamian civilization. It flourished due to abundant rainfall and irrigation facilities built in Mesopotamia and was severely impacted when the Indian Ocean monsoon moved south and changed rainfall patterns. By 2000 B.C., its agricultural economy had collapsed, and today it has become a barren land. Fagan believes this is "the first time the entire city has collapsed in the face of an environmental disaster" (Fagan 2004).

The modern world cities of London, New York, Tokyo, and Paris will also be exposed to unprecedented climate change threats.

The four significant civilizations, civilizations created in the ancient days of humankind, and the world cities in each era, have been discussed. However, human civilizations and world cities are changing with the times. One of the causes of this transition is climate change, and one of the driving forces of its evolution is social and technological innovation.

1.3 Recent Extreme Weather and Economic Loss

The World Meteorological Organization (WMO) defines abnormal weather as "a phenomenon in which the average temperature and precipitation are significantly biased from the average year, and the deviation occurs only once in 25 years or more." In recent years, the frequency of abnormal weather has increased across the globe. There are ongoing changes that could have a significant impact on the global climate. To gain a better understanding of the progress of ordinary weather, we have compiled the news materials of each institution based on the WMO report and have also shown typical abnormal weather that has occurred in recent years (since 2015).

1.3.1 High Temperature

The WMO published the annual meteorological status report every year and compiled the annual meteorology using the "hottest year in history," since 2011. The 2019 report stated that "currently, the average temperature of the Earth has been

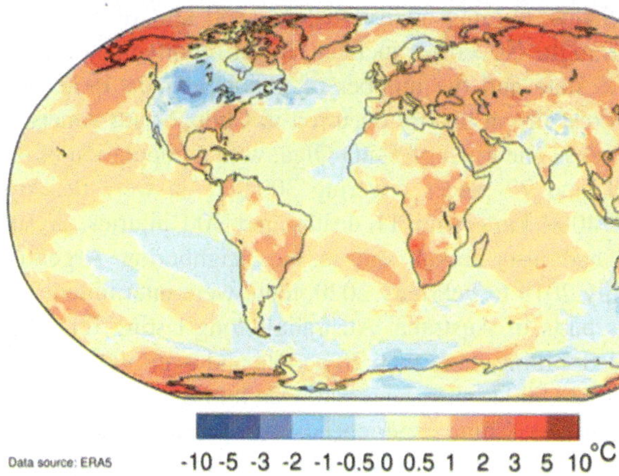

Fig. 1.6 Surface-air temperature anomaly for 2019 for the 1981–2010 average. (Source: European Centre for Medium-range Weather Forecasts (ECMW F) ER A5 data, Copernicus Climate Change Service)

recorded to increase by 1.1 °C compared to before the Industrial Revolution" (WMO 2017). Damage due to high temperatures is occurring all over the world. Figure 1.6 shows a surface-air temperature anomaly for 2019 based on the 1981–2010 average. The overall warmth of the year is evident, although there have been unusual variations in temperature throughout the world. Most areas were warmer than the recent average (1981–2010), while 2019 has been one of the top three in record in Africa since at least 1950 (WMO 2017).

Among the high-temperature reports from many countries, temperatures over 45.6 °C were recorded in late June 2019, the highest ever in French meteorological observation history. This is a higher temperature than the heatwave generated in France in 2003 (World Weather Attribution 2019). It also broke record highs in the history of the Arctic and Antarctica. On January 20, 2020, a temperature of 38 °C was the highest temperature in the Arctic Circle recorded in Verkhoyansk, Sakha Republic of Siberia, Russia. The European Union's Copernicus Climate Change Service (CAMS) reported that the average temperature in the Arctic was ~10 °C above average for March, April, and May of this year (BBC 2020). On the other hand, Antarctic ice is melting faster than has been predicted.

Droughts: Drought damage was reported in California, USA, for all of 2015. While in Vietnam, 2016 was the worst drought for this country in 90 years. In 2018, northern Argentina and its surroundings experienced drought from January to March. The food crisis caused by the drought induced severe economic damage to the area.

Forest fires: High temperatures and drought damage have triggered many forest fires. In 2015, the forest fire damage caused by a drought in California, USA, was reported. In July to September and November 2018, another forest fire broke out in the Western United States, burning approximately 1860 km^2 of forest, and there were also 85 deaths from the November incident. It was reported that the forest fires were the worst in the state and the number of fatalities since 1933. In January 23, 2017, the worst forest fire in history occurred in Chile, South America, lasting approximately two weeks. The town of Santa Olga, with a population of 4000, was completely destroyed, and 77 places covering 300,000 ha were destroyed. The burned area spanned 4000 km^2, with 11 deaths and 300 injuries. A substantial historical forest fire had also broken out in the neighboring Argentina, from mid-January. From July 2019 to February 2020, there were intermittent large-scale forest fires in various parts of Australia. Australia's long-lasting forest fires have been the worst recorded, killing more than 28 people and over 500 million animals and plants. More than 170,000 km^2 of land had been burnt.

Heavy rain and typhoons: China experienced widespread damage from 2015 to 2019 due to typhoons and heavy rains in the summer. Above all, heavy rains impact southern China each year, typically causing hundreds of deaths each year. In Japan, the heavy rains due to the rainy season and typhoons also occurred in July 2018 (typhoon No.7 and seasonal rain front), at Kyushu in 2019 (typhoon Nos. 15 and 19), and in July 2020 (storm rain front). From October 24 to 26, 2019, 99 people were killed in various parts of Japan (based on data from January 10, 2020), and the typhoon destroyed approximately 3200 houses. Other regions of Asia (such as India and Thailand) have also reported on damage caused by heavy rain. Outside the Asian region, massive rain damage occurred in 2017 from southwestern Colombia to Peru, in 2018 to parts of Nigeria, and in 2019 in an area spanning northern to western Africa.

Floods: Rivers inundated due to heavy rains result in the incidence of flood events. The heavy rain leads to floods that impacted southern Africa in January 2015, causing hundreds of deaths. From late June to early July 2018, heavy rain led to the death of 237 people mainly in Okayama, Hiroshima, and Ehime prefectures in Japan (data as of January 9, 2019) and destroyed approximately 7000 homes. From December 1 to 31, 2019, a massive flood caused by heavy rain occurred in the Jakarta metropolitan area of Indonesia. A total of 103 residential areas were flooded nationwide, and 9 people were killed.

1.3.2 Seawater Temperature/Sea Level Rise

Global warming is predicted to bring about an increase in seawater temperatures and sea level rise; the warming seawater creates a thermal expansion causing sea levels to rise. Sea level rise is also due to the decrease in ice sheets in Greenland and Antarctica. Greenland loses an average of 200 Gt of ice annually, and Antarctica loses an annual average of 118 Gt, according to a recent study published by National Aeronautics and Space Administration (NASA) and Washington University in Science magazine. It is feared that this has led to a sea level rise of more than

1.3 cm in the last 16 years alone, affecting many cities and infrastructures facing the sea (Earther 2020)

1.3.3 Economic Loss Due to Abnormal Weather

Extreme weather is usually based on local historical weather data and is the most common 10% of the average for the past, according to the National Centers for Environmental Information (NOAA). Extreme climates such as heatwaves, droughts, storms, floods, gusts, and snowstorms are frequent within the global range. Climate change has a tremendous negative impact on several global problems, such as reduced labor productivity, the spread of infectious diseases, heatwaves, and pollution.

According to a survey by the World Meteorological Organization (WMO), the global economic loss due to abnormal weather in 2017 was 300 billion US dollars, much higher than the loss of approximately 330 billion dollars in 2016. Table 1.1 compares the climate threats and extreme weather events between 1990 and 2016, in terms of the scale of economic loss.

Between 1980 and 2015, the average annual number of abnormal damaging weather events was 582, and the economic loss was US $110 billion, affecting the livelihood of approximately 170 million people. For developed countries, the leading impact of extreme weather damage is economic loss.

A larger number of unnecessary deaths occur in developing countries than their developed counterparts. Moreover, economic loss in developing countries account

Table 1.1 Overview of the economic impacts of climate change and extreme weather from 1990 to 2016. (Source: WMO 2017)

Year	Number of weather disasters (cases)	Economic loss (US$ billion)
1990	412	65.6
1992	390	110.6
1994	458	67.5
1996	469	85.1
1998	501	142.5
2000	519	60.0
2002	451	98.8
2004	424	144.0
2006	606	69.4
2008	524	145.8
2010	625	115.7
2012	719	161.6
2014	726	97.1
2016	797	129.4

(Amounts are in real terms adjusted by the consumer price index in 2016)

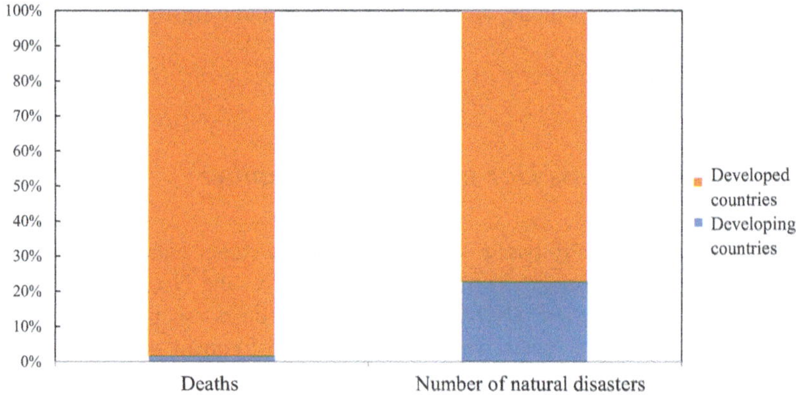

Fig. 1.7 1990–2016 natural disasters in developed and developing countries. (Source: based on the EM-DAT 2017)

for a more significant proportion of GDP. Figure 1.7 shows the number of occurrences of natural disasters and fatalities in developed and developing countries for natural disasters, with more than 5000 deaths since 1900. Although it is not possible to select natural disaster-prone regions, weak infrastructure and disaster prevention measures are likely to increase the scale of social and economic impacts and human casualties caused by disasters. In the future, if the natural environment continues to deteriorate, it is expected that developing countries will suffer more significant damage.

These extreme weather conditions around the world are a sign of human destruction. The issue of climate change is unavoidable and urgently requires mitigation for the survival of human beings and all lives on this planet.

1.4 Warning from IPCC

From the formation, development, and decline of ancient civilizations, civilizations have a profound relationship with climate and the environment. Since the first industrial revolution, particularly in recent years, the temperature rise and abnormal climate have attracted widespread attention, triggering fear and anxiety. The main conclusions of the fifth IPCC report (AR5) WG I "Climate Change 2013-The Physical Science Basis" (IPCC 2013) are summarized as follows.

1.4.1 Causes of Global Warming

Human activity is likely to be the dominant factor in the warming observed since the mid-twentieth century (>95% likely). Atmospheric CO_2, methane (CH_4), and nitrous oxide (N_2O) concentrations have increased to unprecedented levels over the last 800,000 years. In terms of the causes of climate change, a few changes were made compared with previous reports. It was announced that anthropogenic activities are very likely (95%) to be the cause of the warming observed since the mid-twentieth century. It also states that atmospheric CO_2, CH_4, and N_2O have increased to unprecedented levels over the last 800,000 years. It is stated that the changes in total CO_2 emissions and global mean surface temperatures are roughly linear. In other words, the upper limit of temperature rise is determined by the upper limit of total accumulated emissions. For lower temperature increases, a lower cumulative emission is required.

It is stated that the upper limit of cumulative CO_2 emissions with the highest probability (over 66%) is 790 GtC. This is the level that has maintained the global average temperature rise pre-industrialization within 2 °C. By 2011, approximately 515 GtC have already been emitted.

1.4.2 Current State of Global Warming (Observed Facts)

There is no doubt that global warming will occur. From 1880 to 2012, the global surface temperature has risen by 0.85 °C (Fig. 1.8). From 1992 to 2005, the water temperature in the deep ocean below 3000 m is likely to have increased.

In other words, it has been scientifically observed that anthropogenic CO_2 emissions have continued to increase, atmospheric CO_2 concentrations have risen, and global mean temperatures have been rising over the past 100 years.

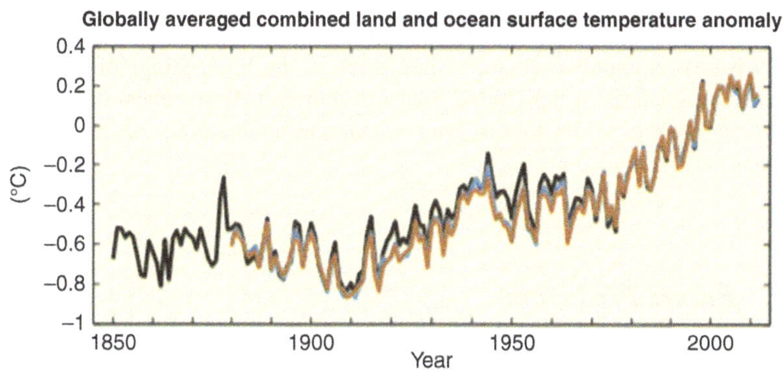

Fig. 1.8 How much did the temperature of the Earth rise? (Source: IPCC 2014)

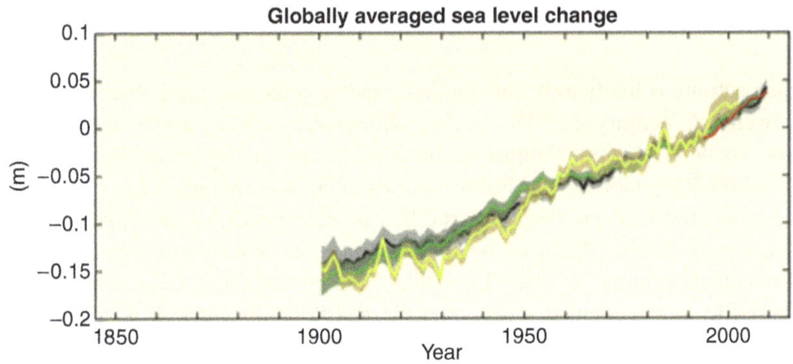

Fig. 1.9 How high is the sea level of the Earth? (Source: IPCC 2014)

1.4.3 Sea Level

The world average sea level rose by 0.19 m for the 1901–2010 period (Fig 1.9). The IPCC has also stated that the world's average sea level will rise in the twenty-first century by 0.26–0.82 m at the end of this century, compared to 1986–2005.

On September 25, 2019, the IPCC released the "IPCC Special Report on The Ocean and Cryosphere in A Changing Climate, SROCC" (IPCC 2019). More than 100 scientists from 36 countries consulted approximately 7000 scientific publications to evaluate the latest scientific literature on the ocean and cryosphere in terms of climate change. The IPCC report proposed that under high-emission scenarios, an average of one-third of the planet's glaciers will be lost by 2100. This loss is not solely constrained to the North and South Poles; the report predicts that the ice volume of smaller glaciers found in Europe, East Africa, the tropical Andes, and Indonesia will diminish by more than 80%. By the end of this century, if carbon emissions continue unabated, the biomass in the world's oceans may be reduced by 15%, and the maximum fishing potential of fisheries may decline by up to 24%.

Until recently, it was believed that sea levels would rise by less than a meter by 2100. However, it is now predicted that it could be about twice as high, according to a study based on expert evaluations published in the Proceedings of the National Academy of Sciences of the United States of America. Researchers have indicated that this could lead to the loss of land, impacting hundreds of millions of people (Jonathan 2019).

1.4.4 Future Predictions

The projected change in global surface temperature by the end of this century is likely to be between 0.3 and 4.8 °C (Fig. 1.10), while the projected rise in sea level is

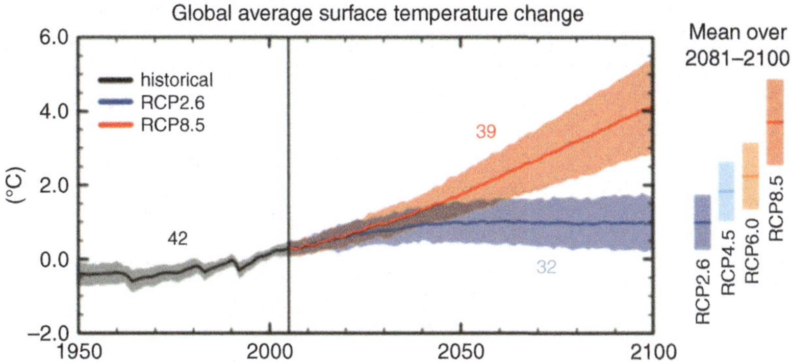

Fig. 1.10 What will happen to the temperature of the Earth? (Source: IPCC 2014)

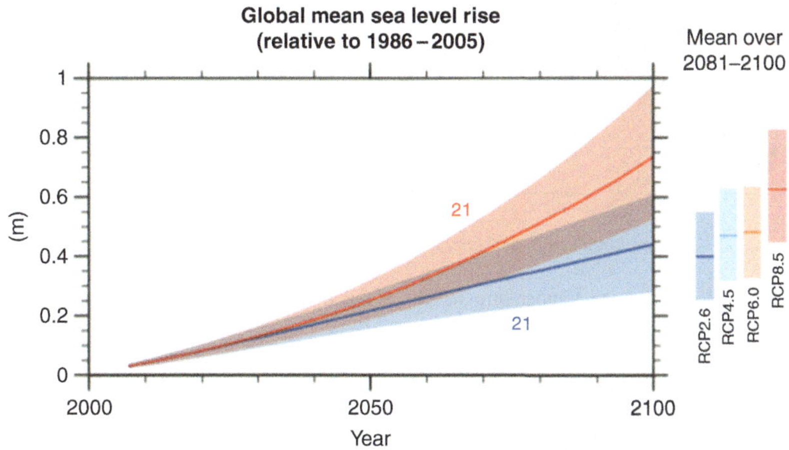

Fig. 1.11 What will happen to the sea level of the Earth? (Source: IPCC 2014)

likely to be in the range of 0.26–0.82 m (Fig. 1.11).The total cumulative CO_2 emissions and changes in the global surface temperature are proportional. The final rise in temperature is related to the range of cumulative emissions (Fig. 1.12). The greater the volume of emissions in the next few decades, the greater the emissions reductions necessary.

To prevent severe disasters from global warming, the 2015 Paris Agreement proposed a goal to control temperature rise to within 2 and 1.5 °C as much as possible (i.e., pre-industrial revolution levels), by the end of this century. For this reason, in the second half of the twenty-first century, there needs to be near-zero GHG emissions. To achieve this goal, each country was required to "propose a voluntary emission reduction target (nationally determined contribution, NDC) to the United Nations" and "take domestic emission reduction measures to achieve this goal." The NDC of the Japanese government is to reduce GHG emissions by 26% of

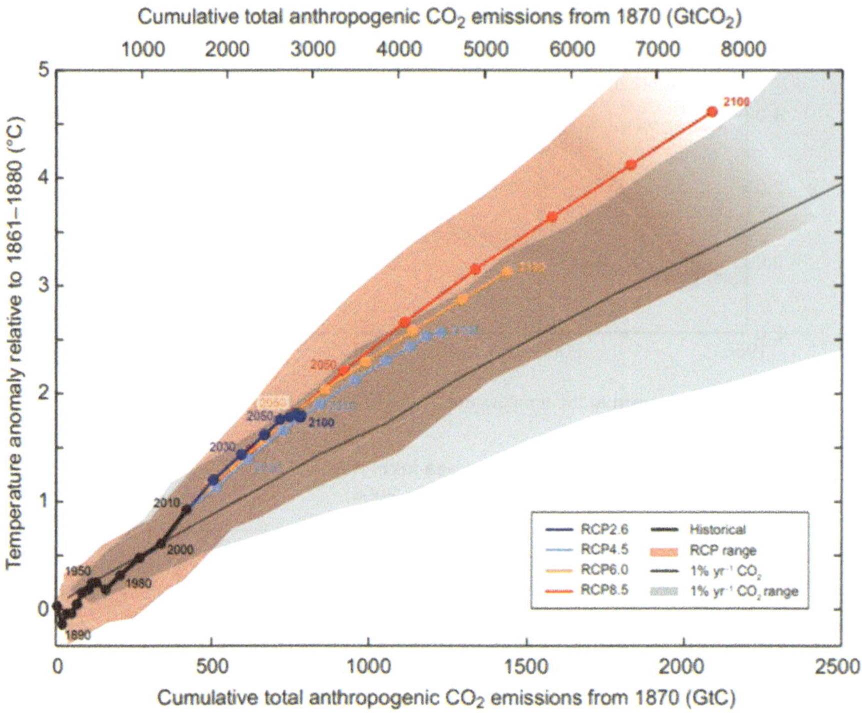

Fig. 1.12 Global average temperature rise as a function of cumulative total carbon dioxide emissions. (Source: IPCC 2014)

2013 levels by 2030. China's NDC aims to reduce emissions per unit of GDP by 60–65% of 2005 levels.

At present, there are at least two problems with NDCs in various countries. First, the NDC is not mandatory target unlike the Kyoto Protocol. As such, whether countries are achieving their NDC is unknown. Second, even if all countries achieve the NDC target, they cannot achieve the 1.5 or 2 °C target of the Paris Agreement. According to the numerical simulations by Tsinghua University research team, under current NDC scenarios, the global average temperature rises by 2100 is predicted to be approximately 3.11 °C. There is only a 6.4% probability that the global temperature rise will be controlled within 1.5 °C and a 15.4% probability that it will be controlled within 2 °C. There is a 32.9% probability that the temperature rise will exceed 4 °C. The prerequisite for achieving the 2 °C temperature target is that the annual reduction rate of energy consumption per unit of GDP is maintained at approximately 2.5%, and the annual reduction rate of carbon emission intensity per unit of energy consumption will exceed 2% in 2020 and increase to approximately 4% and 5% in 2030 and 2040, respectively. The annual reduction rate of global carbon emission intensity per unit of GDP needs to increase from approximately 5% in 2020 to 6.6% in 2040. The global transition to a green, low-carbon, and efficient energy system must be accelerated; the risk is whether the energy system can achieve low-carbon energy. The proportion of fossil energy in the global primary energy needs to decline significantly from the current 85.6% to 57.8% in 2050, while the proportion of coal usage needs to reduce from the current 29% to 12.8% in 2050. The proportion of non-fossil power generation in electricity needs to increase rapidly, from approximately 32% at present to 68% in 2050, while the proportion of fossil energy power generation equipped with carbon capture and storage (CCS) in 2050 needs to reach about 60%. The proportion of electricity as the final energy source also needs to increase from 17.7% (current) to 34.3% in 2050 (Kang 2017).

On October 8, 2018, the IPCC released the Special Report "Global Warming of 1.5 °C" in Incheon, South Korea (IPCC 2018). The report predicts that if the Earth warms by 2 °C (the Paris Agreement goal), there will be twice as many people facing water shortages as a warming to 1.5 °C. The additional warming will also plunge more than 1.5 billion people into extreme heat, and hundreds of millions more people will face infectious diseases such as malaria. The report stated that if the global greenhouse effect develops at the current rate, between 2030 and 2052, the world's average temperature will be 1.5 °C higher than pre-industrial revolution levels. The report listed the impacts of global warming on the climate, including rising sea levels and impacts on ecosystems. The report pointed out that if the global average temperature rise is to be controlled within 1.5 °C, CO_2 emissions must be reduced to 55% of 2010 levels by 2030, and actual CO_2 emissions must be reduced to zero by 2050.

The report pointed out that the current global average temperature has risen by approximately 1 °C from pre-industrial revolution levels. If it continues to increase at the current rate, the global average temperature will rise by 0.2 °C every 10 years. If the global average temperature rises by 2 °C, the risk of heavy rains and tropical depressions in East Asia and North America will significantly increase. Even if the global average temperature rises by 1.5 °C, in 2100, the global sea level will increase in the range of 26–77 cm. There will be a 70–90% reduction in the area of coral reefs; however, once the global temperature rises by 2 °C, coral reefs will almost entirely disappear. The report also details the required global energy source composition when the global average temperature rise is controlled at 1.5 °C. Renewable energy needs to account for 70–85% of all power generation in 2050, coal power generation must be eliminated, and natural gas power generation must use CO_2 recovery and storage technologies (IPCC 2018).

The three trends of increasing carbon emissions, deteriorating air pollution coupled with the weakening of the natural climate cycle, promote the acceleration of climate change in the next 20 years. It is predicted that the frequency of extreme weather events will far exceed expectations. As shown in Fig. 1.13, the temperature will likely rise by more than 1.5 °C in 2030 as opposed to by 2040 (Xu et al. 2018), as predicted in the "Global Warming of 1.5 °C" Special Report (IPCC 2018).

According to the "Emissions Gap Report 2019" released by the United Nations Environment Programme (UNEP) in 2019, GHGs produced by human activities have increased by an average of 1.5% per year over the past 10 years. In 2018,

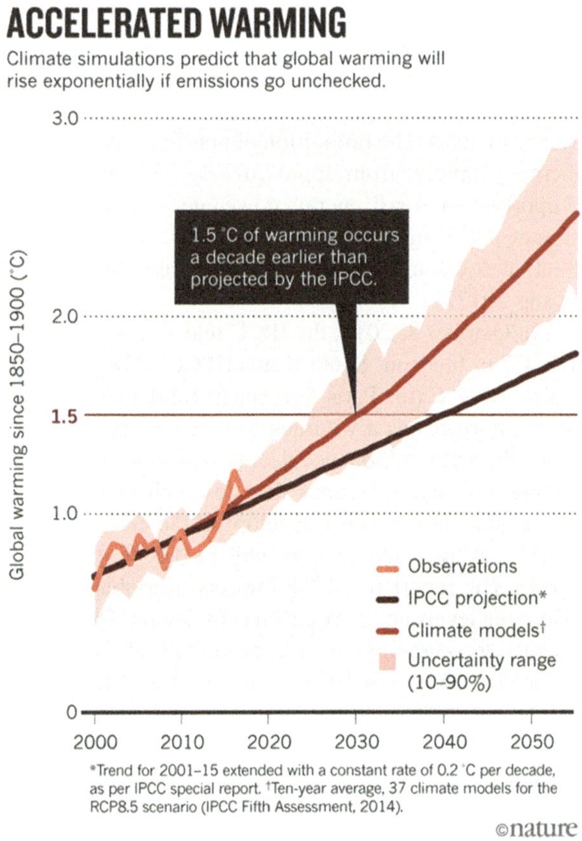

Fig. 1.13 Global temperature rise forecast. (Source: Xu et al. 2018)

carbon emissions reached the highest in history at 55.3 billion tons (calculated as CO_2). According to the current emission reduction targets pledged by countries in the Paris Agreement, emissions in 2030 need to be 56 billion tons. If the emissions continue to increase, the temperature will rise by 3.4–3.9 °C at the end of this century, having destructive effects. Even if countries that have joined the Paris Agreement achieve their emission reduction targets, the temperature will still rise by 3.2 °C (UNEP 2019).

1.5 Uncertainties in Climate Change Issues

1.5.1 What Is Uncertainty?

Uncertainty is a concept that appears in philosophy, statistics, economics, finance, insurance, psychology, and sociology. Uncertainty refers to the unpredictability of the development of things based on the essence and the randomness characterized by

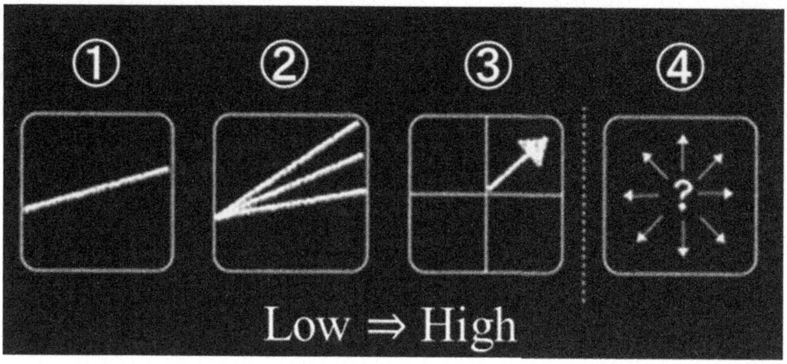

Fig. 1.14 Four types of uncertainty

variables. Uncertainty limits our predictions and observations. Put merely, and uncertainty refers to the inability to accurately know the outcome of a particular event or a confident decision in advance. In other words, as long as there is more than one possible outcome of an event or decision, uncertainty will arise. For example, in economics, uncertainty means that the distribution range and states of economic conditions, such as future gains and losses, cannot be known with certainty. As shown in Fig. 1.14, uncertainty can be divided into the following four types (Courtney et al. 1997):

(a) The future or results are predictable and deterministic. The future can be foreseeable.
(b) There are several limited possibilities in the future or results. Although the future cannot be fully predicted, it can be limited to a specific range of options.
(c) The future or result can only know the approximate direction. It is impossible to predict the extent of the future to a specific range of choices, but the future changes can be predicted with a certain probability and fluctuation range.
(d) The future or results are entirely unpredictable. The uncertainty spans various fields, and there is no basis for predicting the future. The issue of climate change is a typical issue of uncertainty.

1.5.2 Uncertainty in Climate Change Issues

Climate change is a typical interdisciplinary problem. While the climate is one of the tasks in meteorology, it requires research from the perspectives from international cooperation and lifestyle changes and the sustainability of the Earth to clarify the situation and causes, to construe and implement countermeasures. It also needs the knowledge coping with uncertainty from atmospheric chemistry, oceanography, biology, energy land use engineering, agriculture and forestry and economics, law and politics, international relations, cultural anthropology and social psychology, and logic.

Concerning the issue of climate change, there is uncertainty in scientific knowledge (category of cognitive science) as well as uncertainty in human decision-making, including policy-making (category of design science). Uncertainty is one of the essential factors that complicate measures against global warming (Yamaji 2017). Uncertainties in scientific knowledge include uncertainty in recognizing climate phenomena and uncertainty in future projections.

1.5.2.1 Uncertainty in Recognizing Phenomena

In the climate change issue, it is a science that human-made carbon dioxide emissions continue to increase, that atmospheric carbon dioxide levels are rising, and that global mean temperatures have risen over the past 100 years. It is a fact that has been observed for a long time, and it is almost sure that it corresponds to the above uncertainties. However, human society has not fully elucidated the causal relationship, global system, and the temporal and spatial effects on the social system and human system in the climate change problem.

Including the situation of climate change in the past 2000 years, such as the Medieval Warm Period appearing in the North Atlantic from the tenth to the fourteenth century AD, and the beginning of the fifteenth century, the global climate entered a cold period "Little Ice Age" (Fig. 1.3 can also be seen whether China called it the "Ming and Qing Little Ice Age") existed, and the scope and extent of existing climate change remain with uncertainty (IPCC 2007). There are many uncertainties influencing factors (such as the correlation relationship between atmospheric temperature-greenhouse gas concentration-greenhouse gas emissions, the different understandings of the greenhouse effect mechanism, the contribution of water vapor to the greenhouse effect and warming, and others).

The complexity of climate system change determines that human understanding of climate change is inevitable. Many research conclusions on climate change are not final. Many issues need to be further studied. For example, as shown in Table 1.2, the results of the evaluation of the causes of climate change seen in the IPCC report are also uncertain.

"Possibility" is a term that expresses uncertainty quantitatively and is expressed probabilistically based on observation, statistical analysis of model results, and expert judgment.

The most comprehensive evaluation of scientific knowledge on global warming may be the IPCC report. IPCC generally started in 1990, separated every 5 years. According to Table 1.2, we can find that it has a different comment on climate change and the influence of human actives, from FAR (1990), including "The temperature would raise," "Anthropogenic greenhouse gases can cause climate change" to the AR5(2013) "Extremely high possibility" (over 95%), and "There is no doubt that global warming will occur." IPCC's understanding of the causes and extent of climate change is deepening. However, it is expected that damage will

Table 1.2 Assessment of the impact of human activities on global warming in the IPCC assessment report

Reports	Years	Evaluation of the impact of human activities on global warming
First Assessment Report 1990 (FAR)	1990	"The temperature would raise" Anthropogenic greenhouse gases can cause climate change.
Second Assessment Report: Climate change 1995 (SAR)	1995	"Impact is reflected in global climate" Distinct human influences are visible in the world climate.
Third Assessment Report: Climate change 2001(TAR)	2001	"High possibility" (Over 66%) It is high possibility that most of the warming observed in the last 50 years is due to increasing greenhouse gas concentrations.
Fourth Assessment Report: Climate change 2007 (AR4)	2007	"Quiet high possibility" (Over 90%) There is no doubt about global warm.ng will occur. It is quiet possibility that the global warming since the mid-20th century is alomst because of the increased concentrations of anthropogenic greenhouse gases.
Fifth Assessment Report (AR5)	2013–2014	"Extremely high possibility" (Over 95%) There is no doubt about global warming will occur. It is extremely possibility that the main reason of global warming since the mid-20th century is the humankind activities.

Source: Based on IPCC AR5 WGI TSBox TS.1

become apparent over a long period of several generations after greenhouse gas emissions and that the damage will continue for a long time. The causal relationship is not fully understood, the probability of damage is not exact, and it is a highly uncertain problem.

1.5.2.2 Uncertainty in Future Forecast Evaluation

It can be seen from Figs. 1.10–1.12 that future prediction results have a large degree of dispersion. It is due to the complexity of the climate system and the limitations of assessment techniques which make the current assessment results of climate change mostly uncertain. This uncertainty is mainly due to the following aspects:

(a) The uncertainty in the understanding of climate change phenomena

Due to the limited understanding of the carbon cycle, the physical and chemical processes of greenhouse gases and aerosols, including the uncertainties in the greenhouse gas and aerosol emissions data, the carbon dioxide concentration in the atmosphere is transformed into a "radiation force" for the climate system. At times, there is much uncertainty. Besides, the emission of greenhouse gases and aerosol is restricted by many factors such as the population, economy,

Table 1.3 Influencing factors of uncertainty in climate change assessment and prediction

Climate change scenarios	Climate model
	Scenario design
	Model application technology
Evaluation model	Model structure
	Model parameters
	Model input data
Evaluation process	Land-air-sea coupling technology
	Human impact
Countermeasures effect	Policy diversity
	Non-uniqueness of result

and social development of various countries, which makes it difficult to accurately predict the concentration of greenhouse gases in the atmosphere in the future. At the same time, the relationship between temperature rise and the impact is far from clear, which significantly affects the accuracy of the assessment results.

(b) The deficiencies of the climate model and the prediction technology itself

To predict global and regional climate changes in the next 50–100 years, it must rely on complex global land-air-sea coupling models and high-resolution regional climate models. However, the current climate models' descriptions of clouds, oceans, and polar ice caps are still incomplete, and the models cannot handle the effects of clouds and ocean circulation, as well as regional precipitation changes.

Significant variability can occur in climate prediction due to the parameter uncertainties of the equations solved by the physics-based climate model and the structural uncertainties of the climate model. As a result, there is a great deal of uncertainty in decision-making.

The factors affecting the uncertainty of climate change assessment and prediction are shown in Table 1.3. Among them, the imperfection of the climate model itself is one of the fundamental reasons for its uncertainty. Numerical experiments generally obtain future climate change scenarios (computer experiments) using climate models (in fact, in addition to climate models, it also includes economic, energy, environmental policy models, land-use change models, impact models, which are an integrated model). The global climate model (GCM) developed by many research institutions in various countries, relying on the advancement of high-level computer technology, has a particular ability to simulate global, hemispherical climate conditions, and this ability is advancing rapidly. Although different climate models can give a more consistent future climate change trend, the results of climate scenarios output by different climate models are quite different. This is equivalent to the uncertainty type 2 or type 3 shown in Fig. 1.14, that is, the approximate direction can be determined, but there are several different results. Due to the many natural factors that affect climate change, coupled with the interaction and feedback within the atmosphere-sea-land-ice and snow systems, it constitutes the complexity, diversity, and possibility of computational analysis of climate change. The current physical laws of climate phenomena have not been fully explained, let alone a mathematical equation to describe climate phenomena (physical phenomena). Also, greenhouse gas emission prediction is a necessary input condition of the climate model, and the uncertainty of its scenario setting will inevitably affect the output of the climate model.

In terms of predicting global and national future climate change scenarios, large-scale climate models suitable for use by various countries are still under development. The climate models used so far cannot accurately construct scenarios of future climate change in different countries and the world. It becomes a significant constraint that forbids the counties to take further research and countermeasures.

There is significant uncertainty in the realization of the "1.5 °C" target of the Paris Agreement. There are significant uncertainties in "who," "when," "where," "how," and "how much reduction of greenhouse gases can be achieved" to achieve the "1.5 °C" target. It is because there are uncertainties in temperature relations-concentration-amount of reduction-reduction measures (time, place, ways).

1.5.3 Uncertainty in Decision-Making Relationship with Uncertainty

1.5.3.1 Skepticism to Global Warming

As mentioned in Sect. 1.2, climate change has changed drastically since the temperature changed from cold to dry and wet, at least since 10,000 B.C. Although global warming has become more severe in recent years, it is assumed that current global warming is only a part of the natural cycle of global climate change and due to the uncertainty of climate change mentioned above. Some scientists have claimed "Skepticism to Global Warming." There are two main types of "Skepticism to Global Warming."

One comes from scientific uncertainty. It is considered to belong to the category of scientific debate and policy debate. Specifically, it includes "discussion/question on scientific knowledge of global warming," "skepticism on the cause of global warming," "discussion on the carbon cycle," "criticism for dispersion of prediction results," and "impact of global warming" discussion.

The other is systematic skepticism and denial of propaganda activities. It is thought that this is mainly due to the interests generated by measures against global warming, for example, "denial of CO2 cause," "criticism against IPCC," and "criticism against global warming countermeasures." At times, we oppose global warming countermeasures such as the introduction of new energy due to scientific uncertainties or variations in future prediction results of global warming. To this end, in the "Merchants of Doubt" by US science historian Naomi Oreskes and others, how some elite scientists have come to terms with individual companies and political groups and how to organize systematic skepticism and denial is described in detail (Naomi 2011).

The former can be attributed to uncertainty, and it is hoped that the skepticism will gradually disappear as science and technology advances reduce uncertainty. Especially in the problem of social science policy theory, such as global warming countermeasures, unlike the problem of natural science in which there is only one correct answer, there can be multiple answers, which may increase uncertainty. Absent. Although the IPCC is the "mainstream" and most "authoritative" institution in the field of climate change, the science is not a "majority" logic, but sometimes "truth" in the minority. Therefore, it is considered that it is nature the science should have the existence of a healthy "skepticism."

1.5.3.2 Decision-Making with Uncertainty

Decision theory is divided into three categories: decision-making under certainty, decision-making under risk, and decision-making under uncertainty. There are more difficult uncertainties in combating global warming. Policy-making is the act of choosing the future, but there is also uncertainty in the decision itself.

Taking global climate countermeasures needs to give answers to (1) where, (2) when, (3) who, (4) what, and (5) how.

It is currently believed that global warming, which is mainly caused by the greenhouse effect caused by human production and living activities, requires a time interval of several generations to change. This change can allow today's human society to take various measures under instability. To slow down the global warming and adapt, without the catastrophic consequences of the decline of an ancient civilization. Civilized society does not appear to be powerless in the face of climate change, nor can it adapt to climate change indefinitely.

We have to make decisions with uncertainty. There are two essential principles for decision-making under uncertainty: one is the principle, and the other is flexibility.

(a) Principle

"United Nations Framework Convention on Climate Change": We must achieve the ultimate goal of stabilizing the concentration of greenhouse gases in the atmosphere at a level that does not cause dangerous anthropogenic interference with the climate system.

Key conclusions from IPCC Report AR5: It is highly likely that human activity was the predominant factor in the warming observed since the mid-twentieth century (possibly more significant than 95%). The changes in total accumulated CO_2 emissions and global mean surface temperatures are roughly linear.

"Paris Agreement" Goal: The Paris Agreement commits to keeping the global temperature increase to less than 2 °C above pre-Industrial Revolution levels while making efforts to keep it under 1.5 °C higher. It should make virtually global greenhouse gas emissions zero in the second half of the twenty-first century.

Therefore, humankind must actively work to reduce CO_2 emissions in its production activities and daily activities. It is the basic principle.

(b) Flexibility

The issue of global warming is caused by the survival of humankind itself and is the most significant and fundamental problem for humankind. However, the issues that human society faces are diverse, including short-term and long-term, local and global, and socioeconomic environments, and are not limited to global warming issues. The 1997 Kyoto Protocol imposes numerical targets that are legally binding and, at the same time, introduced flexible mechanisms such as emissions trading and CDM (Clean Development Mechanism). It can be said that the first step toward measures to deal with the uncertainty of climate change issues has been started.

Therefore, at the same time as reducing CO_2 emissions, various methods and policies are needed for the sustainable development of human society in different social contexts. Moreover, a social system that can incorporate innovation flexibly, and policy design that can integrate economic system and technical system is required. To realize the Paris Agreement target, it is necessary to strengthen the synergy effect of the economy, environment, and society more efficiently through the cooperation of multiple countries. It was considered policy flexibility to address the uncertainty of the problem.

Under the abovementioned multiple uncertainties, we are trying to make policy choices that significantly affect the future of humankind. The policy must be phased in response to advances in scientific knowledge while maintaining the flexibility to deal with uncertainty. The best mix of principle and flexibility is the basic concept advocating the "East Asian Low Carbon Community" concept described in Chap. 3.

1.6 Low Carbon Society and Global Sustainability

1.6.1 Prerequisites for a Sustainable Human Society

The concept of sustainability is considered to have been formed with the birth, evolution, and development of human society. Forestry (maximum allowable cut) and fisheries (maximum sustainable yield) are the oldest ecological categories. Alternatively, the "theory that man is an integral part of nature" and "the way from what is beneath abstraction" have been used as an oriental philosophy that human beings and wildlife must ultimately obey the laws of nature (Zhou 2013).

The idea of sustainability has received worldwide attention in all fields since the publication of the "Our Common Future" report by the United Nations World Commission on Environment and Development in 1987. In this study, the concept of sustainable development was strongly emphasized; this is a development that meets the needs of the present generation without compromising the needs of future generations.

The minimum requirements for a sustainable human society are in line with the three Herman Daly principles (Herman 1990):

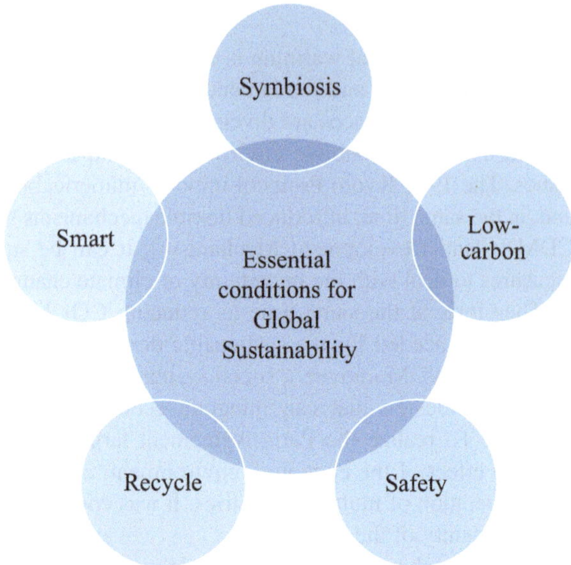

Fig. 1.15 Essential conditions for human society to achieve sustainable development

(a) The sustainable use of renewable resources requires that consumption is not greater than the regeneration rate of resources.
(b) The sustainable use of non-renewable resources requires that the rate of consumption is not greater than the pace at which renewable substitutes are established.
(c) The sustainable pace of pollution and waste requires that production is not greater than the pace at which natural systems are able to absorb, recycle, or neutralize these pollutants.

Based on the twentieth-century development model of mass production, mass consumption, mass disposal, and mass pollution, ultimately leading to the manifestation of global warming due to large-scale CO_2 emission, are symbolic of not meeting the premise of these three principles.

As an alternative, we propose five sustainability principles (G-SUS Five Principles), as shown in Fig. 1.15:

(a) Recycling: maximize resource utilization
(b) Low carbon: minimization of environmental load
(c) Symbiosis: humanity and nature exist in harmony
(d) Safety: a safe and secure society
(e) Smart: optimization of social-economic technological systems (i.e., minimize cost, maximize utility)

From the beginning of the twenty-first century, the development of the science of sustainability as an academic discipline expanded to the spatial scale of nations,

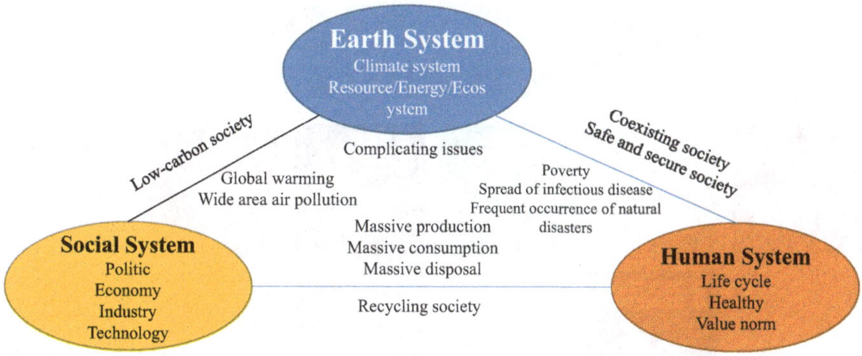

Fig. 1.16 Basic scheme of sustainability science. (Source: based on Integrated Research System for Sustainability Science (IR3S))

regions, cities, companies, and products or in fields such as society, economy, environment, and institutions. With the varied understanding and definition of sustainability depending on the theorist, the amount of accumulated research is enormous. As such, the science of sustainability is still in development as a supra-disciplinary educational system.

As shown in Fig. 1.16, we considered the Earth to consist of resources/energy, natural ecosystems (the basis of human survival), economic systems, political systems, industrial structures, and technical systems. A sustainable society may be realized by re-building the human system consisting of lifestyles, health, safety, security, and value norms and repairing their mutual relationships (IR3S).

1.6.2 Sustainable Development Goals

The SDGs at the United Nations Summit in September 2015 aim to end poverty, protect the planet, and ensure peace and prosperity for all. These are the goals set by the 193 member states of the United Nations to achieve over 15 years (2016–2030). As shown in Fig. 1.17, there are 17 significant goals and 169 concrete targets to achieve.

The primary aims of the SDGs are linked to the following priority areas:

1. Energy Security Issues (7th in SDGs): this aims to address issues associated with the rapid increase in energy demand, the potential depletion of fossil fuels, the promotion of energy conservation and renewable energy diffusion, risks for nuclear power accidents, and spent nuclear fuel.
2. Resource/Recycling Issues (12th in SDGs): this aims to address the depletion of metal resources, health/environmental problems from by illegal recycling of electric/electronic devices, health/environmental problems associated with the mining of natural resources.

Fig. 1.17 Sustainable Development Goals proposed by the United Nations in 2015

3. Climate Change (13th in SDGs): this aims to promote international cooperation and respond to climate change, acknowledging that they are interconnected at the root of social systems and must be systematically resolved through strategic policy development.

These are global issues that require the development and implementation of policies that are based on mutually beneficial international cooperation.

The rapid progress of global warming and the higher frequencies of abnormal weather highlight the urgency of establishing multidimensional cooperation and international institutional design to realize a low-carbon society for the long term. The Paris Agreement entered into force in 2016, and the US withdrawal from the Agreement in 2017 has strengthened this urgency. The Chinese, Japanese, and Korean nations (i.e., CJK countries) have set its own voluntary goal (INDC); this is not sufficient to address this global issue. Cooperation toward wide-area, low-carbon societies, including by Japan, China, and South Korea, is essential. While Japan possesses the world's highest level of low-carbon technology patents, it lags in innovation by entering the vast Asian market and cannot fully contribute to the creation of a green economy and the expected environmental pollution problem. Therefore, it is essential to position the required significant reduction in CO_2 as a strategic issue shared by all of East Asia and build a wide-area, circulating, low-carbon society created by actors across different dimensions.

1.6.3 Low-Carbon and East Asian Sustainability

The East Asian region is characterized by the world's most extraordinary diversity in terms of politics, economy, technology, culture, religion, and ethnicity. On the environmental side, we are faced with a wide range of common issues such as economic development (overcoming poverty), addressing pollution, global warming, reduced biodiversity, ozone depletion, and trans-boundary air pollution. In regions where rich and developing countries live alongside each other, such as East Asia, all countries share the burden of environmental degradation. As such, the need for inter-regional cooperation in resolving environmental problems is becoming increasingly apparent; these problems do not heed physical or geopolitical boundaries. There is no other problem with of greater significance than air pollution, acid rain, and global warming. East Asia faces a trilemma of poverty, pollution, and global environmental issues. Ensuring the security of strategic energy resources based on mutually beneficial international cooperation in East Asia and building policy development scenarios and roadmaps for that purpose are essential in their own right.

A community is a union of people who share the same interests and goals. The issue of global warming is a serious issue for a community that shares the common interests and goals of humankind. This includes climate mitigation and adaptation, impact, and the reduction of environmental damage. Due to the urgency of climate change issues, the commonality, and necessity of realizing a low-carbon society, the add-on benefits of carbon abatement countermeasures, and characteristics of CO_2 itself, it is vital that the "East Asian Low-Carbon Community" concept is materialized.

References

BBC (2020) Recorded 38 degrees in Siberia or highest possibility in the Arctic. https://www.bbc.com/japanese/53147673. Accessed 31 July 2020 (in Japanese)

Cabinet Office (2015) Fifth science and technology basic plan. https://www8.cao.go.jp/cstp/kihonkeikaku/index5.html. Accessed 1 July 2020

China Meteorological Administration (2002) The sudden change in climate and the decline of ancient civilizations four thousand years ago. http://www.cma.gov.cn/kppd/kppdqxsj/kppdqhbh/201212/t20121213_196305.html. Accessed 1 July 2020

Courtney H, Kirkland J, Viguerie P (1997) Strategy under uncertainty. Harv Bus Rev 75(6):66–79

Earther G (2020) NASA space lasers offer' fantastically detailed' look at the World's ice loss. https://earther.gizmodo.com/laser-satellites-offer-fantastically-detailed-look-at-t-1843197751. Accessed 31 July 2020

EM-DAT (2017) International Disaster Database 2017. https://www.emdat.be/. Accessed 6 May 2020

Fagan BM (2004) The long summer: how climate changed civilization, New York

Fukui News (2019) Relationship between climate change and ancient civilization, 120 points are showing the rise and fall of Egypt. https://www.hokurikushinkansen-navi.jp/pc/news/article.php?id=NEWS0000020693. Accessed 1 July 2020

Gordon CV (1950) The Urban revolution. Town Plan Rev 21:3–17
Herman D (1990) Towards some operational principles of sustainable development. Ecol Econom 2:1–6
IPCC (2007) Fourth assessment report. https://www.ipcc.ch/site/assets/uploads/2018/03/ar4_wg2_full_report.pdf. Accessed 6 May 2020
IPCC (2018) Global warming of 1.5 °C. Special report.
IPCC (2013) Climate change 2013: the physical science basis. https://www.ipcc.ch/report/ar5/wg1/ 2020/05/06
IPCC (2014) AR5 synthesis report: climate change 2014 https://www.env.go.jp/earth/ipcc/5th/pdf/ar5_wg1_overview_presentation.pdf. Accessed 31 July 2020 (in Japanese)
IPCC (2019) Special report on the ocean and cryosphere in a changing climate. https://www.ipcc.ch/srocc/. Accessed 6 May 2020
Jonathan W (2019) Rising sea levels pose threat to homes of 300m people – study. https://www.theguardian.com/environment/2019/oct/29/rising-sea-levels-pose-threat-to-homes-of-300m-people-study. Accessed 6 May 2020
Kajita H, Kawahata H, Wang K, Zheng H, Yang S, Ohkouchi N, Utsunomiya M, Zhou B, Zheng B (2018) Frequent and abrupt cold episodes around 4.2 ka in the Yangtze-delta: the collapse of the earliest rice cultivating civilization, Japan Geoscience Union meeting, 2018
Kang C (2017) Key scientific issues and theoretical research framework for power
Liu Y (2009) Climate change and the rise and fall of historical dynasties. Science in China: Earth Sciences.
Ministry of the Environment (1995) Quality of the environment in Japan 1995. https://www.env.go.jp/en/wpaper/1995/eae240000000000.html. Accessed 6 May 2020
Naomi O, Conway EM, Jacques T (2011) Merchants of doubt: how a handful of scientists obscured the truth on issues from Tobacco smoke to global warming. Le Pommier, France
UNEP (2019) Emissions gap report 2019. Global progress report on climate action. https://www.unenvironment.org/interactive/emissions-gap-report/2019/. Accessed 6 May 2020
Wang S (2005) 2200-2000BC climate change and the decline of ancient civilization. Advances in Natural Science 15(19):1094–1098
Weiss H, Bradley RS (2001) What drives societal collapses? Science 291:609–610
Weiss H, Courtyard MA, Wetterstron W et al (1993) The genesis and collapse of third millennium north Mesopotamian civilization. Science 261(20):995–1004
World Meteorological Organization (2017) 2017 is set to be in the top three hottest years, with record-breaking extreme weather. https://public.wmo.int/en/media/press-release/2017-set-be-top-three-hottest-years-record-breaking-extreme-weather. Accessed 31 July 2020
World Weather Attribution (2019) The human contribution to record-breaking June 2019 heatwave in France. https://www.worldweatherattribution.org/human-contribution-to-record-breaking-june-2019-heatwave-in-france/. Accessed 31 July 2020.
Xu Y., Ramanathan V. & Victor D. 2018. Global warming will happen faster than we think. https://www.nature.com/articles/d41586-018-07586-5. Accessed on 2020/05/06.
Yamaji K (2017) A long-term strategy for global warming countermeasures. Science Council open symposium, 2017, Sep 27. http://www.scj.go.jp/ja/event/pdf2/170927-1.pdf. Accessed 1 Aug 2020.
Zhou W (2013) Introduction to sustainability science. Law and Culture Publishing House, Kyoto
Zhu K (1973) Preliminary research on China's climate change in the past five thousand years. Sci China 2:1973

Chapter 2
Climate Change Strategy and Emission Reduction Roadmap for China, Japan, and South Korea

Xuepeng Qian, Yishu Ling, and Weisheng Zhou

Abstract In order to limit the increase in global temperatures to below 2 °C, as outlined in the Paris Agreement, measures to reduce carbon emissions from large emitters in East Asia—China, Japan, and South Korea—have attracted global attention. Differences among the three countries in terms of the environment, economy, and energy mean that each is at a different stage of emission reduction. In this chapter, we take into account environmental, economic, and energy characteristics in China, Japan, and South Korea and assess their emission reduction strategies. While China is simultaneously grappling with poverty eradication, environmental protection, and responding to climate change, Japan and South Korea are combating climate change with leading technologies and high levels of energy efficiency. Although there are many differences among the three countries, all advocate realizing low-carbon societies and achieving sustainable development. Moreover, ensuring energy security, developing low-carbon technologies, cultivating low-carbon industries, and adapting to climate change are challenges faced by all three countries. We analyzed targets and countermeasures for Intended Nationally Determined Contributions (INDC) under the Paris Agreement. In the process of achieving emission reduction targets, the respective efforts of each country are indispensable and contribute to international cooperation.

X. Qian
College of Asia Pacific Studies, Ritsumeikan Asia Pacific University, Beppu, Oita, Japan
e-mail: qianxp@apu.ac.jp

Y. Ling
Graduate School of Policy Science, Ritsumeikan University, Ibaraki, Osaka, Japan
e-mail: ps0290ih@ed.ritsumei.ac.jp

W. Zhou (✉)
College of Policy Science, Ritsumeikan University, Ibaraki, Osaka, Japan
e-mail: zhou@sps.ritsumei.ac.jp

W. Zhou et al. (eds.), *East Asian Low-Carbon Community*,
https://doi.org/10.1007/978-981-33-4339-9_2

2.1 Introduction

Although the Paris Agreement was adopted in 2015, there is still no clear consensus as to whether developed countries should take the lead in reducing emissions. In 2019, the United States withdrew from the Paris Agreement on the grounds that it affected its own interests. Against this background, the emission reduction actions of Asian countries are worthy of attention. Among the top 12 emitters of carbon dioxide, two-thirds are in Asia (Fig. 2.1). China, Japan, and South Korea ranked first, fifth, and seventh globally, accounting for about 33.44% of emissions in 2018 (IEA 2019). Therefore, climate change countermeasures taken in China, Japan, and South Korea will be critical for global emission reductions.

One difficulty in negotiating emission reductions is the need to simultaneously balance interests among developed countries, among developing countries, and between developed and developing countries. China, Japan, and South Korea are developing, developed, and semi-developed countries, respectively; thus cooperation among them could serve as a pilot for global cooperation (Fig. 2.2). Moreover, China, Japan, and South Korea are all large energy consumers reliant on fossil fuels. Transition from a high energy consumption society to a low-carbon society, called decarbonization, is a pressing issue. In this chapter, we analyze the current situation in these three countries from the perspectives of economy, energy, and the environment. Then, we clarify the limits of economic growth and environmental impact when building a decarbonized society among these countries. Finally, we develop a

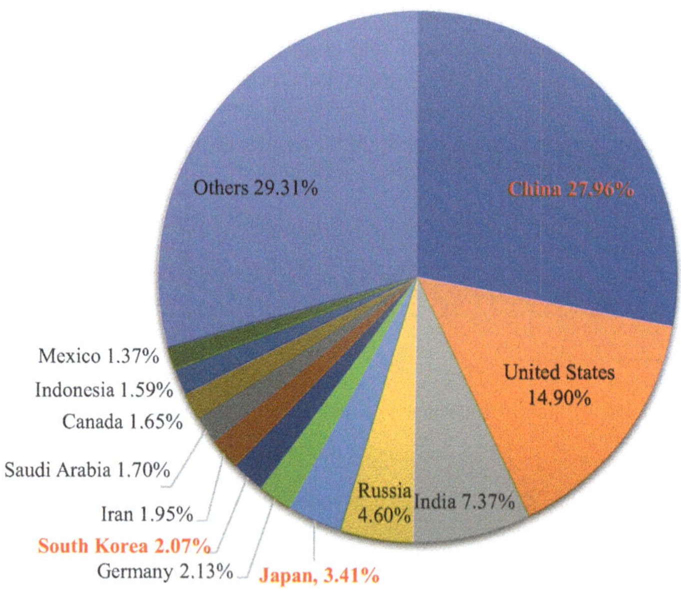

Fig. 2.1 Proportion of global carbon dioxide emissions contributed by the largest emitters in 2018. Source: IEA (2019)

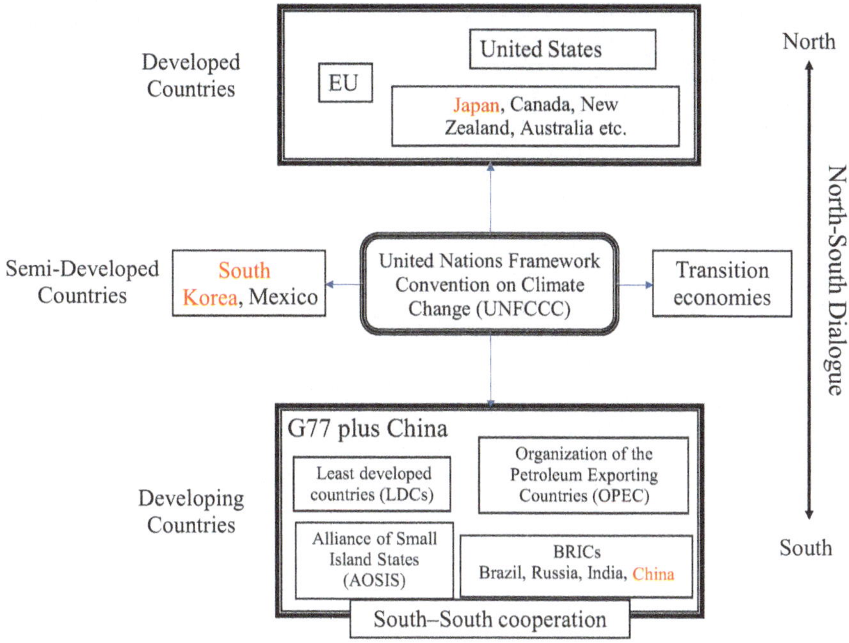

Fig. 2.2 Position of the focal countries in the United Nations Framework Convention on Climate Change (UNFCCC). Source: Qian (2012)

reduction roadmap based on INDC targets and countermeasures for China, Japan, and South Korea.

2.2 Economic Prospects in China, Japan, and South Korea

China, Japan, and South Korea have experienced rapid economic development since the twentieth century. However, after 1990, rates of development in Japan and South Korea slowed and were particularly affected by two financial crises (Fig. 2.3). Although the growth rate in China was also affected by these crises, it remained positive (Fig. 2.4). In 1999, China's gross domestic product (GDP) purchasing power parity (PPP) surpassed that of Japan for the first time and has been the highest of the three nations since. Except for a negative growth rate in 1998 in South Korea, the growth rates in China, South Korea, and Japan are generally at high, medium, and low levels, respectively, which corresponds with the developmental stage of each country.

The development trajectories of China, Japan, and South Korea show three similar curves; that of China currently resembles the historical trajectory of Japan and South Korea. However, after more than 30 years of rapid growth, China remains in the early stages of development, and labor productivity is low (Fig. 2.5); high

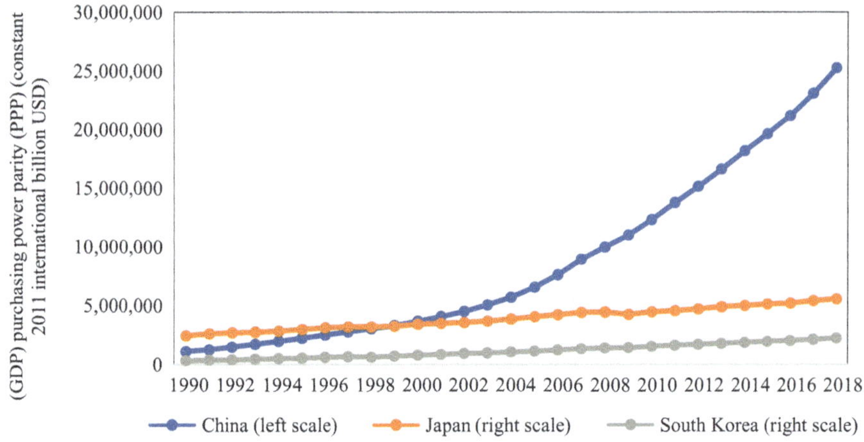

Fig. 2.3 Gross domestic product (GDP) purchasing power parity (PPP) (constant 2011 international billion USD) in 1990–2018 in China, Japan, and South Korea. Source: IMF (2019)

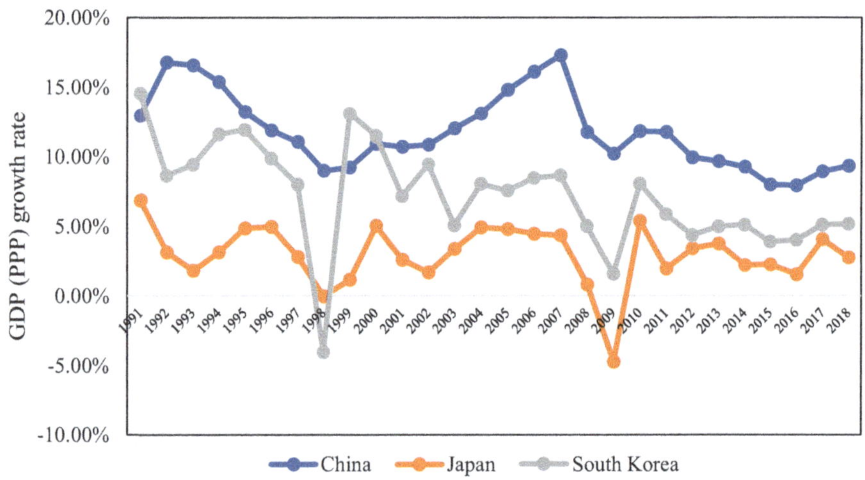

Fig. 2.4 Growth rate of gross domestic product (GDP) purchasing power parity (PPP) (constant 2011 international billion USD) in 1991–2018 in China, Japan, and South Korea. Source: IMF (2019)

growth has only begun to translate into increased productivity in the last 10 years. In the last 5 years, labor productivity in South Korea has approached that of Japan, whereas labor productivity in China was approximately 50% that of Japan in 2019.

Urbanization rate provides another means to understand the developmental stages of the three countries (Fig. 2.6). In 1990, both Japan and South Korea had reached an urbanization rate of approximately 80%, while China remained at 26%. Since 2010, the urbanization rate in Japan has stalled at 91% and that of South Korea at 81%. In

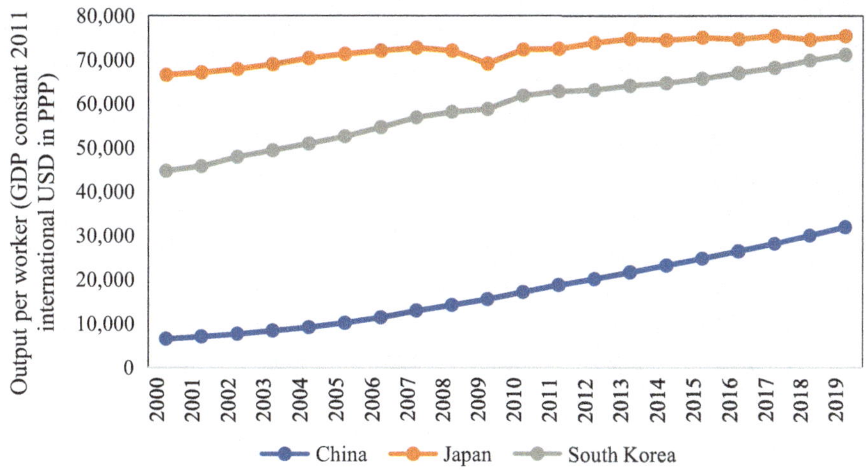

Fig. 2.5 Labor productivity in China, Japan, and South Korea. Source: ILOSTAT (2020)

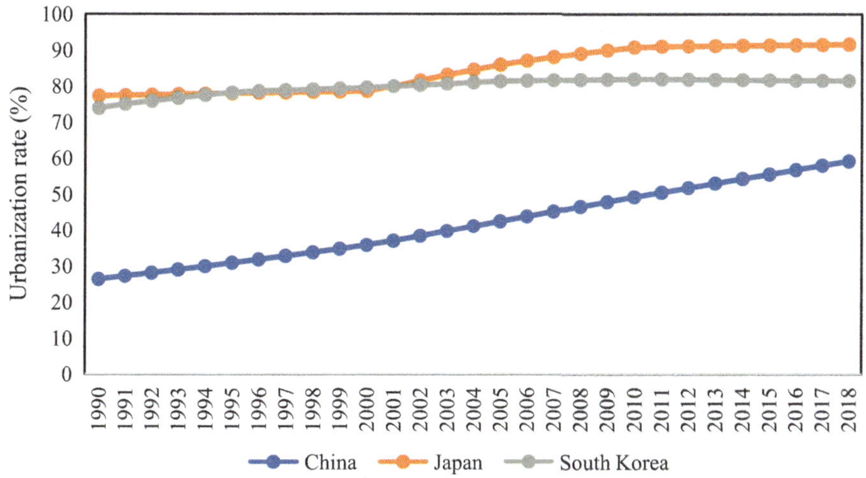

Fig. 2.6 Urbanization rate in China, Japan, and South Korea. Source: UN (2019a, b)

contrast, China continues to exhibit 1% annual growth, which if maintained would lead to 70% urbanization by 2030.

The economic situation can be analyzed from the perspective of industrial structure. In China, the proportion of primary industry (e.g., agriculture) continues to decline, the status of secondary industry (e.g., manufacturing) as the largest sector has been consolidated, and the share of tertiary industry (e.g., services) has increased significantly (Fig. 2.7).

The pattern of industrial structure in Japan has not changed from tertiary, secondary, and primary industry, in descending order (Fig. 2.8). The proportion of

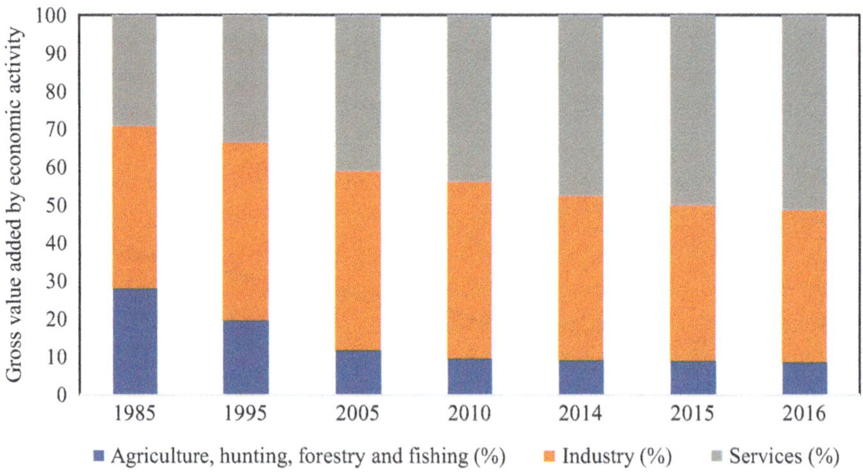

Fig. 2.7 Gross value added by economic activity type in China. Source: UN (2019a, b)

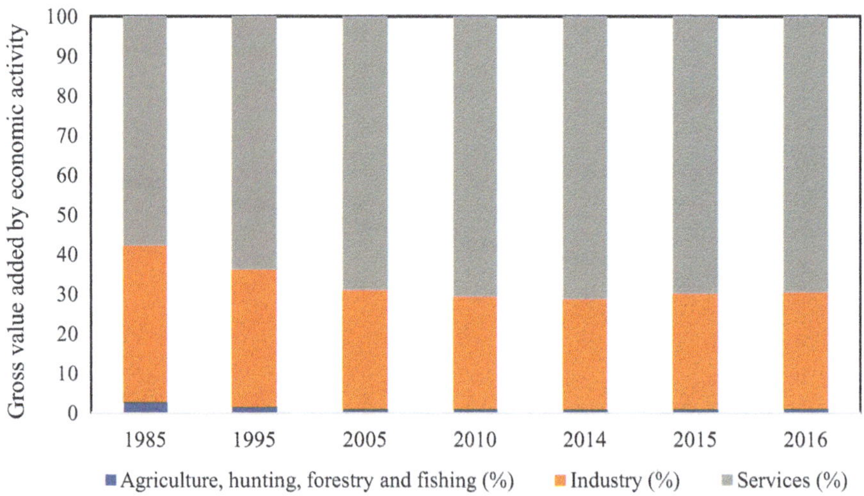

Fig. 2.8 Gross value added by economic activity type in Japan. Source: UN (2019a, b)

primary industry has continued to decline, falling to 1.15% in 2016. The proportion of tertiary industry has continued to rise, reaching 69.31% in 2016. As early as the 1960s, Japan began to propose the concept of a post-industrial and smart society. The contribution of secondary industry to GDP has also continued to decline, falling to 29.52% by 2016, which is the lowest of all three countries.

After the 1980s, South Korea was at a post-industrial stage, implementing an industrial policy of technological departure. The industrial structure is technology-intensive and requires upgrading (Fig. 2.9). The proportion of primary industry

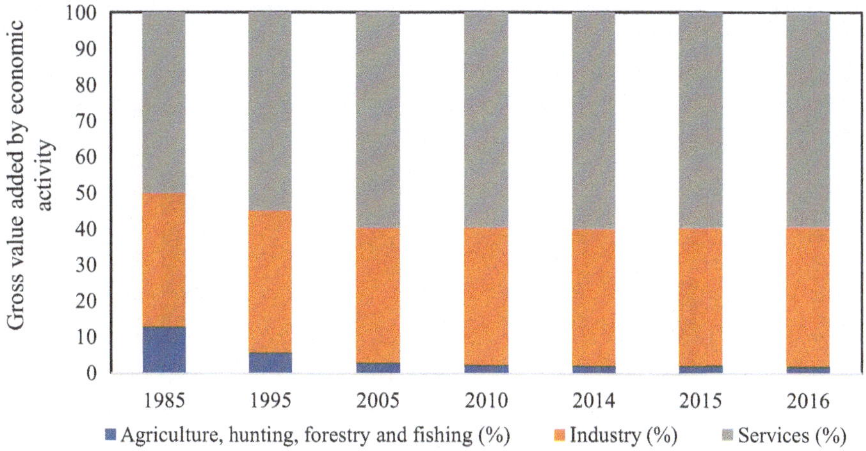

Fig. 2.9 Gross value added by economic activity type in South Korea. Source: UN (2019a, b)

quickly dropped, the proportion of secondary industry has remained almost unchanged, and the proportion of tertiary industry continues to rise.

According to Petty-Clark's law, Japan has entered the stage of post-industrialization, while South Korea is transitioning from industrialization to post-industrialization. China is still in the process of industrialization. The characteristics of economic development in Japan are effective economic policies led by the government at all stages; a shift in industrial structure from technology-intensive to knowledge-intensive; a "trade-oriented nation" strategy that makes full use of foreign countries to promote export-oriented economic growth; and an emphasis on technology introduction and research and development (Fig. 2.10). The development characteristics of the Korean economy are government-led resource allocation; an export-oriented development strategy; a "science and technology-oriented nation" strategy that centers development on technology-intensive industries to promote export-oriented economic growth; and attaching importance to foreign investment and technology. In economic development in China, the government has always played an active role in promoting effective macroeconomic control. In the process of economic globalization, China has seized processing links in the value chain and has begun competing in international markets. China has also introduced large amounts of capital and technology from Europe, the United States, Japan, South Korea, and elsewhere to accelerate its development.

We analyzed the economic status of the three countries in terms of GDP, productivity, urbanization rate, and industrial structure. We infer that as a developing country, China will remain in the transitional stage of industrialization and urbanization for the next decade. Japan will continue to promote industrial informatization and will remain ahead of the other two countries. South Korea will strive to transform into a post-industrial society. However, due to the impact of COVID-19, rates of economic development in China, Japan, and South Korea will be greatly reduced.

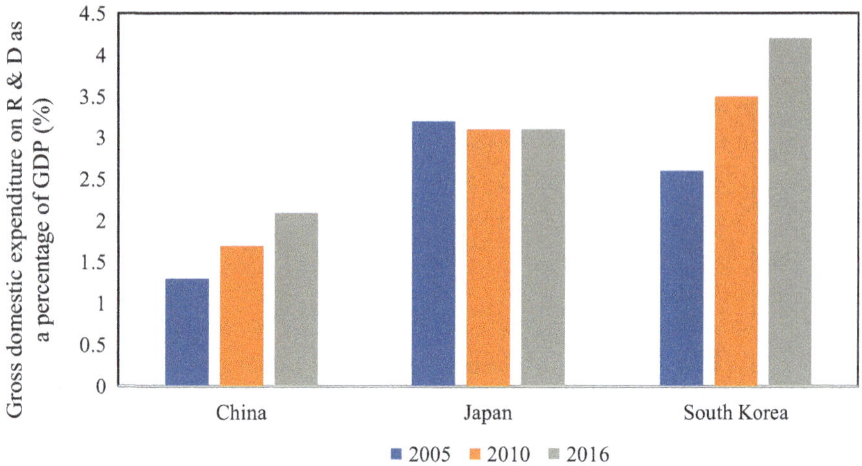

Fig. 2.10 Gross domestic expenditure on research and development (R&D) as a percentage of gross domestic product (GDP) in China, Japan, and South Korea. Source: UN (2019a, b)

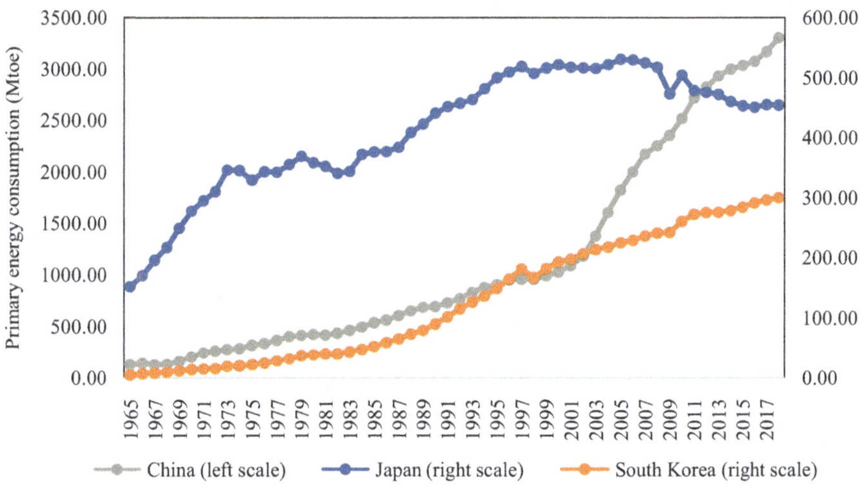

Fig. 2.11 Primary energy consumption in 1965–2018 in China, Japan, and South Korea. Source: BP Statistics (2019)

2.3 Energy Consumption in China, Japan, and South Korea

China, Japan, and South Korea are big energy consumers (Fig. 2.11). After the 1973 oil crisis, primary energy consumption in Japan developed at a slower rate than before. Primary energy consumption also decreased during the 1979 oil crisis, the Asian Financial Crisis, the financial crisis of 2007–2008, and particularly after the Tohoku earthquake. With increases in population and economic reform beginning in

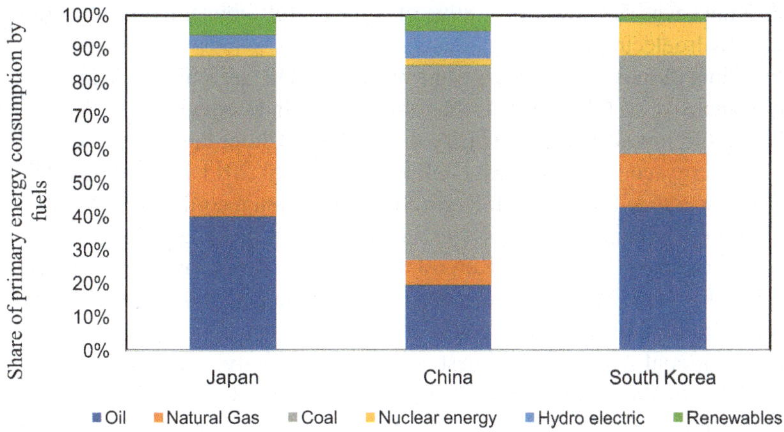

Fig. 2.12 Share of primary energy consumption by fuel in 2018. Source: BP Statistics (2019)

1978, primary energy consumption in China has accelerated since the late 1970s; it became the biggest energy consumer in Asia in 1977. Since 2001, when China became a member of the World Trade Organization (WTO), energy consumption has increased rapidly. In contrast, rapid growth in South Korea from the 1960s to the 1980s contributed stable and robust growth both in GDP and primary energy consumption. However, the Asian Financial Crisis shrank the economy and primary energy consumption of South Korea in 1998. In the 2000s, South Korea changed its economic development model and maintained strong growth in both GDP and energy consumption.

China, Japan, and South Korea are all reliant on fossil fuels for their development. Unlike Japan and South Korea, which both depend on oil, China depends on coal, accounting for approximately 60% of its total energy consumption in 2018 (Fig. 2.12). In addition to fossil fuels, nuclear is an essential energy source in each country. Japan decreased the proportion of nuclear power in its total energy consumption to 2.44% in 2018, due to the Fukushima Daiichi nuclear disaster. The percentage of nuclear in China is 2.03%, which is the lowest proportion of the three, but at 66.6 million tons of oil equivalent (Mtoe) is the most substantial volume among the three countries. In South Korea, nuclear accounted for approximately 10.00% of total energy consumption in 2018, which is the highest among the three countries and demonstrates a strategic priority. However, the president elected in 2017 aims to phase out nuclear over the next 45 years.

Renewable energy is becoming an increasingly critical energy option that could reduce GHG emissions as an important component of the energy mix. In 2018, annual renewable energy consumption in China was 415.6 Mtoe (12.58% of overall energy consumption), including 272.1 Mtoe of hydroelectricity. In Japan, annual renewable energy consumption was 43.7 Mtoe (9.62% of overall consumption), including 18.3 Mtoe of hydroelectricity. Annual renewable energy consumption in

South Korea was 5.7 Mtoe or 1.89% of total energy consumption, including 0.7 Mtoe of hydroelectricity.

Prevailing trends in the energy intensity of GDP are compared in Fig. 2.13. Energy intensity of GDP in 2018 has sharply declined compared to that in 1990. Japan has the lowest energy intensity of GDP and China the highest. From 2013, primary energy consumption per GDP PPP (constant 2011 international USD) in China is lower than that in South Korea. In 2018, China recorded the lowest level, at 0.145 toe/USD compared to 0.158 toe/USD in South Korea. Gaps in energy intensity among China, Japan, and South Korea are narrowing. For instance, the gap in energy intensity of nominal GDP between Japan and China in 1990 was 0.874 toe/USD, narrowing to 0.187 toe/USD in 2018.

In China, coal remains the largest source of electricity, accounting for approximately 70% of the nation's electricity in 2018 (Fig. 2.14). Due to high levels of PM2.5 pollution nationwide and pressure to achieve emission reduction targets, China, as the world's biggest GHG emitter, is expected to act rapidly to transfer to cleaner energy. As part of its 13th Five-Year Plan, a total of 150 GW of new coal capacity has been canceled or postponed until at least 2020 (National Energy Administration 2017). Increasingly strict controls on total coal capacity and power plant emissions are expected to result in the retirement of up to 20 GW of older plant capacity and spur technological upgrades to China's remaining 1000 GW of coal power. Other fuels, such as natural gas, nuclear power, and renewables, are expected to make up increasing shares of electricity generation in China in the future.

In 2010, a year before the Tohoku earthquake, the composition of power sources in Japan was coal (27.8%), liquefied natural gas (LNG, 29.0%), oil (8.6%), nuclear (25.1%), hydroelectric (7.3%), and other renewables (2.2%; Fig. 2.15). Due to the temporary closure of nuclear power plants after 2011, the percentage of nuclear declined to 0 in 2014 before slightly increasing to 8.9% in 2018 (BP 2019). Oil and LNG generation increased to compensate for the decline in nuclear. Renewables also compensated for some of the decline in nuclear power, increasing from 2.2% in 2010. Annual renewable electricity generation was 168.35 TWh, including 90.67 TWh of hydroelectricity, contributing approximately 16.73% of electricity generation in 2018 (BP 2019).

Like Japan, South Korea is a major energy importer. Electricity generation mainly comes from conventional thermal power. South Korea began its transition to cleaner energy with a 2017 power supply plan. Over 60 coal power plants are in operation in South Korea, one-third of which are due for retrofitting. The country has placed a heavy emphasis on nuclear power generation, accounting for nearly 30% of electricity. Despite the 2011 Fukushima Daiichi nuclear disaster in Japan, the South Korean government planned for significant expansion of its nuclear power industry, aiming to increase the share of electricity generation accounted for by nuclear to 60% by 2035. However, this was cancelled in 2017, leading to promotion of renewable energy in South Korea. In 2018, annual renewable electricity generation was 19.00 TWh, including 7.29 TWh of hydroelectricity, contributing approximately 3.28% of electricity generation (Fig. 2.16).

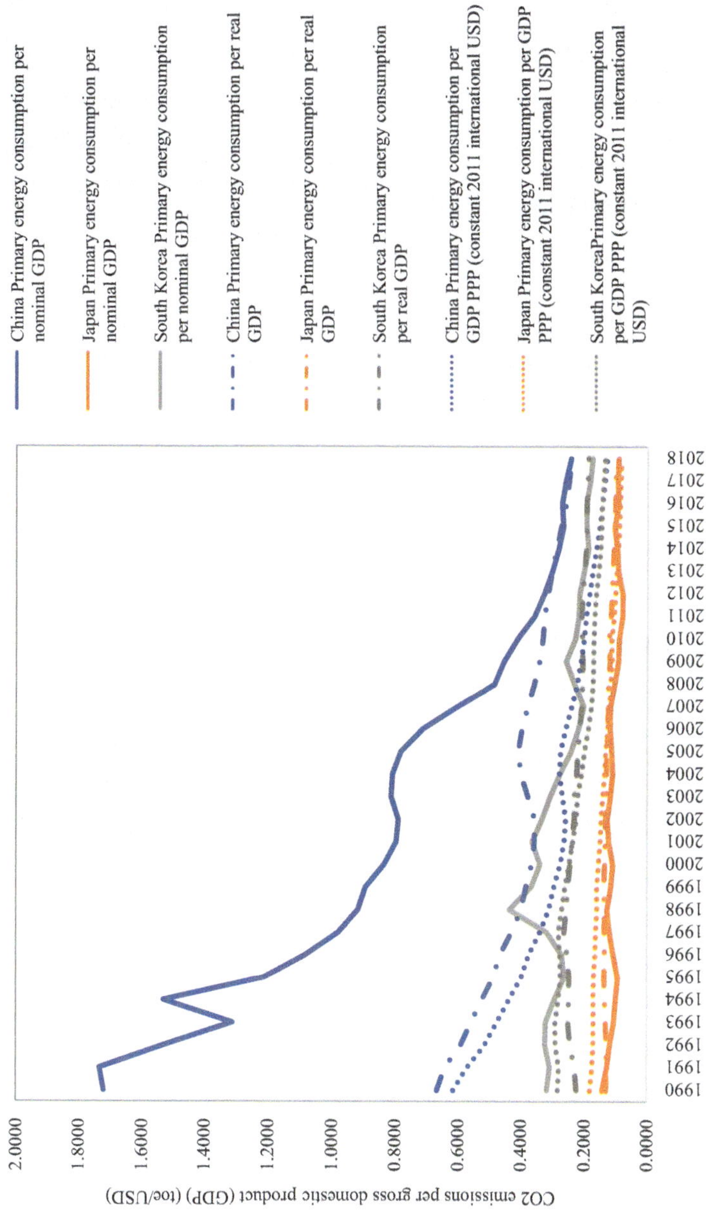

Fig. 2.13 CO_2 emissions per gross domestic product (GDP) in 1990–2018 in China, Japan, and South Korea. Source: BP (2019), UN (2019a, b)

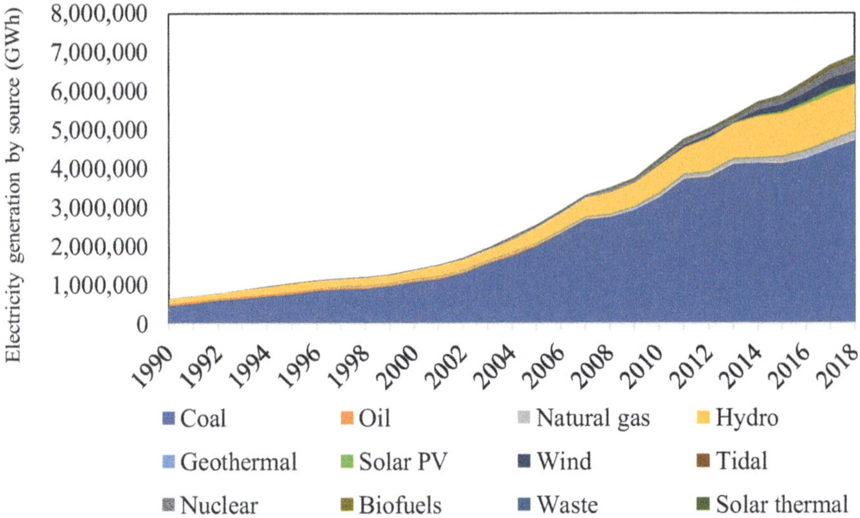

Fig. 2.14 Electricity generation by source in China in 1990–2018 (GWh). Source: IEA (2019)

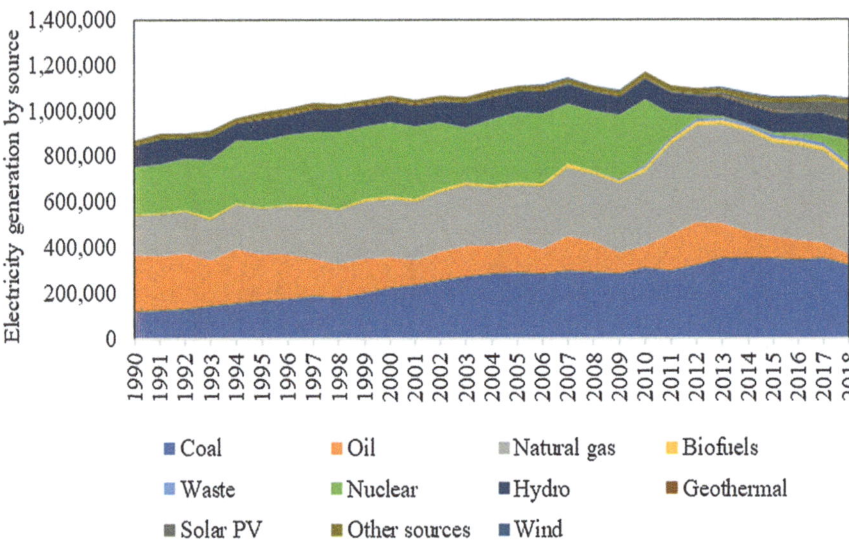

Fig. 2.15 Electricity generation by source in Japan in 1990–2018 (GWh). Source: IEA (2019)

Between 2000 and 2018, global renewable energy power capacity increased from 842 GW to 2350 GW (IRENA 2019). Renewable electricity generation in China, Japan, and South Korea has grown more rapidly since the mid-2000s. China expanded its renewable energy sector with unprecedented speed from 2006, and there are apparent increases in both Japan and South Korea from the late 2000s (Fig. 2.17).

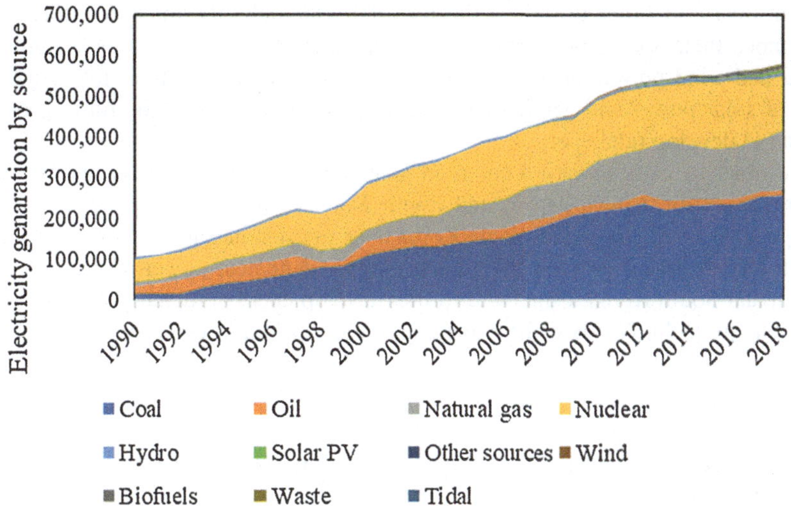

Fig. 2.16 Electricity generation by source in South Korea in 1990–2018 (GWh). Source: BP (2019)

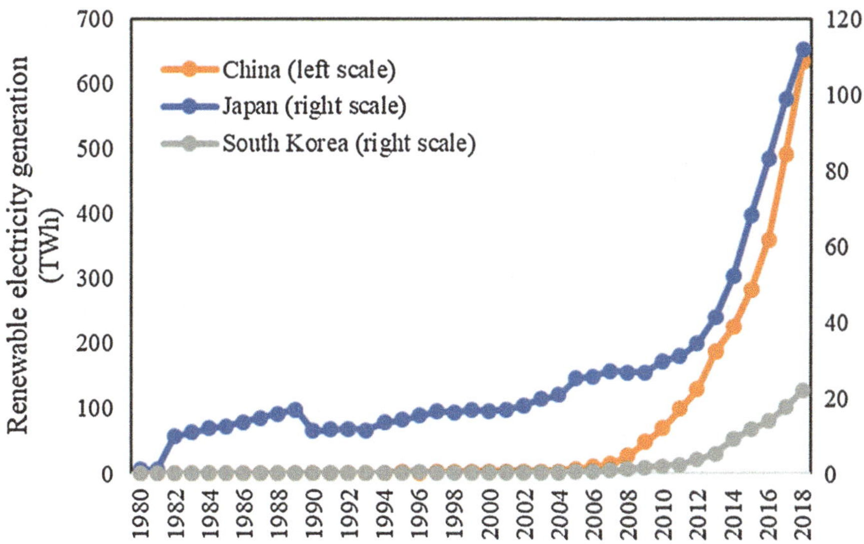

Fig. 2.17 Renewable electricity generation in China, Japan, and South Korea. Source: BP (2019)

Through analyzing the energy status of China, Japan, and Korea in terms of energy consumption, energy use efficiency, and power generation, it is clear that development in all three countries is dependent on fossil fuel. China is the largest coal consumer, while Japan and South Korea are more dependent on oil. All three countries are major energy importers and face similar crises in energy security. In recent years, in order to reduce the use of fossil fuels, the three countries have expanded their use of renewable energy. China has achieved remarkable results in solar and wind power. Japan has exploited its geographical features to expand the use of solar power and geothermal energy. South Korea has made full use of the ocean to develop tidal energy.

In order to work toward a common goal of a low-carbon society and improve energy security, China, Japan, and South Korea will further expand the use of renewable energy. Japan and South Korea hope to use renewable energy to generate electricity to fill the gap left by reducing levels of nuclear power. Compared with Japan and South Korea, China will further develop energy efficiency and introduce new energy technologies while expanding the use of renewable energy.

2.4 Climate Change Strategy: Building Decarbonized Societies

2.4.1 Domestic Strategies

In order to mitigate climate change and achieve green economic development, China, Japan, and South Korea have made concerted efforts to move toward decarbonization.

In Japan, a global warming countermeasure plan has been drawn up to guide government actions. Cross-sectoral policies, basic policies, and promotion of international cooperation are being encouraged, and efforts are being made to achieve goals under a cooperative system. In the industrial sector, energy conservation and management of facilities and equipment are important, including the introduction of equipment with high energy-saving performance and the use of energy management systems. In the business sector, it is mandatory to comply with energy-saving standards for new buildings. Similar standards will be introduced for existing buildings. Efforts are being made to increase the stock of high-efficiency lighting, such as LEDs, to 100% by 2030 and to improve energy-saving performance through a top runner system. In the household sector, measures include introduction of 5.3 million household fuel cells and a top runner system. Dissemination of next-generation vehicles such as electric vehicle (EV) and fuel cell vehicle (FCV) and improvement of fuel are measures being taken in the transportation sector. After Japan was awarded the "Fossil Award" the second time by the international environmental NGO "Climate Action Network" in COP 25, Minister of Economy, Trade and Industry responded as the closure of the coal-fired power plant, which is not energy efficient in 2020. There are about 140 coal-fired power plants in Japan, of which 114 are outdated and inefficient. Therefore, in the energy sector, conversion from thermal power generation to renewable energy is the main issue but countermeasure. The launch of a regional carbon trading market will also play a key role in moving toward sustainable development.

There have been four stages of climate change strategy in China. The first is the observation stage (before 1997). From the 1990s, the adverse effects of climate change began to be recognized globally. A climate change response agency was established in China, charged with addressing overall climate change issues. The second stage is the learning stage (1997–2005). China, regarded as a developing country, had no emission reduction obligations under the Kyoto Protocol. Nevertheless, since 1989, it has changed its economic structure of heavy industry, promoted the use of renewable energy, and improved energy efficiency. In 1998, the National Climate Change Countermeasures Coordination Subcommittee, consisting of representatives of relevant ministries, was established as a place to study climate change policy in China. Above all, the importance of realizing a low-carbon society was recognized. Subsequently, China began to implement measures against climate change. The third stage is the cooperation stage (2005–2013). Policy has been strengthened since 2005 by demonstrating the importance of climate change at the 17th National People's Congress. On June 12, 2007, the *Notice of the State Council on the Diet's climate change countermeasures and the establishment of energy-saving emission reduction sub-assemblies* was published, with an emphasis on climate change issues. Based on this notification, the Countermeasures for Climate Change Leaders Group was established to determine critical strategies, policies, and measures for climate change issues, operations, research, and international cooperation. Interdepartmental coordination of international policy negotiation became important. Both government officials and NGOs began to seek public participation. Since 2006, China has published annual national level reports on progress in addressing climate change issues. By 2007, China had approved 1000 clean development mechanism (CDM) projects, 155 of which were registered as Chinese projects and contribute to a reduction of 92 million tons of CO_2 each year. These actions and policies have resulted in the avoidance of some CO_2 emissions. The fourth stage is the stage of seeking dominance (after 2013). In 2014, the United States and China agreed to strengthen greenhouse gas reduction efforts. The United States withdrew from the Paris Agreement in 2019, while China proposed the Belt and Road Initiative in 2017. China promised to invest in clean energy, build scientific cooperation, and support other countries in tackling climate change. As China seeks to lead the world in combating global climate change on behalf of the United States, the role of controlling energy and the environment is emphasized. China is the world's largest greenhouse gas emitter but is continuing its efforts to reduce coal use and expand renewable energy. In 2017, it abandoned plans to construct 103 new coal-fired power plants, and the National Energy Agency announced the investment of more than 360 billion USD in renewable energy by 2020. China also plans to expand a carbon trading scheme originally piloted in seven provinces and cities to cover an additional area by 2020.

There are three main climate change strategies in South Korea. The first is a national carbon trading scheme. When launched in 2015, more than 525 companies participated in this emission trading scheme, covering approximately 68% of national greenhouse gas emissions. The 525 participating companies have their own reduction targets. The Ministry of the Environment's Certification Committee reviews emission reports annually, and emissions are verified by third parties. There are plans to link to carbon markets in the European Union and China, in order to improve cost efficiency and effectiveness, increase market liquidity, and promote bilateral and multilateral cooperation. The second strategy is a carbon tax and the development of a green information technology (IT) industry. Currently, the carbon tax is not obligatory, but the government is considering carbon tax policy in preparation for future carbon markets. Korean companies are currently focusing on the development of new technologies and are implementing eco-campaigns with the aim of reducing greenhouse gas emissions using new technologies. Green IT is one such solution. The Ministry of Administration has set up a calculation center for green energy conservation and is developing a comprehensive plan to promote it. The final strategy is promotion and utilization of renewable energy. The eighth power supply plan emphasized reducing dependence on coal-fired power. New equipment is prohibited in principle. By 2022, ten old facilities (3.4 GW) will be abolished. As of 2034, coal-fired power plants that have been in operation for more than 30 years will be ceased. Besides, in the same year, it was announced that dependence on nuclear power plants would be reduced and the focus moved to LNG and renewable energy generation. In South Korea, nuclear power is the main power source, accounting for 30% of total power generation; thus a move away from it will be a "great change in energy policy." By 2020, South Korea plans to build more than 10 GW of solar power generation facilities (EIA 2020).

In summary, four important trends have emerged in the three countries:

1. Green manufacturing has begun to replace traditional manufacturing.
2. Renewable energy and new energy sources are developing rapidly due to coal phase-out plans.
3. Investment in environmental protection-related fields continues to increase.
4. Regional or national carbon markets are developing rapidly.

2.4.2 *International Cooperation in Energy and the Environment*

The history of international environmental cooperation among China, Japan, and South Korea is reviewed from 1999 to the present. Since 2008, the situation has changed from unidirectional support to building strategic, mutually beneficial relationships. Cooperation must account for risks.

2.4.2.1 Cooperation Among China, Japan, and South Korea in 1999–2008

Environmental cooperation in the East Asia region was promoted by the Tripartite Environmental Ministers Meeting (TEMM) in Seoul in 1999 (Trilateral Cooperation

Secretariat (TCS) 2019). Cooperation among the three countries with regard to the environment started under the framework of ASEAN+3. The TEMM became a platform of the highest level for each country to promote regional environmental management. During this period, the three countries had already established a mechanism for regular meetings of leaders—ministerial meetings—and a consultation mechanism for international affairs. The sixth 10+3 Leaders' Meeting adopted the East Asia Research Group Final Report, *Promoting the Evolution of the 10+3 Leaders' Meeting to the East Asia Summit (EAS)* in 2002. As a result, China, Japan, and South Korea began to explore the construction of an East Asian Community under the framework of regional cooperation, including in energy and the environment. Also, during this period, Japan continued to support China through launching large-scale, long-term investments and transferal of technology via official development assistance (ODA) and Japan International Cooperation Agency (JICA) projects.

2.4.2.2 Cooperation Among China, Japan, and South Korea After 2008

Since 2008, the China-Japan-South Korea trilateral summit has been held annually, marking a new stage in international cooperation. In September 2011, the TCS was established in Seoul to promote peace and common prosperity among the three countries. Trilateral cooperation attaches great importance to the promotion of coordination in environmental protection and renewable energy (TCS 2019). During this period, the summit adopted several documents detailing measures to strengthen cooperation in environmental protection in areas such as environmental education, climate change, and biodiversity protection. A consensus was reached that the three countries should promote sustainable growth by developing green energy and improving energy efficiency. Based on this consensus, the East Asian Low-Carbon Growth Partnership was developed to promote low-carbon growth in countries of the EAS in 2012 (Ministry of Foreign Affairs of Japan (MOFA) 2015). Dialogue emphasized the importance of national low-carbon growth strategies, technology, market mechanisms, and cooperation among various stakeholders (MOFA 2015). Furthermore, an East Asian Low-Carbon Community framework was advocated, aimed at achieving a regional low-carbon society (Zhou et al. 2010).

Japan decided in 2005 to terminate its annual loan program by 2008. This was by far the biggest part of its ODA to China and demonstrated a shift from unidirectional support to building a strategic, mutually beneficial relationship based on market mechanisms. The CDM promoted rapid growth in environmental and energy cooperation. The first Japan-China Third Country Market Cooperation Forum was held in Beijing in 2018. By strengthening cooperation between China and Japan in third-country markets, a pathway to realize a tripartite, or multilateral, cooperation model is provided. The third-county market cooperation approach provides a way for participating countries to share benefits more equally.

In summary, cooperation among China, Japan, and South Korea in energy and the environment has a long history and has changed from unidirectional support to mutually beneficial cooperation by employing market-oriented approaches.

2.5 Reduction Roadmaps Based on INDC Targets and Countermeasures in China, Japan, and South Korea

According to INDC targets (Table 2.1), by 2018 China had lowered carbon dioxide emissions per unit of nominal GDP by 60–65% from 2005 levels (Fig. 2.13). However, emissions per unit of real GDP and GDP PPP still need to be improved. Levels in China in 2018 were roughly equivalent to those in Japan in 1980; thus there is immense potential for cooperation to reduce emissions. It is estimated that there is a theoretical reduction potential between China and South Korea and Japan and South Korea. This is the basis of cooperation among the three countries to reduce emissions. Mitigation potential in South Korea is limited due to its industrial structure, with a large share of manufacturing and high energy efficiency in major industries (UNFCCC 2019) leading to an expected 30% reduction in emissions by 2030. This will be supplemented by trading on the international carbon market. Climate change policy directions in the three countries (Table 2.2) show that development of innovative low-carbon technology, transfer of existing technology, and international cooperation are vital to achieve regional emission reductions.

2.5.1 Transition from Coal to Clean Energy

According to their INDCs, all three countries will proactively increase the share of non-fossil fuels in primary energy; coal currently dominates their energy structures.

There are nearly 2500 coal-fired power plants in China, with a total installed capacity of almost 970 GW by the end of 2017, an annual increase of approximately 2%. The coal power industry in China faces problems of overcapacity, and arrested economic growth has led to a decline in power demand. The Chinese government is seeking to continue economic restructuring and adopt more consumption-led and environmentally sustainable growth, which has further decreased oil and coal demand. Natural gas and renewable energy consumption is growing rapidly. According to the energy development plan of the 13th Five-Year Plan, the total installed capacity of coal-fired power should be kept below 1.1 billion kW by 2020. After 2020, coal consumption should be reduced and adopt a long-term downward trend.

There are more than 100 coal-fired power plants in Japan. After the Fukushima nuclear power plant accident, Japan increased investment in new coal-fired power

Table 2.1 The INDC targets and countermeasures of China, Japan, and South Korea

	China	Japan	South Korea
Target	• To lower carbon dioxide emissions per unit of GDP by 60% to 65% from the 2005 level • To achieve the peaking of carbon dioxide emissions around 2030 and making best efforts to peak early • To increase the share of non-fossil fuels in primary energy consumption to around 20% • To increase the forest stock volume by around 4.5 billion cubic meters on the 2005 level	• At the level of a reduction of 26.0% by fiscal year (FY) 2030 (1.042 billion t-CO2 eq) compared to FY 2013, 25.4% reduction compared to FY 2005 • Aiming at a 35% increase in energy efficiency by 2030 • The ratio of renewable energy within the total electric power generated will be around 22–24%, and the ratio of nuclear power will be around 20–22% (solar is expected to increase 7× and wind/geothermal energy 4× from the current level) in 2030	• 37% reduction from the business-as-usual (BAU, 850.6MtCO2eq) level by 2030 • One-third of total reductions depend on international carbon markets • Other sectors will offset the industrial sector mitigation
Scope (sectors and gases)	Sectors: 1. Energy 2. Industrial processes and product use 3. Agriculture 4. Land use, land-use change, and forestry (LULUCF) 5. Waste Gases: CO_2, CH_4, N_2O, HFCs, PFCs, SF_6, and NF_3	Sectors: 1. Energy (a) Fuel combustion (b) Fugitive emissions from fuels (c) CO2 transport and storage 2. Industrial processes and product use 3. Agriculture 4. LULUCF 5. Waste Gases: CO_2, CH_4, N_2O, HFCs, PFCs, SF_6, and NF_3	Sectors: 1. Energy 2. Industrial processes and product use 3. Agriculture 4. Waste Gases: CO_2, CH_4, N_2O, HFCs, PFCs, SF_6, and NF_3
Main countermeasures	• Implementing national and regional strategies • Building low-carbon energy system and energy efficient and low-carbon industrial system	• Energy policies and the energy mix • Developing the Plan for Global Warming Countermeasures • Removals by LULUCF	• Industrial sector: the GHG and Energy Target Management System (TMS); a nationwide emissions trading scheme • Building sector:

(continued)

Table 2.1 (continued)

	China	Japan	South Korea
	• Controlling emissions from building and transportation sectors • Increasing carbon sinks • Enhancing support in terms of science and technology • Promoting carbon emission trading market and international cooperation	• JCM and other international contributions	managing energy efficiency • Transport sector: expanding infrastructure; strengthen the average emission standard • MRV system to monitor business with large amounts of GHG in above sectors • International contributions

Source: UNFCCC (2019)

Table 2.2 Climate change policy directions in the three countries

	China	Japan	South Korea
Coal-fired power generation	Reduction	Increase	Reduction
Renewable energy	Supportive	Supportive	Supportive
Carbon pricing	• Environmental protection tax • Nationwide ETS panel with pilot ETS	• Tax for climate change mitigation • Two metropolitan ETS	• Coal consumption tax • Nationwide ETS
Energy efficiency and energy conservation	• Potential for improvement	Leading level among developed countries with limited potential for improvement	High energy efficiency of major industries with relatively limited potential for improvement

Source: UNFCCC (2019)

plants. By 2050, the total installed capacity of coal-fired power plants is expected to remain at around 20 GW. Future energy policy in Japan will promote an energy composition as it was before the Fukushima nuclear power plant accident, using coal and nuclear energy as the basic power resources. Combined with INDC renewable energy development targets, energy structure has become the basis for increasing construction of coal-fired power plants in Japan. The fifth basic energy plan implemented by the Japanese government in July 2018 continues this trend.

In South Korea, coal has been the main driver of carbon dioxide emissions for two decades. Coal-fired power plants are the largest single contributor to domestic greenhouse gas emissions. Due to strong support from the current Korean government, future trends in energy structure include reducing the number of nuclear power plants and supporting renewable energy and LNG. The current government plans to phase out coal and nuclear power while expanding renewable energy to 20% of total power generation by 2030.

2.5.2 Improvement of Energy Efficiency

In China, while clear and enormous expectations are placed on the introduction of non-fossil fuels, energy conservation has yet to be fully addressed. In contrast, Japan is a leader among developed countries, with limited potential for further improvement. In South Korea, there is high energy efficiency in major industries with relatively limited potential for improvement. Both Japan and South Korea continue to improve domestic efficiency in iron and steel industries and cooperate with other countries, including China, to improve the energy efficiency of others.

Considering the mature utilization of thermal power in China, Japan, and South Korea and the high energy efficiency of Japan and South Korea, international cooperation among the three countries is advantageous. Japan is the greatest investor in overseas financing for the development of coal-fired power plants, followed by South Korea and China, which together contribute the most globally to coal power generation assistance.

2.5.3 Promoting International Cooperation

Based on potential and historical coordination among China, Japan, and South Korea, international cooperation is key to realize regional emission reductions. The Paris Agreement recognizes carbon trading mechanisms at national and regional levels and provides political and policy support. China, Japan, and South Korea are developing domestic markets, aiming to introduce new carbon market mechanisms by 2020. It is timely to lay foundations now for connection to regional carbon markets.

In addition to carbon trading mechanisms, development of innovative low-carbon technology and transfer of existing technology are expected via international cooperation to achieve ecologically sensitive design in energy and materials cycles. Considering history among the three countries, market-oriented mechanisms under the East Asian Low-Carbon Community framework are expected to be an effective approach to promote international cooperation.

2.6 Conclusion

In international climate change negotiations, China, Japan, and South Korea are positioned as developing, developed, and semi-developed countries, respectively. As the main emitters globally, it is indispensable to analyze emission reduction strategies in these three countries in order to promote global emission reductions. In this chapter, we analyzed the economic status of the three countries in terms of GDP, productivity, urbanization rate, and industrial structure. We analyzed the energy status of China, Japan, and Korea in terms of energy consumption, energy use efficiency, and power generation. Although at different developmental stages, the countries are following similar trajectories. However, development in China and South Korea in both the economy and energy should not simply repeat the pathway of Japan. Based on INDC targets and countermeasures for climate change, climate change policy dictates a transition from coal to clean energy, improvement of energy efficiency, and promoting international cooperation under the East Asian Low-Carbon Community framework.

References

BP (2019) BP statistical review of world energy 2019 [eBook], 68th edn. pp. 8–9, 51–53. https://www.bp.com/content/dam/bp/business-sites/en/global/corporate/pdfs/energy-economics/statistical-review/bp-stats-review-2019-full-report.pdf. Accessed 1 Apr 2020

EIA (2020) South Korea is one of the world's largest nuclear power producers. https://www.eia.gov/todayinenergy/detail.php?id=44916. Accessed 29 Nov 2020

IEA (2019) World energy outlook 2019. https://www.iea.org/reports/world-energy-outlook-2019. Accessed 18 Apr 2020

ILOSTAT (2020) Labor productivity. https://ilostat.ilo.org/topics/labour-productivity. Accessed 12 June 2020

IMF (2019) IMF annual report. https://www.imf.org/external/pubs/ft/ar/2019/eng/assets/pdf/imf-annual-report-2019.pdf. Accessed 18 Apr 2020

IRENA (2019) Renewable energy capacity statistics 2019. https://www.irena.org/-/media/Files/IRENA/Agency/Publication/2019/Mar/IRENA_RE_Capacity_Statistics_2019.pdf. Accessed 18 Apr 2020

Ministry of Foreign Affairs of Japan (2015) East Asia low carbon growth partnership dialogue. https://www.mofa.go.jp/policy/environment/warm/cop/ealcgpd_1204/index.html. Accessed 18 Apr 2020

National Energy Administration "The 13th five-year plan for energy development" (2017). https://policy.asiapacificenergy.org/sites/default/files/%E8%83%BD%E6%BA%90%E5%8F%91%E5%B1%95%E2%80%9C%E5%8D%81%E4%B8%89%E4%BA%94%E2%80%9D%E8%A7%84%E5%88%92pdf.pdf (in Chinese). Accessed 15 Aug 2019

Qian B (2012) A study about the domestic and foreign policy of china under climate change. Master thesis, Ritsumeikan University

Trilateral Cooperation Secretaiat (2019) Trilateral cooperation. https://www.tcs-asia.org/en/cooperation/dashboard.php. Accessed 18 Apr 2020

UNFCCC (2019) Annual report. https://unfccc.int/annualreport. Accessed 14 Apr 2020

United Nations (2019a) GVA by kind of economic activity. http://data.un.org. Accessed 10 June 2020

United Nations (2019b) Gross domestic expenditure on R & D. http://data.un.org. Accessed 1 June 2020

Zhou W, Nakagami K, Su X, Ren H (2010) Development of policy frame and evaluation model of the "East Asia Low Carbon Community" concept. Environ Technol 39:536–542

Chapter 3
Concept and Framework of the East Asian Low-Carbon Community

Weisheng Zhou

Abstract Achieving a low-carbon society is a common goal among developed and developing countries. Despite the uncertainties related to addressing global climate change, it is essential to tackle this issue in order to ensure a sustainable decarbonized society. To solve this global problem, local efforts coordinated at a national level are indispensable. The realization of a wide-area low-carbon society through cross-border multilateral cooperation and policy integration is also required. Realization of a wide-area low-carbon society that transcends national borders will create a sustainable and vibrant international society in harmony with economic development, pollution control, and society, where measures are taken to mitigate and adapt to global climate change. This chapter presents an analysis of characteristics such as the uncertainty of the climate change problem and also describes the concept of the East Asian Low-Carbon Community and its multilayered structure centered on Japan, China, and South Korea.

3.1 Introduction

Despite the many uncertainties associated with climate change, the realization of a low-carbon society is an indispensable and necessary condition beyond the achievement of a sustainable decarbonized society. In order to address global climate change, local efforts by individual countries are indispensable, while simultaneously, realization of a wide-area low-carbon society through cross-border multilateral cooperation and policy integration is desired. East Asia is one of the most diverse regions in the world, with multiple countries at different levels of development. It is facing challenges such as climate change and environmental destruction and resource and environmental constraints associated with economic growth. These problems are particularly concentrated and acute in East Asia.

W. Zhou (✉)
College of Policy Science, Ritsumeikan University, Ibaraki,, Osaka, Japan
e-mail: zhou@sps.ritsumei.ac.jp

W. Zhou et al. (eds.), *East Asian Low-Carbon Community*,
https://doi.org/10.1007/978-981-33-4339-9_3

The realization of a low-carbon society involves the creation of a sustainable and vibrant international society in which the economy, environment, and society are in harmony and global climate change is addressed. Elemental issues of significance for this are the development of innovative low-carbon technologies and the transfer of existing technologies, the creation of low-carbon economic and industrial systems, and the transformation of low-carbon social systems such as life cycles, and ecological energy and material cycles, through international collaboration. Through the design and piloting of model projects, the study will demonstrate the feasibility of a low-carbon society, present a road map that embodies the process of transition to a sustainable low-carbon society, and guide construction of a low-carbon society in Asia. Policy proposals and empirical research will contribute to the creation of a model of cooperation for embodying Japan-China strategic reciprocal relationships.

This chapter presents an analysis of major international climate change agreements, such as the Kyoto Protocol and the Paris Agreement, their significance and limitations, and the climate measures taken by China, the world's largest carbon dioxide emitter. Primary energy consumption and CO_2 emissions of Japan, South Korea, and China account for approximately 30% of the global total; thus the concept and multilayered framework of an East Asian Low-Carbon Community are introduced.

3.2 Major International Climate Change Agreements

3.2.1 Major Agreements Under the Framework Convention on Climate Change

Table 3.1 shows the main steps in international negotiations on climate change. The first international United Nations Conference on the Human Environment was held in 1972 in Stockholm, Sweden. Negotiations were held in Berlin in 1995 following the adoption of the Framework Convention on Climate Change in 1992. The Berlin Mandate was agreed at the First Conference of Parties (COP1) and the Kyoto Protocol in 1997, and finally the Paris Agreement was adopted in 2015. Other agreements related to climate change have also been made.

The United Nations Framework Convention on Climate Change (UNFCCC) aims to stabilize atmospheric greenhouse gas concentrations at levels that would not have dangerous impacts on the climate system. Different country types are defined based on historical and current greenhouse gas emissions, leading to "common but differentiated responsibilities" through "considering the right to promote sustainable development." Annex 1 countries include members of the Organisation for Economic Co-operation and Development (OECD), other developed countries, and economies in transition, such as Russia and countries in Eastern Europe. Developing countries are classified as Non-Annex 1 and include China and India. In particular, the contents of Article 4 of the Convention differ between developed and developing

Table 3.1 International climate change negotiations

Year	Event
1972	First UN Environment Conference in Stockholm forming the United Nations Environment Program (UNEP)
1988	Establishment of the International Panel on Climate Change (IPCC)
1992	UN Framework Convention on Climate Change (UNFCCC) signed and ratified by the United States
1995	Parties to the UNFCCC meet in Berlin (the First Conference of Parties (COP1) to the UNFCCC) to outline specific emissions targets
1997	Kyoto Protocol adopted: first global treaty on reducing greenhouse gas emissions. Developed countries accept binding targets for emissions reduction for 2012 and 2020
2001	The United States rejects the Kyoto Protocol
2003	Formulation of operational rules for sink clean development mechanism (CDM), CDM board activity report, and discussion on future global warming prevention framework at COP9 EU implements EU Emission Trading Scheme
2004	Russia ratifies Kyoto Protocol, meeting threshold of signatories to enter into force
2005	Kyoto Protocol enters into force
2007	Bali Action Plan launches parallel negotiations under Framework Convention
2008	Adaptation Fund launched at COP14
2009	World leaders negotiate Copenhagen Accord at COP15
2010	Green Climate Fund established under the Agreement at COP16
2011	New target set for 2015 in Durban Platform at COP17
2012	End of the first commitment period for the Kyoto Protocol Canada withdraws from Kyoto Protocol
2014	Green Climate Fund established under the Agreement at COP16
2015	Adoption of the Paris Agreement by 196 parties to the UNFCCC
2016	Paris Agreement enters into force
2017	Concrete climate action commitments in COP22 The United States withdraws from Paris Agreement
2018	Commitment to further operationalize the Paris Agreement
2019	The US Trump administration officially notifies the United Nations of its withdrawal from the Paris Agreement (official withdrawal in November 2020)

countries. The most significant difference is that only developed countries set (voluntary) targets to return greenhouse gas emissions to 1990 levels by 2000 (UN 1992). However, the targets were not mandated, and ultimately were not met.

3.2.2 Significance and Limitations of the Kyoto Protocol

COP1 was held in Berlin in 1995. The Protocol and other legal instruments applied to developed countries since 2000 were agreed in 1997 under the Berlin Mandate, which prevented the imposing of new obligations on developing countries and became the basis for creating the Kyoto Protocol.

The Kyoto Protocol set legally binding individual quantitative targets for greenhouse gas emissions in developed countries. Although the climate change framework treaty, which is the parent treaty of the Kyoto Protocol, does not mandate greenhouse gas emission targets, it sets a voluntary target of returning emissions to 1990 levels by 2000, which was not achieved. Perhaps because of this, negotiations for the Kyoto Protocol considered whether the treaty should have legally binding goals and ultimately agreed that mandates would be required. The Kyoto Protocol set a goal of reducing greenhouse gas emissions by 5% from 1990 levels by 2008–2012 (first commitment period) for all developed countries. Obligations were not to be introduced for developing countries, including China and India.

Considering the scientific uncertainties of climate change issues, international cooperation is essential to achieve emission reduction goals efficiently through the use of market mechanisms and to provide flexibility to contribute to the sustainable development of developing countries. "Kyoto mechanisms," such as emissions trading, and the clean development mechanism, were introduced.

In 1997 when the Kyoto Protocol was introduced, the world's largest emitter of greenhouse gases was the United States (by total annual emissions, per capita emissions, and cumulative emissions). It eventually withdrew from the Protocol, stating the following reasons: negative impacts on the US economy, unfairness due to nonparticipation of developing countries, and uncertainty of climate change. The United States accounts for approximately one-fourth of global and 40% of developed countries' CO_2 emissions. The Kyoto Protocol set a goal of reducing greenhouse gas emissions from developed countries by 5%, but reductions in the United States reached approximately half that before its withdrawal.

The Kyoto Protocol covers a wide range of issues, including measures to protect and mitigate the risks of US withdrawal and nonparticipation, "participation" of developing countries, how to share burdens among countries, policy measures such as the role of economic instruments, and forms of consensus building. It was historically significant in providing the first legally binding quantitative targets for greenhouse gas reduction and introducing flexible Kyoto mechanisms.

3.2.3 Significance and Limitations of the Paris Agreement

Regarding emission reductions from 2013 to 2020 (the second commitment period of the Kyoto Protocol), each country submitted its 2020 target based on the Cancun Agreement adopted at COP16, held in Cancun, Mexico, in December 2010. Some countries, including Japan, did not participate. The international community agreed on a long-term goal of "keeping the temperature rise below 2 degrees Celsius compared to before industrialization." However, it was not possible to reduce emissions to the extent that this target was attainable. Therefore, negotiations began in 2012 to create an international treaty in which all countries, both developed and developing, commit to reduction targets after 2020. As a result, COP21 agreed to the Paris Agreement as an alternative to the Kyoto Protocol in December 2015.

The Paris Agreement commits to keeping the global temperature increase to less than 2 °C above pre-Industrial Revolution levels while making efforts to keep it under 1.5 °C higher. In order to achieve this goal, almost no greenhouse gas emissions can be permitted globally in the latter half of the twenty-first century. Thus, it will be necessary to introduce measures in both developed and developing countries in 2020. All countries participate in "deciding voluntary reduction target (NDC: nationally determined contribution) and sending it to the United Nations every five years" and "taking domestic measures for reduction to achieve it" (Mandatory (Article 4 (2), Article 4 (9))).

While the Kyoto Protocol set targets only for developed countries, the Paris Agreement covers almost all countries, including China and India. After submitting targets, implementation of domestic measures is obligatory. However, while the Kyoto Protocol obliged developed countries to achieve legally binding targets, the Paris Agreement is not obligatory, and only voluntary target NDCs are required to be submitted. While the United States emits approximately 15% of global carbon dioxide, it has actively supported efforts of developing countries to combat rising temperatures and sea level and extreme weather, providing funds and technology. The withdrawal of the United States will undoubtedly increase the challenge of achieving the goals of the Paris Agreement, creating a limitation.

3.2.4 History of China's Climate Change Framework

Since China is now the world's largest emitter of CO_2 and the largest developing (emerging) country, it has significant responsibility and an important position in the climate framework. The transition of China's climate measures can be divided into the following four stages:

Stage 1 (until 1997 COP3 Kyoto Conference): Resistance (or observation)

As of 1997, China's CO_2 emissions were 2959 million tons (equivalent to 13% of global emissions, 56% of those of the United States). CO_2 emissions per capita were 2.4 tons (equivalent to 63% of global average, 12% of that of the United States). In China, climate change was classified as a problem in the distant future, and it was felt that developed countries should take the initiative to reduce it (the developed country responsibility theory). Overcoming poverty was prioritized in domestic policy.

Stage 2 (until 1997–2005 Kyoto Protocol came into effect): Learning

Future abnormal weather changes were forecast in the IPCC report. Frequency of abnormal weather events increased in China, such as drought in the Yellow River, flood in the Yangtze River, and worsening desertification. The radicalization of destruction increased scientific awareness of climate change. Furthermore, most measures against global warming, especially CO_2 reduction measures such as energy saving and new energy measures, also had domestic relevance, leading to synergistic benefits from CO_2 reduction. While reducing CO_2, energy structures can be improved, clean and renewable energy introduced, energy security increased, industrial structure and energy saving and utilization efficiency improved, and pollution issues such as acid rain addressed. Synergistic benefits including significant afforestation, limiting population growth, and realization of a low-carbon society are goals that should be pursued by developing and developed countries alike, since their importance in the future is being gradually recognized.

Stage 3 (2005–2015: Adopted Paris Agreement): Cooperation

In 2006, carbon dioxide emissions from China (6004 Mt) surpassed those of the United States (5602 Mt) for the first time, and China became the largest global emitter. However, per capita emissions in 2018 were 6.8 tons, only 43% of the United States' 15.8 tons. China began to be actively involved in the climate framework (international cooperation on CO_2 reduction and domestic actions) while still adhering to the principles of the developed country responsibility theory and "not complying with legal reduction obligations." In particular, in China, it is striving to realize a low-carbon economic society, which is also called a "low-carbon boom."

At COP15 in Copenhagen, Denmark, in 2009, the Chinese government revealed a domestic legally binding voluntary goal: "By the year 2020, CO_2 emissions per GDP will be 40–45% reduced compared to 2005." With this public announcement, countries such as India, Brazil, and Indonesia also declared voluntary goals. Although evaluation of the outcomes of COP15 is divided, the most significant achievement was the announcement of voluntary goals by major developing countries such as China and India.

Most measures against global climate change, especially CO_2 reduction, are also domestic measures and have synergistic effects. China recognized that developing countries should also seek to achieve a low-carbon society as a prerequisite for sustainability. A tendency to use "external pressure" and "internal pressure" of "climate change" emerges as a driving force for national development.

Stage 4 (2015–present): Traction

China achieved its voluntary CO_2 emission reduction target presented at COP15 2 years early. At the end of 2017, CO_2 emissions per GDP decreased by 46% compared to 2005, exceeding the target of a 40–45% reduction by 2020. This climate action did not come at the expense of economic growth; the Chinese economy grew by 1.48 times in 2005–2015, while carbon emissions per GDP decreased by 38.6% (UNFCCC 2018). Besides, as described in Chap. 15, in 2011, China piloted seven carbon emissions trading scheme projects regulating the power generation, steel, and cement manufacturing sectors. By the end of 2017, 200 million tons of emissions were being traded. The National Development and Reform Commission (NDRC) launched a national CO_2 emissions trading system for the power generation sector in December 2017. Under this system, if a company's emissions exceed their allowance, an additional allowance can be purchased from a company with low emissions. The scheme already includes close to 1700 companies with total emissions of over 3 billion tons and is the largest in the world. The scheme will be expanded to other industries in the future. In the field of new energy, as described in Chap. 19, the technology development and introduction of new energy sources in China are world-leading.

Sino–US relations enabled the finalization of the Paris Agreement adopted at COP21 (2015). The US president at the time, Barack Obama, and Chinese president Xi Jinping succeeded in finding a compromise, leading to successful development of the Agreement. Conversely, the current US administration officially notified the United Nations of its withdrawal from the Paris Agreement in November 2019 and has distanced itself from technical and financial cooperation. On the contrary, China promptly clarified its commitment to the Agreement; at the 2017 Davos Conference, President Xi said, "All parties to the Paris Agreement should promote this without turning against it. It is our responsibility to future generations" (Xi 2017).

At the G20 Osaka Summit in 2019, the heads of 19 countries (excluding the United States, which had already announced its withdrawal from the Paris Agreement) reiterated their commitment to fully implement the Paris Climate Convention. During the summit, China, France, and the United Nations successfully held a tripartite meeting on climate change, in which China announced its commitment to fulfill its obligations in the Paris Agreement, accelerate the mobilization of resources to address climate change and promote sustainable development, and cooperate to collectively respond to climate change. The establishment of a low-carbon community is a specific strategy for a collective response to climate change.

China will not replace the role of the United States following its withdrawal from the Kyoto Protocol and the Paris Agreement, but due to China's economic development, responsibility for increasing CO_2 emissions, and frequent abnormal weather, it will actively drive CO_2 emission reductions and global climate measures.

3.3 Concept of East Asian Low-Carbon Community

3.3.1 Background

3.3.1.1 Characteristics and Issues of East Asia

The East Asian region is characterized by an enormous diversity of politics, economy, technology, culture, religion, and ethnicity. In an environmental context, common issues are faced, such as economic development (overcoming poverty), addressing pollution, global climate change, biodiversity loss, ozone depletion, and transboundary air pollution. In regions like East Asia where developed and developing countries coexist, all countries share the burden of environmental degradation, and the need for regional cooperation in resolving environmental problems is becoming increasingly apparent. Problems such as air pollution, acid rain, and climate change do not respect physical or geopolitical boundaries. East Asia faces a trilemma of poverty, pollution, and global environmental issues. Ensuring the security of strategic energy resources based on mutually beneficial international cooperation in East Asia and building policy development scenarios and roadmaps are essential.

The Paris Agreement adopted at COP21 in December 2015 stipulates the goal of "2 °C control, 1.5 °C effort," and the reduction of greenhouse gas emissions from anthropogenic activities to virtually zero in the latter half of this century. By 2030, Japan will reduce CO_2 emissions by 26% compared to 2013, China will reduce CO_2 emissions per GDP by 60–65%, and South Korea will reduce CO_2 emissions by 37% compared to business as usual (BAU). Achieving a low-carbon society to reduce global climate change is a goal in common between developed and developing countries. However, Japan has already achieved the world's highest levels of energy saving and efficiency. Reducing CO_2 further will be costly, and additional dramatic reductions are impossible.

On the other hand, China, with sizeable CO_2 emissions, has a high, cost-effective reduction potential, but has limited capacity. Once the co-benefits of economic growth, pollution control, and low carbonization become apparent, there will be a strong incentive for low-carbon policy development. Other developing countries such as Brazil, India, and Mexico face these common challenges. Tables 3.2, 3.3, and 3.4 show climate measures in Japan, China, and Korea, respectively.

3.3.1.2 Technical Gap and Potential in China, Japan, and South Korea

Figure 3.1 compares CO_2 emissions (CO_2 emission intensity) per real GDP (purchasing power parity (PPP) conversion) among the three countries. From 1971 to 2014, the CO_2 emission intensity of Japan, China, and South Korea differs from each other. The causes of this difference include energy structure, industrial structure, technology level, lifestyle, and exchange rate. For example, if the efficiency of

Table 3.2 Climate change policies in Japan

Year	Event
1997	Launch of the Global Warming Prevention Headquarters
1998	Release of "General principles of the countermeasures for global warming" and "Act on Promotion of Global Warming Countermeasure"
2002	Plan for Global-Warming Prevention revealed Promulgation of the revised "Act on Promotion of Global Warming Countermeasures"
2005	Kyoto Protocol comes into effect. Revision of the "Act on Promotion of Global Warming Countermeasures" comes into effect, and headquarters reestablished in the Cabinet Cabinet decision on Kyoto Protocol target achievement plan
2007	Proposition of Cool Earth 50 (also known as Cool Earth)
2008	Announcement of Cool Earth Promotion Program
2009	Proposal of "Hatoyama Initiative" on Global Warming
2013	Enclosed Policy on Global Warming Countermeasures
2015	Action Policy for Global Warming Countermeasures Based on Paris Agreement Submission of Japan's nationally determined contribution towards post-2020 GHG emission reductions: reduction of 26.0% by fiscal year (FY) 2030 compared to FY 2013 (25.4% reduction compared to FY 2005) (approximately 1.042 billion t-CO_2eq. as 2030 emissions)
2018	Long-term growth strategy based on the Paris Agreement

Table 3.3 Climate change policies in China

Year	Event
1990	Set up National Leading Group on Climate Change, Energy Conservation and Emissions Reduction
1998	State Council adjusted original climate change coordination group and established "National Climate Change Coordination Group" led by former National Development Planning Commission with participation of 13 departments
2006	Established China Clean Development Mechanism Fund and management center
2007	Launched China's National Climate Change Program
2008	Enacted China's Policies and Actions for Addressing Climate Change
2011	China's Policies and Actions for Addressing Climate Change Working plan of GHG emission control for the 12th Five-Year Plan (FYP) Launch of seven pilot carbon emissions trading schemes
2014	China's National Plan on Climate Change (2014–2020)
2015	Submission of China's nationally determined contribution, promising to achieve peak carbon dioxide emissions by 2030 or as soon as possible. By 2030, CO_2 emissions per unit of GDP should fall to 60–65% of 2005 levels
2016	Working plan of GHG emission control for the 13th FYP
2017	Launch of national carbon emissions trading market
2018	China announced that its carbon emission intensity has fallen 45.8% from 2005, achieving its target 2 years ahead of schedule

coal-fired power generation in China can be improved to be higher than that of Japan, the amount of CO_2 reduction would be equivalent to approximately half of Japan's annual emissions. This would also significantly reduce coal consumption in

Table 3.4 Climate change policies in South Korea

Year	Event
1998	Composed Global Warming Countermeasures Committee
1999	First comprehensive countermeasures in response to Climate Change Convention (1999–2001)
2001	Established Framework Committee for Climate Change Treaties
2002	Second comprehensive countermeasures in response to Climate Change Convention (2002–2004)
2005	Third comprehensive countermeasures in response to Climate Change Convention (2005–2007)
2008	Fourth comprehensive countermeasures in response to Climate Change Convention (2008–2013)
2010	Framework Act on Low Carbon, Green Growth, creating legislative framework for mid- and long-term emissions reduction targets
2015	Submission of South Korea's nationally determined contribution (NDC): emission reduction of 37% from the business as usual (BAU) level by 2030
2018	Released draft 2030 roadmap for achieving NDC
2019	Third Energy Master Plan up to 2040 adopted

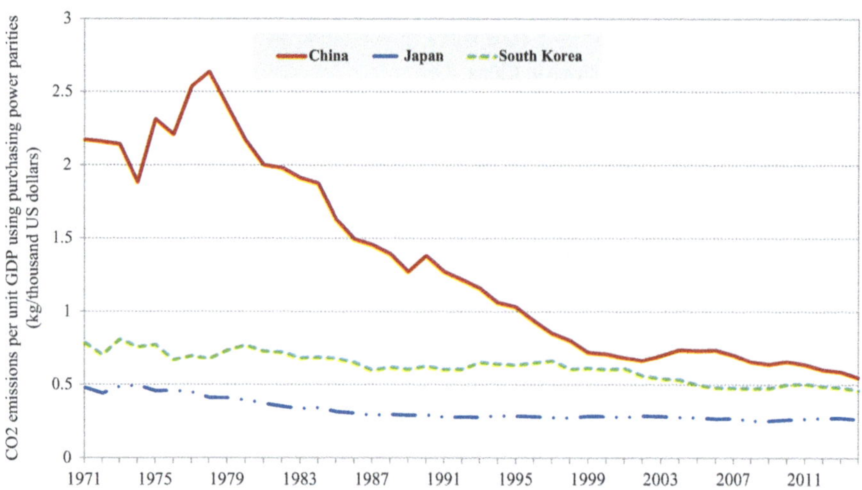

Fig. 3.1 Comparison of CO_2 emissions per unit GDP using purchasing power parities of Japan, China, and South Korea. Source: 2017 EDMC Handbook of Energy & Economic Statistics

China, thereby reducing air pollutants, improving economic efficiency, and promoting industrial development and economic growth in Japan.

However, there is an expiry date for technology, and as shown in Fig. 3.1, the difference in technology between China and Japan shrinks rapidly. Nevertheless, the technological gap among the three countries remains significant; industrial technology should be expanded internationally through industry–government–academia collaboration.

3.3.2 Characteristics of Climate Change and the "CO_2 Problem"

3.3.2.1 Urgency of the Problem

The Paris Agreement established a global long-term goal to control global average temperature rise since the Industrial Revolution to within 2 °C and to strive to control it to 1.5 °C. To achieve this goal, we should "reach the global peak of greenhouse gas emissions as soon as possible" and "achieve a balance between anthropogenic emissions and removal of greenhouse gas sources in the second half of this century," i.e., achieve net-zero global greenhouse gas emissions by the second half of this century.

The concentration of atmospheric carbon dioxide measured in Hawaii on May 3, 2019, exceeded 415 ppm, the highest value in 800,000 years. Abnormal weather events, estimated to be caused by climate change, are occurring worldwide. Measures to combat climate change and prevent the scale of anthropogenic activities from exceeding the limits of the Earth are urgently needed. Action must be taken as quickly as possible; there is no time for debate. The commitments of the Paris Agreement should be fulfilled and the mobilization of resources to tackle climate change and promote sustainable development accelerated. The establishment of a low-carbon community is a specific strategy for a collective response to climate change.

3.3.2.2 Goal Commonality

Realizing a low-carbon society is a common goal of developed and developing countries. Reducing CO_2 and balancing environmental cycles and growth will contribute to achieving sustainable development, as prioritized through the UN Sustainable Development Goals (SDGs) adopted by the UN General Assembly in September 2015. Climate change will affect the world's ability to achieve the SDGs.

3.3.2.3 Specificity of Measures

Technological breakthroughs will be critical for realizing a wide-area low-carbon society while ensuring global economic development. Countermeasure policies consist of the following five pillars:

1. Promotion of global energy conservation
2. Substantial introduction of clean energy (such as conversion between fossil fuels, development and introduction of renewable energy and nuclear power)
3. Development of innovative environmental technology (such as CO_2 separation, collection, storage, recycling)
4. Expansion of CO_2 absorption sources (such as afforestation)

5. Development of innovative energy-related technologies for the next generation (such as space solar power generation, fusion technology)

Based on the characteristics of the CO_2 problem, countermeasure technologies include "no-regret countermeasures" (such as energy saving, afforestation), "minimum-regret countermeasures" (such as fuel conversion, introduction of new energy), and "specialized countermeasures" (such as carbon dioxide capture and storage, CCS). Excluding specialized measures, the technologies contribute to domestic and local improvements in addition to CO_2 reduction effects, such as improved economic efficiency, addressing pollution, and increasing energy security, and are therefore co-beneficial measures. Thus, regardless of whether global climate change occurs or is caused by CO_2, the realization of a low-carbon society is desirable.

3.3.2.4 Difference in Reduction Costs

Reduction costs vary greatly depending on location, time, quantity, and stakeholders. For example, according to the Research Institute of Innovative Technology for the Earth (RITE), the estimated cost of Japan's 2030 NDC marginal CO_2 abatement reduction is very high. The range of marginal reduction costs among national commitment drafts is extensive. In contrast, the NDC target that China is currently submitting has a marginal reduction cost of almost zero and can be achieved without a cost burden (Fig. 3.2) (Akimoto et al. 2017). Cooperation with countries that have small marginal reduction costs of climate change countermeasures is essential for achievement of Paris Agreement reduction targets.

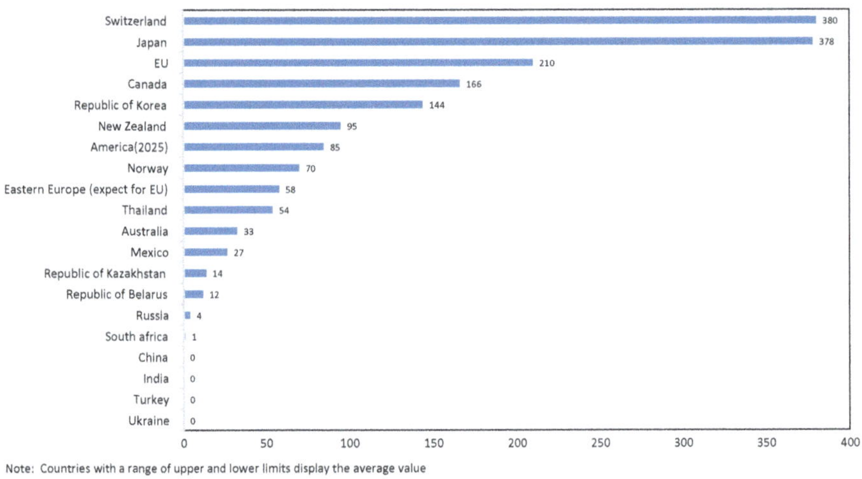

Fig. 3.2 Estimated carbon dioxide marginal abatement cost (US dollars/t-CO2). Note: Average values are displayed for countries for which a range of upper and lower limits are available

3.3.2.5 Specificity of Carbon Dioxide

Greenhouse gases targeted for emission reduction under the Kyoto Protocol include carbon dioxide (CO_2), methane (CH_4), nitrous oxide (N_2O), hydrofluorocarbons (HFCs), perfluorocarbons (PFCs), and sulfur hexafluoride (SF6). The IPCC Fifth Assessment Report estimates that carbon dioxide contributes more than 60% of artificially emitted greenhouse gases (IPCC 2013). Carbon dioxide has the same impact on global climate change, regardless of its source or location. There are not many victims, and it is non-point source pollution, so anyone can be a perpetrator and is also a victim. Damage becomes apparent over a long period of several generations; the damage will continue for a long time, so an international response is urgently required.

3.3.2.6 Scientific Uncertainty of Climate Change Issues

Warming of the climate system has been evident since 1950, and there have been many changes that have not been seen in previous decades to hundreds of years. The IPCC Fifth Assessment Report concludes that human activity is highly likely (95% or greater) to have been the dominant factor in the warming observed since the mid-twentieth century (IPCC 2013). However, the causal relationship of climate change is not fully understood, and the probability of damage is not clear; thus it represents a highly uncertain problem. Therefore, it is necessary to prioritize measures and roadmaps for greenhouse gas reduction that have significant reduction effects, low costs, and low risks and contribute to sustainability. It is crucial to build a wide-area low-carbon society that crosses national borders. The Kyoto Mechanism defined by the Kyoto Protocol is a combination of principles (actual realization of reduction target) and flexibility (efficient realization of reduction target), considering the scientific uncertainty of climate change.

The tenacity and uncertainty of the climate change problem, the characteristics of CO_2 (similar effects regardless of emission location), and the beneficial effects of CO_2 countermeasures will contribute to global sustainability. Also, the beneficial effects will contribute to the development and transfer of innovative technology, the transformation of economic and social systems, and strategic innovation to build an internationally reciprocal, wide-area, low-carbon society that transcends national borders. This will allow the achievement of sustainable development in developing countries. This wide-area low-carbon society is named the East Asian Low-Carbon Community, centered on cooperation among Japan, China, and South Korea. The concept was first proposed in 2008, and subsequent research has sought to bring about its existence (Zhou 2008; Zhou et al. 2010, 2014).

3.3.3 *Framework for the East Asian Low-Carbon Community*

In the climate framework, Japan, South Korea, and China are developed, middle-income, and developing (emerging) countries, respectively. East Asia needs to simultaneously address global issues of climate change and local issues of economic development and pollution control. The concept of the East Asian Low-Carbon Community, as shown in Fig. 3.3, consists of time, space, policies, and outcomes. It has a multilayered structure on four axes (Zhou et al. 2010).

3.3.3.1 First Axis: Generational Equity

There is controversy over balancing intergenerational equity, or sustainable development, and equity within generations, specifically (1) inter-industry, income-layer, and regional equity issues and (2) international equity issues. One reason for equity disputes between nations is the diversity of nations discussing measures to tackle climate change. Relevant circumstances such as the volume of emissions of a nation compared to its population, the degree of economic development, energy structure, and industrial structure must be taken into account.

Atmospheric greenhouse gas concentrations are a global public good. By maintaining concentrations at a reasonable level, a specific range of temperatures will be conserved, benefiting all nations and people, for multiple generations. From an economic point of view, it is difficult to exclude someone from enjoying the

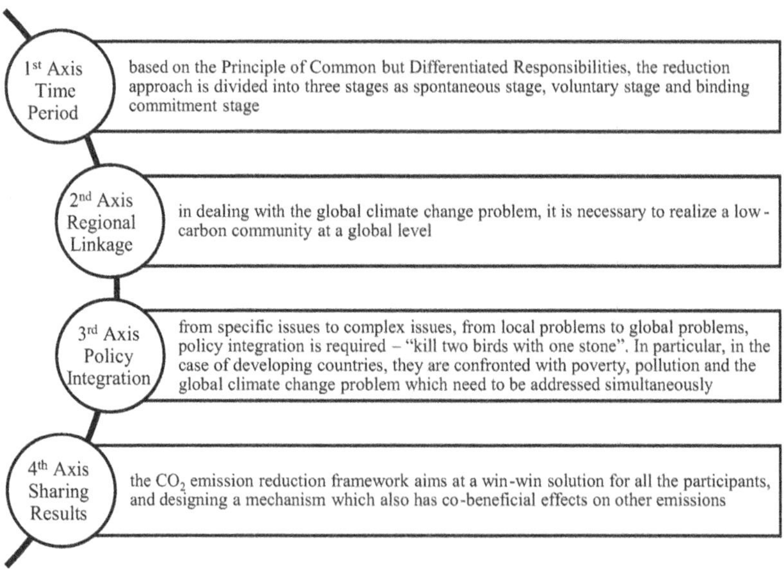

Fig. 3.3 Multilayered structure of the East Asian Low-Carbon Community Initiative. Source: Zhou et al. 2010

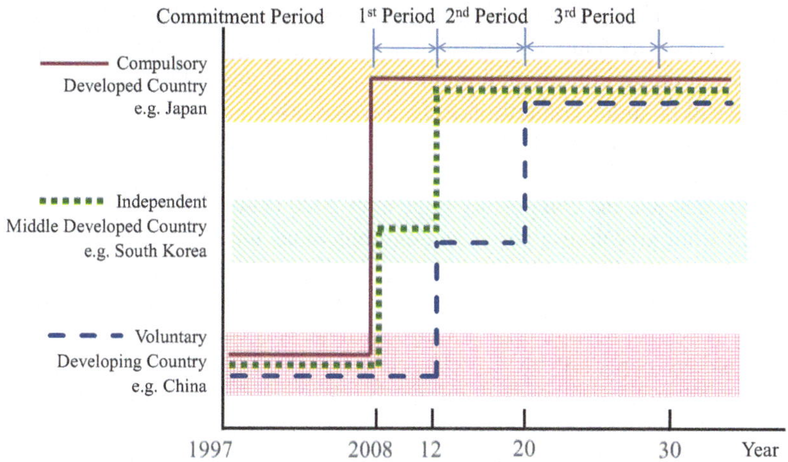

Fig. 3.4 Commitments and targets of different countries under the UNFCCC. Source: Zhou 2004

benefits of greenhouse gas concentrations (non-exclusivity), and the enjoyment of others cannot change the total concentration of greenhouse gases. Since greenhouse gas concentrations are non-competitive, this provides a typical example of a public good, and since it has global spread, it is a global public good. Thus, all humans have universal rights and responsibilities for this global public good that is characterized by non-competitiveness and non-exclusivity. The right to emit CO_2 is equal to all humans; all people have equivalent CO_2 emission rights.

A regulation of equal permissible emission per capita under the total amount is used to equalize per capita emission rights. For example, atmospheric CO_2 concentration in 2100 is the permissible range for global climate change prevention. The permissible amount of per capita emissions divided by the predicted global population gives the total amount of permissible emissions allowing stabilization at the target 450 ppm (approximately twice the amount before the Industrial Revolution) (Zhou 2006).

According to the principle of "common but differentiated responsibilities" stipulated in the Framework Convention on Climate Change, equity must be sought between and among generations. As shown in Fig. 3.4, the world can be divided into three regions: (1) developed countries (such as the United States, Japan), (2) middle-income countries (such as Korea, Mexico), and (3) developing countries (such as China, India). Corresponding to this division, participation in the climate change framework is either compulsory (with legally binding targets, for developed countries such as the United States, EU, Japan) or voluntary (legally binding). Some countries volunteered to set targets, although they were not required to (such as South Korea, Mexico). Developing and emerging countries will take voluntary reduction measures. Division into three stages has been proposed (Zhou 2004, 2006). All countries were in the voluntary stage until 2008. From 2008 to 2012, developed countries and some middle-income countries were in the compulsory stage, and developing countries entered the voluntary stage. From 2013 to 2020, developed countries and some middle-income countries will enter the compulsory stage, and developing countries will enter the voluntary stage. From 2020, almost all countries will be in the compulsory stage (Zhou 2004).

Taking China as an example, as shown in Fig. 3.4, the author predicted that the voluntary stage would last until 2012 and the self-imposed stage would be between 2013 and 2020 and the compulsory stage after 2020 (Zhou 2001, 2004). China was exempted from greenhouse gas reduction obligations until 2012, the first commitment period of the Kyoto Protocol, as stipulated by the Kyoto Convention on the Framework Convention on Climate Change and the Berlin Mandate. Regarding the target for 2013–2020 announced at COP15, China was not legally bound to reduce CO_2 emissions per GDP by 40–45% compared to 2005 by 2020. However, it mandated this voluntary target domestically, incorporating it into implementation of the 12th and 13th Five-Year Plans (2011–2015 and 2016–2020), moving from the voluntary to self-imposed stage. From 2020, China's per capita CO_2 emissions will exceed the global average, and it is expected to have legally binding targets for reduction as in developed countries (Zhou 2001, 2004).

This projection of China's climate response schedule is mostly in line with the Paris Agreement, which will require legally binding commitments from both developed and developing countries from 2020 in order to meet its targets. All countries participated in "deciding voluntary reduction target (NDCs) and sending it to the United Nations every five years" and "taking domestic measures for reduction to achieve it."

Developed countries, including Japan, have sequentially experienced economic development, pollution problems, and global climate change, but China is facing these three challenges simultaneously. Generally, the most urgent environmental problems are domestic problems associated with urbanization, such as air pollution, water pollution, and waste management, and measures against global warming are inevitably of low priority. However, domestic environmental problems facing China and global environmental problems do not conflict. For example, both acid rain and global warming are caused by substances generated by the consumption of fossil fuels, so it is possible to implement measures that contribute to the simultaneous solution of both issues. Since greenhouse gases have a long-term impact, developing countries cannot wait to take effective and appropriate measures; it is necessary to address these three challenges facing China simultaneously.

3.3.3.2 Second Axis: Spatial Cooperation

Solving global climate change is impossible with the efforts of only one country. It is essential to realize a local low-carbon society through urban–rural cooperation, an inter-city low-carbon community through international inter-city cooperation, and a wide-area low-carbon society through transnational cooperation. The term "glocal" has been coined to combine global and local and represents the idea of "think globally, act locally." It is defined as a regional concept, located between "global"

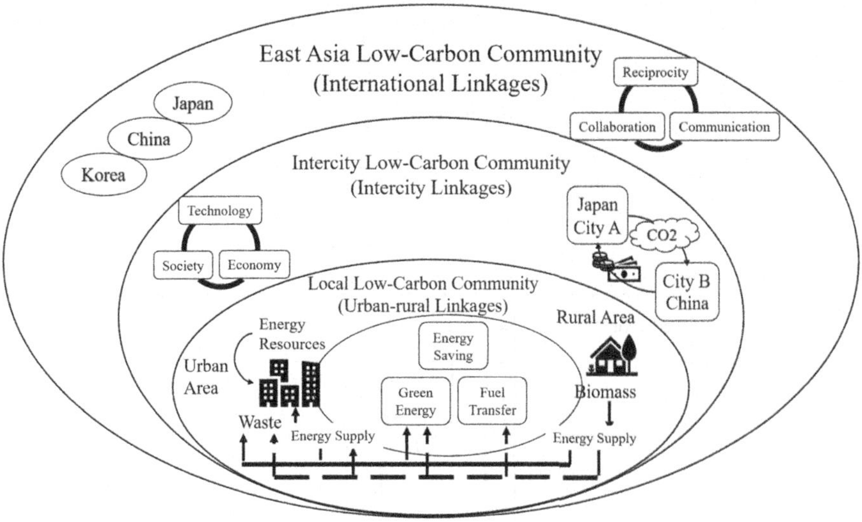

Fig. 3.5 A spatial visualization of the East Asian Low-Carbon Community concept

and "local." This is the spatial category of the "East Asian Low-Carbon Community" concept proposed in this chapter (Fig. 3.5).

Achieving a low-carbon society is a goal shared by developed and developing countries, such as Japan and China. Meanwhile, the Chinese government is mobilizing domestic policies, including the 13th Five-Year Plan, and is working to achieve the goals of the Paris Agreement. China has set a target to reduce primary energy consumption to 5 billion tons of standard coal by 2020. It has an energy-saving rate of 15% and a renewable energy utilization rate in 2020 of 15% of primary energy consumption. In order to achieve these targets, in 2017, a national emissions trading exchange was launched, a regional allocation system for the introduction of renewable energy was introduced, coal consumption has been limited, and natural gas utilization was promoted. China has established measures such as the promotion of electric vehicle development and the reform of electric power liberalization. China has a high reduction potential, and such reductions are cost-effective, but its capacity is limited. Japan has experienced two oil shocks and has accumulated technologies related to energy conservation, high energy efficiency, and renewable energy use. Applying these technologies and expertise in China will contribute to the solution of China's climate change problems and will provide new business opportunities for Japanese companies.

3.3.3.3 Third Axis: Policy Integration

The environmental problems facing society are diversifying and expanding. Many are interrelated and must be solved concurrently. As shown in Fig. 3.6, an integrated multiple-target policy is required to move from individual to complex and from local

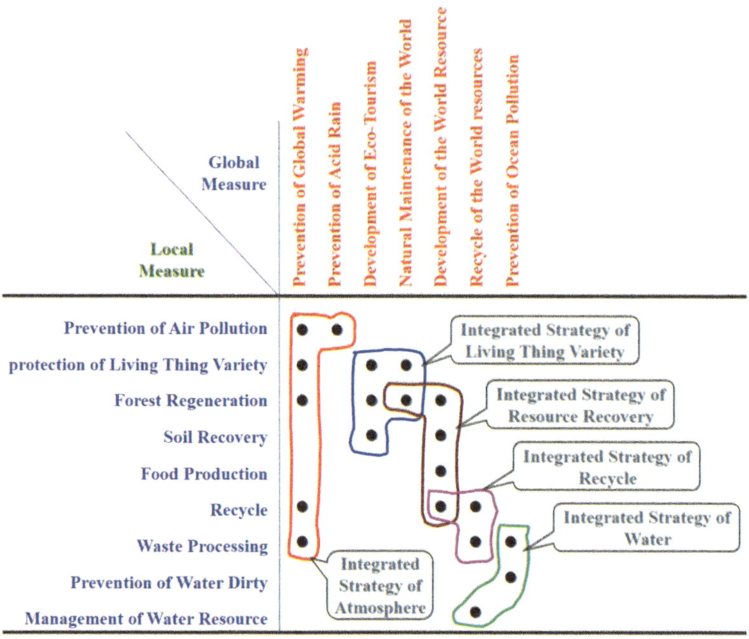

Fig. 3.6 Integration of local and global measures

to global problems. Especially in developing countries, poverty, pollution, and global environmental problems must be simultaneously addressed. For example, the combined economies of Japan, China, and South Korea will increase to account for 25% of the global economy, power consumption will increase to 25%, and CO_2 emissions will increase to 30%. Sulfur oxides (SO_x), dust, and ozone depletion are severe. In addition to global warming prevention measures, integrated atmospheric strategies will take measures to address acid rain and cross-border pollution (air pollution) prevention, biodiversity protection, reforestation, recycling, and waste treatment. It is important to consider measures that integrate local and global environmental measures in the future.

Contemporary social issues have a comprehensive character in which politics, economy, and technology are intricately intertwined over the short- and long-term, at local and global levels. Policies to solve these issues are plural and form a complicated system. From a scientific point of view, the technologies for reducing carbon dioxide discussed above can be divided into the following three types.

The first type is no-regret behavior. Even if climate change is unrelated to carbon dioxide emissions, and even if temperature increases do not occur in the future, there is no regret associated with taking the action. This type of behavior is an inevitable choice to maintain sustainable development, such as tree planting, afforestation, energy saving, and emission reduction. No-regret behavior is the primary strategy to reduce carbon dioxide emissions.

The second type is acts of minimal regret. Such behaviors mainly include the development of new energy such as wind, solar, and biomass. Compared with traditional fossil fuels, the cost of developing new energy is higher, and there are certain risks in development and utilization. However, in order to optimize the energy structure and ensure energy supply sustainability, such actions must be taken, even under risky conditions and even if there is no climate change problem.

The third type is specialization and mainly refers to the capture and storage of anthropogenic carbon dioxide emissions, a method specially devised to deal with climate change. Sequestering this carbon dioxide is only a temporary solution to climate change and has no other benefit. At present, no country has mature technology for this.

Countries across the world are balancing developing economies with overcoming pollution and coping with global climate change, and there is massive potential for international cooperation in the field of environmental energy. The ultimate goal of all countries is to achieve sustainable development. It is proposed here that through bilateral and multilateral cooperation (third-country market cooperation), funds, technology transfer, and capacity improvement can be provided to promote the sharing of benefits and responsibilities across an international wide-area low-carbon community.

Developed countries have sequentially experienced economic development, regional environmental problems (such as regional pollution), and global environmental problems such as global climate change. Developing countries are facing all three simultaneously. Acidification and global warming are prevalent because their main causative substances are due to fossil fuel combustion, and their causes are deeply rooted in modern civilization. It is important to consider measures to integrate local and global environmental measures. Environmental issues are global but cannot be tackled without including local measures. Steady accumulation is essential. In order to realize a low-carbon society, it is crucial to support local measures in order to have developing countries participate.

3.3.3.4 Fourth Axis: Resultant Reciprocity

The framework of a low-carbon community is not a zero-sum game, but a system of win-win games that benefit all parties and co-benefits that contribute to the simultaneous solution of local and global issues. Specific measures must be implemented.

Under the globalized economic system, international competitiveness is a necessary condition for a company to survive. Since a private company owns its technology, technology transfer may reduce international competitiveness on the transferee side and hollow out industry and technology. There is often a lack of appropriate protection of intellectual property rights in developing countries. Many factors hinder technology transfer, such as a lack of systems and financial mechanisms to promote transfer, inadequate systems for promoting technology digestion, and a lack of efficiency. Therefore, market mechanisms should be utilized in order to promote efficient technology transfer and use.

In addition to reducing environmental pollutants such as CO_2 and SO_x, market needs and economic profits of both parties will be considered, selecting projects that share economic and environmental benefits obtained by both parties. With this method, the risk of technology transfer in developed countries and the financial burden for technology transfer in developing countries will be significantly reduced, and the efficiency of energy-intensive fields such as older thermal power plants will be reduced. Remodeling becomes possible, offering many business opportunities to developed countries. In order to promote this type of technology transfer, the two participating governments should have specific policy incentives, such as securing human resources to promote technology transfer and facilitate financial measures (e.g., subsidy system or preferential tax policy). Companies in developed countries will see measures such as the transfer of energy-saving technologies as business opportunities and take proactive measures, for example, by linking the Clean Development Mechanism (CDM) and Energy Service Company (ESCO).

The realization of a low-carbon community concept in East Asia centered on Japan, China, and South Korea will create a sustainable and vibrant international society in which, in addition to measures against global warming, economic development, pollution control, and society are in harmony.

3.4 Conclusion

The realization of the East Asian Low-Carbon Community concept will create a sustainable and vibrant society in which the economy, environment, and society are in harmony, while global climate change countermeasures are promoted (Zhou 2017). In order to materialize this concept, this study has introduced key projects including the development of innovative low-carbon technology and the transfer of existing technology, the presence or absence of new energy and nuclear power generation, low-carbon economy industry, creation and innovation of low-carbon systems such as an East Asian carbon emission trading system, eco-design of energy and material cycles through international collaboration, local low carbonization through collaboration between urban and rural areas, realization of low-carbon recycling through a pilot model project, presenting a roadmap that embodies the transition to a sustainable low-carbon society, making policy recommendations to guide the construction of a low-carbon society in the Asian region, and developing an East Asian low-carbon community integrated evaluation model.

References

Akimoto K, Sano F, Tehrani SB (2017) The analyses on the economic costs for achieving the nationally determined contributions and the expected global emission pathways. Evol Instit Econom Rev 14(1):193–206

IPCC (2013) Climate change 2013: the physical science basis, IPCC AR5 WGI SPM. https://www.ipcc.ch/report/ar5/wg1/. Accessed 23 July 2020

UN (1992) United Nations framework convention on climate change. https://unfccc.int/resource/docs/convkp/conveng.pdf. Accessed 23 July 2020
UNFCCC (2018) China meets 2020 carbon target three years ahead of schedule. https://cop23.unfccc.int/news/china-meets-2020-carbon-target-three-years-ahead-of-schedule. Accessed 23 July 2020
Xi J (2017) Share the responsibility of the times and promote global development. World Economic Forum 2017 annual meeting, Davos, 17 January 2017
Zhou W (2001) Participation problems of developing countries in climate change framework—a case study of China. J Pol Sci 9:1–20
Zhou W (2004) China's participation problem in the climate change framework. J Energ Resour Soc 25(6):1–6
Zhou W (2006) How developing countries can engage in GHG reduction: a case study for China. Sustain Sci 1:115–122
Zhou W (2008) Towards the realization of wide-area low-carbon society: the East Asia Low-Carbon Society. Environ Conserv Eng 37:642–646
Zhou W, Nakagami K, Su X, Ren H (2010) Policy framework and evaluation model of the East Asia Low-Carbon Community. Environ Conserv Eng 39:536–542
Zhou W, Su X, Qian X (2014) Study on the introduction of CO_2 emissions trading system for realizing East Asian Low-Carbon Community. J Pol Sci 8:85–97
Zhou W (2017) East Asia low-carbon community concept and its realization. In: Shindo E, Zhou W (eds) Open the way for East Asian Collaboration. Hanadensha, Tokyo, pp 105–118

Chapter 4
Modeling the East Asian Low-Carbon Community

Xuanming Su and Weisheng Zhou

Abstract In order to realize the East Asian Low-Carbon Community, a comprehensive energy and economic evaluation model must be developed. The model can be used to formulate future emission reduction plans and quantitatively evaluate the economic and environmental impacts of an East Asian Low-Carbon Community, including the effects of economic and technological measures and international cooperation. This chapter develops an integrated energy and economic evaluation model—the Glocal Century Energy Environment Planning (G-CEEP) model. First, we introduce the model framework and explore the modeling method, including the two-level constant elasticity of substitution (CES) production function, technological learning, the macroeconomic model, environmental evaluation, and the objective function. Using the G-CEEP model, we project a business-as-usual (BAU) scenario and a CO_2 mitigation scenario to 2050, for China, Japan, and Korea. Then we evaluate co-benefit effects, considering SO_2 and NO_x. Finally, a policy perspective based on the analysis is provided.

4.1 G-CEEP Model

4.1.1 Model Framework

The G-CEEP model is a large-scale nonlinear integrated planning model that has been used to analyze a low-carbon economy among China, Japan, and Korea within five-year periods from 2005 to 2100 (Su et al. 2010, 2012a). The model consists of

X. Su
Research Institute for Global Change/Research Center for Environmental Modeling and Application/Earth System Model Development and Application Group, Japan Agency for Marine-Earth Science and Technology (JAMSTEC), Yokohama, Japan
e-mail: suxuanming@jamstec.go.jp

W. Zhou (✉)
College of Policy Science, Ritsumeikan University, Ibaraki, Osaka, Japan
e-mail: zhou@sps.ritsumei.ac.jp

W. Zhou et al. (eds.), *East Asian Low-Carbon Community*,
https://doi.org/10.1007/978-981-33-4339-9_4

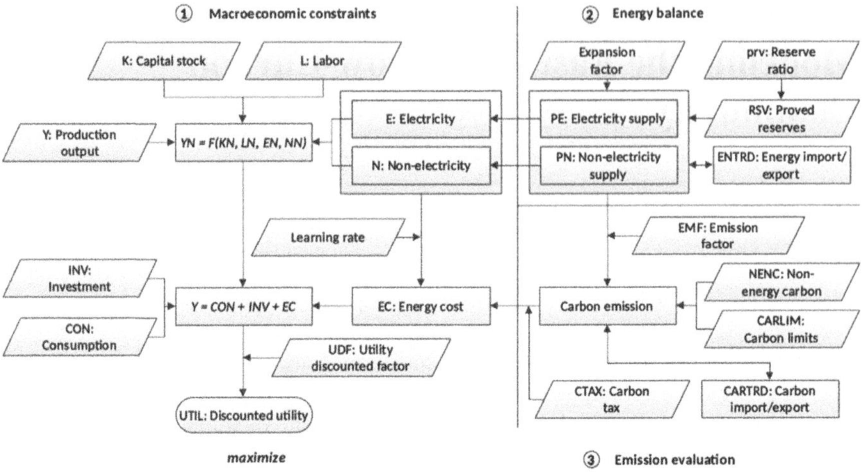

Fig. 4.1 Flow chart of Glocal Century Energy Environment Planning (G-CEEP) model

three sub-models: macroeconomic, energy balance, and environmental. The production output in the macroeconomic sub-model is the sum of consumption, investment, and energy system cost. Investment is determined by the initial investment and annual growth rate. The relationship of the production output, capital stock, population, and consumption of energy for electricity and non-electricity is expressed by a two-level CES production function (Su et al. 2012b). The key constraints in the energy balance sub-model are the energy system cost constraint and energy supply and demand balance. Depletion of fossil fuels, such as coal, oil, and natural gas, and the annual available renewable energy are considered as strict constraints. The environmental sub-model is used to calculate the relative energy and industrial emissions according to the emission factors under specific scenarios. Details are shown in Fig. 4.1.

4.1.2 Model Equations

4.1.2.1 Two-Level CES Production Function

Most macroeconomic energy models or energy models with macroeconomic descriptions are based on the two-level CES production function. The GREEN model nests capital and energy with low substitution elasticity, and this aggregation is combined with labor through a higher elasticity of substitution (Burniaux et al. 1991). The GLOBAL 2100 model uses nested capital and labor against energy (Manne and Richels 1992). We follow the macroeconomic model with a two-level CES production function proposed by Manne and Richels (1992), which adds energy as factor inputs in the second level of the traditional two-input CES production function.

In the G-CEEP model, energy consumption is further divided into that for electricity (EN) and non-electricity (NN). Thus, there are four inputs: capital stock (KN), labor (LN), electricity, and non-electricity. The two-level CES production function can be written as:

$$\mathrm{YN} = \gamma\left[\delta(\delta_1 \mathrm{KN}^{-\rho_1} + (1-\delta_1)\mathrm{LN}^{-\rho_1})^{\frac{\rho}{\rho_1}} + (1-\delta)(\delta_2 \mathrm{EN}^{-\rho_2} + (1-\delta_2)\mathrm{NN}^{-\rho_2})^{\frac{\rho}{\rho_2}}\right]^{-\frac{1}{\rho}} \tag{4.1}$$

where YN is the output; $0 < \delta, \delta_1, \delta_2 < 1$ are distribution parameters; and ρ (ρ_1, ρ_2) ≥ -1 are substitution parameters. The elasticity of substitution is specified as:

$$\sigma = \frac{1}{1+\rho} \tag{4.2}$$

where σ is a constant dependent on the substitution parameter, ρ (ρ_1, ρ_2). When estimated ρ (ρ_1, ρ_2) is less than -1, the elasticity of substitution σ is negative, which is a theoretical impossibility (Prywes 1986). If substitution elasticity $\sigma < 1$, the two related input factors can be interpreted as complements. When $\sigma > 1$ the two related input factors are substitutes.

The G-CEEP model describes the relationship between capital stock, labor, and energy with a two-level CES production function. This introduces nonlinearity to the large-scale optimization model, which increases computational complexity but renders the model more integrated and economically meaningful. In addition to the two-level CES production function, another nonlinearity, technological learning, is introduced.

4.1.2.2 Technological Learning

Technological learning is widely used as a key factor to analyze the decreasing cost of novel technologies. Although the introduction of endogenous technological change will inevitably increase computational complexity due to nonlinearity associated with the learning curve, we have tried to incorporate the technological learning method in order to analyze the prospects of specific technological options and the related potential cost decreases.

The single factor learning curve is expressed as the following (Kahouli-Brahmi 2008):

$$C_t(Q_t) = a_t Q_t^{-\alpha} \tag{4.3}$$

where t is a specific technology, C_t denotes the cost per unit of production, a_t is the cost of the first unit produced, Q_t is cumulative production, and α is a parameter incorporating "learning by doing." The parameter a is determined by a given point on the learning curve, commonly the initial point:

$$a_t = \frac{C_{t0}}{(Q_{t0})^{-\alpha_t}} \tag{4.4}$$

Thus, the progress rate is given by:

$$p_t = 2^{-\alpha_t} \tag{4.5}$$

The learning rate is given by:

$$l_t = 1 - 2^{-\alpha_t} = 1 - p_t \tag{4.6}$$

The progress rate is the rate at which the cost declines when cumulative production doubles and is sometimes considered as the slope of the learning curve.

4.1.2.3 Macroeconomic Model

Application of the macroeconomic sub-model in the G-CEEP model is based on the two-level CES production function with four inputs: capital stock, labor, electricity, and non-electricity. Input values can be obtained by the following equations.

Production output and new production output:

$$Y_{t,r} = Y_{t-1,r} \times (1-\mu)^5 + \mathrm{YN}_{t,r} \tag{4.7}$$

Labor force and new labor force:

$$L_{t,r} = L_{t-1,r} \times (1-\mu)^5 + \mathrm{LN}_{t,r} \tag{4.8}$$

Electricity and new installed electricity:

$$E_{t,r} = E_{t-1,r} \times (1-\mu)^5 + \mathrm{EN}_{t,r} \tag{4.9}$$

Non-electricity and new non-electricity:

$$N_{t,r} = N_{t-1,r} \times (1-\mu)^5 + \mathrm{NN}_{t,r} \tag{4.10}$$

New capital:

$$\mathrm{KN}_{t,r} = 5/2 \times \left[(1-\mu)^5 I_{t,r} + I_{t+1,r}\right] \tag{4.11}$$

Capital and new capital:

$$K_{t,r} = K_{t-1,r} \times (1-\mu)^5 + \mathrm{KN}_{t,r} \tag{4.12}$$

Two-level CES production function:

$$\mathrm{YN}_{t,r} = f(\mathrm{KN}_{t,r}, \mathrm{LN}_{t,r}, \mathrm{EN}_{t,r}, \mathrm{NN}_{t,r}) \tag{4.13}$$

Cost function:

$$Y_{t,n} = C_{t,n} + I_{t,n} + \mathrm{EC}_{t,n} \tag{4.14}$$

Terminal condition:

$$K_T \times (\omega + \mu) \le I_T \tag{4.15}$$

where t, r denotes period t and region r and T is the final period in the model; YN, KN, LN, EN, and NN represent new production output, new capital stock, new labor, new electricity, and new non-electricity, respectively; μ is defined as the annual depreciation rate where its exponent represents the number of years (5) per period; and ω is annual growth rate. New capital depends on annual investment I; $\mathrm{YN}_{t,r}$ equation is the two-level CES production function; C is consumption; and EC is energy cost.

To ensure that the rate of investment is adequate to provide for replacement and net growth of capital stock during subsequent periods, a terminal constraint (Eq. 4.15) is applied at the end of each planning horizon (Svoronos 1985).

4.1.2.4 Environmental Evaluation

The environmental evaluation output is used to calculate emissions of CO_2, SO_2, and NO_x under specific scenarios, through emission factors of different energy sources subjected to relevant environmental policy constraints.

Energy emission factors vary regionally in accordance with sectors of energy consumption, as well as fossil fuel characteristics. Emissions are defined as the volume of energy consumption multiplied by the emission factors.

The CO_2 emissions in a specific period, region, and sector can be estimated using the following equation:

$$E_{CO_2,\mathrm{trec}} = Q_{\mathrm{trec}} \mathrm{EF}_{CO_2,\mathrm{rec}} \tag{4.16}$$

where $E_{CO_2,trec}$ is the volume of CO_2 emissions in a specific period, region, and sector; Q_{trec} is the volume of energy consumption in a specific period, region, and sector; and $EF_{CO_2,rec}$ is the CO_2 emission factor for a specific region and sector.

The limit of annual CO_2 emissions is defined as follows:

$$\begin{aligned} &\sum\nolimits_{e\in ET}(\mathrm{cece}_{t,r,e} \times \mathrm{PE}_{t,r,e}) + \sum\nolimits_{e\in NT}(\mathrm{cecn}_{t,r,e} \times \mathrm{PN}_{t,r,e}) - \mathrm{CIM}_{t,n} \\ &\quad + \mathrm{CEX}_{t,n} \\ &\quad \leq \mathrm{CLIM}_{t,n} + \mathrm{NENC}_{t,n} \end{aligned} \tag{4.17}$$

where $\mathrm{cece}_{t,r,e}$ and $\mathrm{cecn}_{t,r,e}$ are carbon emission coefficients for electricity and non-electricity, respectively; $\mathrm{PE}_{t,r,e}$ and $\mathrm{PN}_{t,r,e}$ are the supply of electricity and non-electricity, respectively; and $\mathrm{CIM}_{t,n}$ and $\mathrm{CEX}_{t,n}$ denote carbon import and export during carbon trading. The constant on the right-hand side, $\mathrm{CLIM}_{t,n}$, is the carbon limit in region r, and $\mathrm{NENC}_{t,n}$ represents non-energy uses of some fossil fuels.

SO_2 emissions are estimated according to the following equation:

$$E_{SO_2,trec} = 2Q_{trec}S_{trec}\alpha_{s,trec}(1 - R_{trc}) \tag{4.18}$$

where $E_{SO_2,trec}$ is the volume of SO_2 emissions in a specific period, region, and sector; Q_{trec} is the volume of energy consumption in a specific period, region, and sector; S_{trec} is the sulfur content in a specific period, region, and sector; $\alpha_{s,trec}$ is the SO_2 emission factor for a specific period, region, and sector; and R_{trc} is the desulfurization rate for a specific period, region, and sector. A coefficient of 2 is used because the atomic weight of SO_2 is twice that of S.

NO_x emissions are calculated using the following equation:

$$E_{NO_x,trec} = Q_{trec} \cdot \mathrm{EF}_{ec} \cdot (1 - \mathrm{RE}_{tc}) \cdot (1 - \mathrm{DE}_{tc} \cdot \mathrm{PR}_{tc}) \tag{4.19}$$

where $E_{NO_x,trec}$ denotes NO_x emissions; Q_{trec} is energy consumption; EF_{ce} is the emission factor of NO_x; RE_{tc} is the efficiency of NO_x emission reductions; DE_{tc} is the efficiency of equipment to remove NO_x; and PR_{tc} is the popularization ratio of equipment to remove NO_x. Here, $\mathrm{DR}_{tc} = \mathrm{DE}_{tc} \cdot \mathrm{PR}_{tc}$ is the rate of removal of NO_x.

4.1.2.5 Objective Function

The objective of this model is to maximize discounted consumption. The output production is the sum of consumption, investment, and energy cost. When considering the impact of environmental taxes on the energy system, environmental taxes are considered to be part of the energy cost; discounted consumption is maximized, and the levels of each decision variable under the optimized state are identified, as shown in the following equation:

$$\mathrm{UTIL} = \sum_{r=1}^{R} \sum_{t=1}^{T} \left(5 \times \prod_{\tau=0}^{t-1} (1 - \mathrm{udr}_\tau)^5 \times \log\left(C_{\mathrm{t,r}}\right) \right) \tag{4.20}$$

where UTIL is discounted consumption; C is annual consumption; and udr is the utility discount rate.

4.2 Scenario Analysis of Carbon Reductions

4.2.1 Macroeconomic and Emission Reduction Assumptions

The G-CEEP model draws on a series of databases related to macroeconomic assumptions, energy production and associated technologies, and environmental emissions data. Energy data for the initial year come from the International Energy Agency (IEA 2011a, b). Energy-related emission factors are also estimated from the IEA (2010).

Population and gross domestic product (GDP) data are adopted from Shared Socioeconomic Pathway (SSP) 2 (O'Neill et al. 2014, Fricko et al. 2016). The SSPs are designed for future climate change assessment and include five representative scenarios. SSP1 is an ideal sustainable scenario with low challenges to both mitigation and adaptation, while SSP2 represents intermediate challenges with moderate greenhouse gas emissions (GHGs) and is considered to be a balanced scenario. Other scenarios indicate different levels of mitigation and adaptation challenges. For example, SSP3 has great challenges to both mitigation and adaptation, whereas adaptation challenges dominate in SSP4 and mitigation challenges in SSP5. Here, we choose SSP2 to represent business-as-usual (BAU), considering moderate socioeconomic assumptions. According to these assumptions, the global population will increase to approximately 9.2 billion by the middle of this century and decline to 8.7 billion by the end of the century, while global GDP will increase to about 400 trillion USD (2005) per year by the end of the century. All data are shown in Fig. 4.2.

We use nationally determined contributions (NDCs) under the Paris Agreement as mid-term emission reduction targets. The latest emission reduction proposals submitted to the United Nations Framework Convention on Climate Change (UNFCCC) show that China hopes to lower CO_2 emissions per unit of GDP by 60 to 65% from 2005 levels by 2030. Japan plans to reduce GHG emissions by 26.0% by fiscal year (FY) 2030 compared to FY 2013. South Korea has pledged to reduce emissions by 37% from BAU levels, which is projected to be 850.6 $MtCO_2$-eq. GHG emissions by 2030 (UNFCCC 2020). We assume the same GHG reduction rates after 2030 compared with BAU scenarios for each region. A relatively detailed introduction can be seen in Chap. 14.

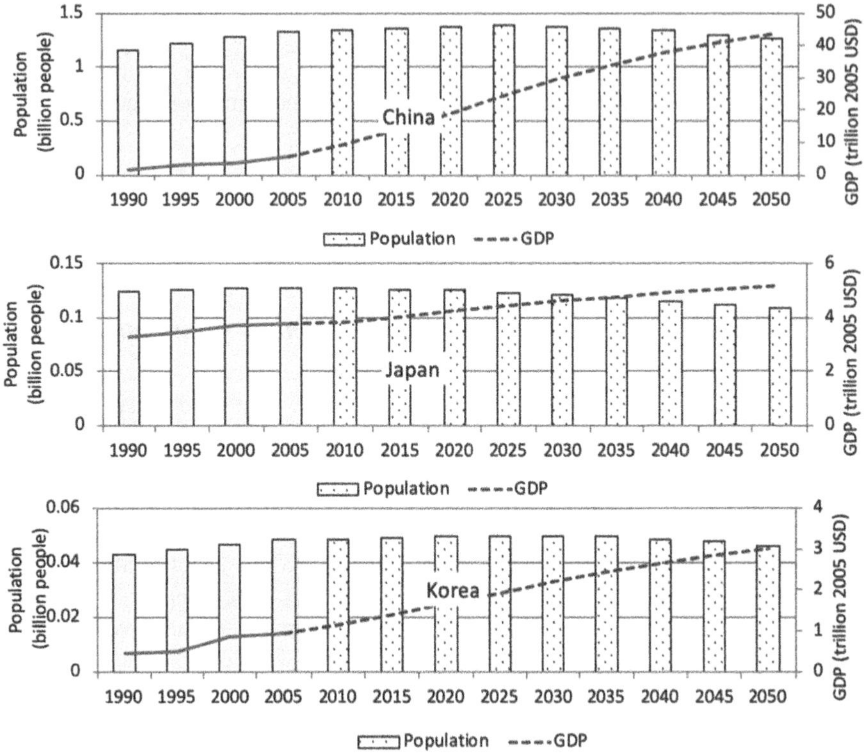

Fig. 4.2 Projections of population and gross domestic product (GDP)

4.2.2 Future Energy Projections and Carbon Mitigation

4.2.2.1 Energy Projections

Using the population and GDP assumptions illustrated above, first, the BAU scenario is generated, which is based on SSP2. As the emission reduction target of Korea in 2030 is based on the BAU scenario, we define the carbon emission constraint for Korea subject to the BAU scenario with reduced carbon emissions. The carbon emission constraints for China and Japan are defined according to historical data. Thus, the target scenario (TAR) is generated.

The primary energy projections in the reference scenarios are given in Fig. 4.3. The primary energy consumptions of the BAU and TAR scenarios are given in Fig. 4.4. The primary energy projections of reference scenarios for China and Korea are expected to grow significantly by the middle of this century, while that of Japan remains fairly stable due to relatively low growth rates of population and GDP. Total primary energy consumption in the BAU scenario will reach 201.7, 23.2, and 15.9 exajoules for China, Japan, and Korea, respectively, in 2030. Coal remains the most important fossil fuel for China and will account for 58.0% of energy consumption in

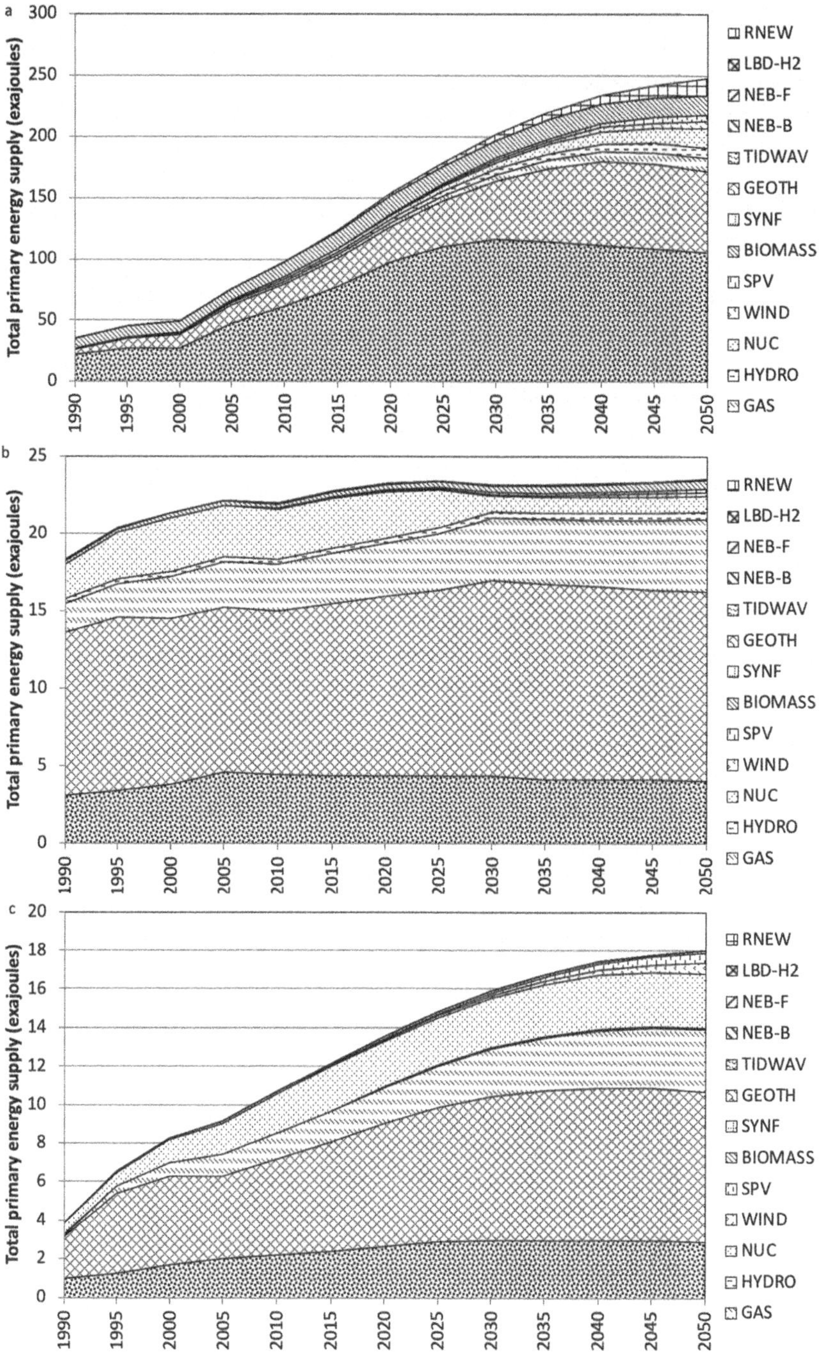

Fig. 4.3 Primary energy projection in reference scenario for (**a**) China, (**b**) Japan, and (**c**) Korea. *GAS* gas-fired electric, *HYDRO* hydroelectric, *NUC* nuclear electric, *WIND* wind electric, *SPV* solar photovoltaics, *BIOMASS* biomass electric, *SYNF* coal-based synthetic fuels, *GEOTH* geothermal electric, *TIDWAV* tide and wave electric, *NEB-B* non-electric backstop from biomass, *NEB-F*

2030 in the BAU scenario. Oil and natural gas will account for 23.1 and 2.8%, respectively. Despite attempts in China to increase nuclear energy, the installed capacity is currently relatively small; thus its share in total primary energy consumption will be 2.8%. Hydropower and other renewable energy consumption will account for 13.4%. Primary energy consumption changes only slightly for Japan in the BAU scenario, because of a relatively low economic growth rate. Coal, oil, and natural gas will account for 19.0, 54.5, and 17.3%, respectively, of total primary energy consumption by 2030. Nuclear will account for 4.5% with 1.0 EJ in 2030. Hydropower and other renewable energy will account for approximately 4.7%. The total primary energy consumption of Korea will increase by 17.5% from 2020 to 2030 in the BAU scenario, and coal, oil, and natural gas will account for 18.6, 46.9, and 15.7%, respectively, in 2030. Nuclear power will account for 15.9% and hydropower and other renewable energy approximately 2.9%.

Carbon emissions are reduced through switching carbon-intensive fossil fuels for fuels with lower carbon intensity or carbon-free fuels. In 2030, coal consumption in China will decrease from 58.0% in the BAU scenario to 45.2% in the TAR scenario. Oil will increase from 23.1% in BAU to 26.0% in TAR. Natural gas will increase from 2.8% in BAU to 3.5% in TAR. Nuclear will shift from 2.8% in BAU to 3.0% in TAR, and hydropower and renewable energy will shift from 13.4% in BAU to 22.2% in TAR. In Japan, energy sources are apt to shift to cleaner fossil fuel (natural gas) or renewable energy due to the wide use of various emission abatement technologies. In 2030, coal consumption in Japan will decrease from 4.4 EJ in the BAU scenario to 3.5 EJ in the TAR scenario. Oil will decrease from 54.5% in BAU to 41.4% in TAR. Natural gas will increase from 17.3% in BAU to 23.3% in TAR. Nuclear will shift from 4.5% in BAU to 5.6% in TAR, and hydropower and renewable energy will shift from 4.7% in BAU to 9.0% in TAR. In 2030, coal consumption in Korea will decrease from 3.0 EJ in the BAU scenario to 1.5 EJ in the TAR scenario. Oil will decrease from 46.9% in BAU to 34.3% in TAR. Natural gas will increase from 15.7% in BAU to 21.4% in TAR. Nuclear will shift from 15.9% in BAU to 21.5% in TAR. Hydropower and renewable energy will shift from 2.9% in BAU to 10.2% in TAR.

4.2.2.2 GDP Losses

GDP losses from the BAU scenarios are given in Fig. 4.5, with 1.8, 1.5, and 2.4% losses for China, Japan, and Korea, respectively, in the TAR scenario in 2030. The GDP loss caused by carbon reductions is affected by total energy cost, which is considered as a proportion of GDP. In China, despite reducing emission intensity to 65% of 2005 levels, the total carbon emission reduced from the BAU scenario in

---

Fig. 4.3 (continued) non-electric backstop from coal, *LBD-H2* hydrogen from carbon free source, *RNEW* other renewables

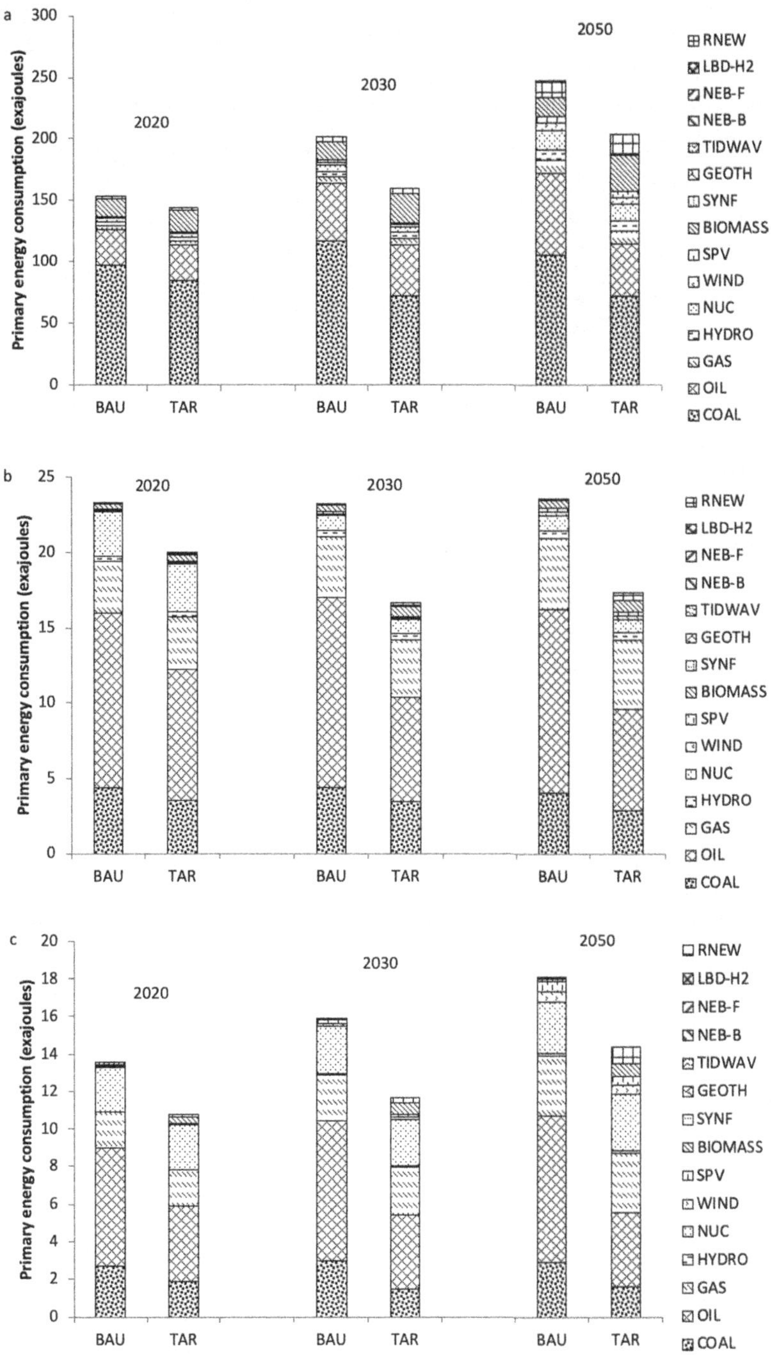

Fig. 4.4 Primary energy consumption in business-as-usual (REF) and target (TAR) scenarios for (**a**) China, (**b**) Japan, and (**c**) Korea

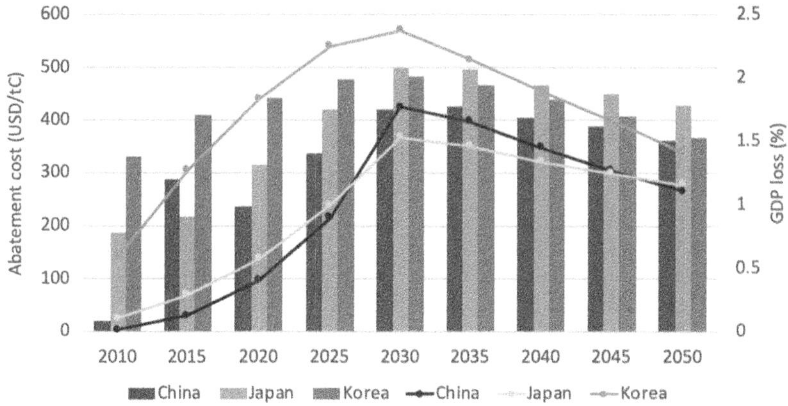

Fig. 4.5 Gross domestic product (GDP) loss from reference scenario if target scenario is implemented

Table 4.1 Major indicators for China

Indicators	2020	Scenarios in 2030	
		BAU	TAR
Population (millions)	1.38	1.38	1.38
GDP (billions in 2005 USD)	19,111.5	29,709.7	29,183.4
Total primary energy consumption (TPEC, MTOE)	3658.3	4816.0	3819.7
Energy intensity (TOE per thousand USD in 2005)	0.19	0.16	0.13
Coal share in TPEC (%)	63.5%	58.0%	45.2%
Oil share in TPEC (%)	18.8%	23.1%	26.0%
Natural gas share in TPEC (%)	2.5%	2.8%	3.5%
Nuclear share in TPEC (%)	2.0%	2.8%	3.0%
Carbon emissions (billion tons carbon)	3.0	3.9	2.6
Carbon intensity (ton carbon per thousand USD in 2005)	0.16	0.13	0.09

2030 is 20.7%. The energy cost is relatively high in Korea, and GDP loss will be up to 2.4% in 2030. Korea has an ambitious carbon abatement target, and thus its energy costs as a proportion of GDP will be high (den Elzen et al. 2011).

4.2.2.3 Major Indicators

Major indicators are summarized in Tables 4.1, 4.2, and 4.3. The energy intensity of China will decrease from 0.19 TOE per thousand USD (2005 USD, similarly hereinafter) in 2020 to 0.16 TOE per thousand USD in 2030 in the BAU scenario and decline to 0.13 TOE per thousand USD in 2030 in TAR. There is a 65% reduction in carbon intensity in China in the TAR scenario from 2005 levels. Energy intensity in Japan will shift from 0.13 TOE per thousand USD in 2020 to 0.12 TOE per thousand USD in 2030 in BAU and decrease to 0.09 TOE per thousand USD in

Table 4.2 Major indicators for Japan

Indicators	2020	Scenarios in 2030	
		BAU	TAR
Population (millions)	0.125	0.120	0.120
GDP (billions in 2005 USD)	4240.2	4628.2	4557.5
Total primary energy consumption (TPEC, MTOE)	556.0	553.5	397.8
Energy intensity (TOE per thousand USD in 2005)	0.13	0.12	0.09
Coal share in TPEC (%)	18.8%	19.0%	20.7%
Oil share in TPEC (%)	49.8%	54.5%	41.4%
Natural gas share in TPEC (%)	14.7%	17.3%	23.3%
Nuclear share in TPEC (%)	12.9%	4.5%	5.6%
Carbon emissions (billion tons carbon)	0.38	0.41	0.27
Carbon intensity (ton carbon per thousand USD in 2005)	0.09	0.09	0.06

Table 4.3 Major indicators for Korea

Indicators	2020	Scenarios in 2030	
		BAU	TAR
Population (millions)	0.049	0.050	0.050
GDP (billions in 2005 USD)	1651.6	2197.1	2144.9
Total primary energy Consumption (TPEC, MTOE)	323.2	379.6	277.8
Energy intensity (TOE per thousand USD in 2005)	0.20	0.17	0.13
Coal share in TPEC (%)	19.8%	18.6%	12.7%
Oil share in TPEC (%)	46.7%	46.9%	34.3%
Natural gas share in TPEC (%)	13.9%	15.7%	21.4%
Nuclear share in TPEC (%)	17.6%	15.9%	21.5%
Carbon emissions (billion tons carbon)	0.22	0.25	0.15
Carbon intensity (ton carbon per thousand USD in 2005)	0.13	0.12	0.07

TAR in 2030. Compared to 2020, carbon intensity in Japan has no significant change in 2030 for BAU, while it declines to 0.06 tC per thousand USD in TAR. For Korea, energy intensity will shift from 0.20 TOE per thousand USD in 2020 to 0.17 TOE per thousand USD in 2030 in BAU and decline to 0.13 TOE per thousand USD in TAR in 2030. Carbon intensity decreases from 0.13 tC per thousand USD in 2020 to 0.12 tC per thousand USD in 2030 in BAU and decreases to 0.07 tC per thousand USD in TAR in 2030.

4.3 Co-benefit Effects on SO_2 and NO_x Emissions

4.3.1 Background Information

Transition to a low-carbon community is a common goal in both developed and developing countries. Using an energy model is an efficient way to analyze this challenging issue. Air pollutants, such as SO_2 and NO_x, are released in various industrial processes. These form acid rain and have significant impacts on human health. As both these air pollutants, as well as carbon emissions, derive from fossil fuel consumption, carbon abatement strategies usually achieve two aims simultaneously: carbon emissions are reduced and air pollutants are also alleviated. In order to measure the degree of alleviation of SO_2 and NO_x emissions under different abatement strategies, we use the G-CEEP model to evaluate the effects of each abatement strategy on SO_2 and NO_x emissions in China, Japan, and Korea.

Efforts to reduce CO_2 emissions will also reduce SO_2 and NO_x. The emission sources are fossil fuels, although emission volumes from consuming one unit of heat value vary for different fuels. SO_2 usually comes from consumption of a variety of coal products, as both energy and non-energy sources, including derived coal and coke. The consumption of oil products is the second major source of SO_2 emissions. A small amount of SO_2 comes from the consumption of natural gas and biomass. NO_x mainly comes from the consumption of oil products, especially gasoline and other light fractions of oil, medium distillates (diesel, light fuel oil), and heavy fuel oil. The consumption of coal and natural gas releases a small amount of NO_x, though this is non-negligible if a large volume of energy is consumed, such as in China where coal is the major energy source.

Reducing SO_2 and NO_x emissions usually relies on abatement technologies or related environmental policies. Nurrohim and Sakugawa (2004) discussed the possible impacts of greenhouse gas abatement measures on NO_x and SO_2 emissions from manufacturing industries in Hiroshima Prefecture. Their predictions showed that NO_x and SO_2 emissions might decrease from 18.9 and 21.5 kt in 1990 to 15.1 and 14.2 kt in 2010, respectively, if Japanese energy policy and targets in voluntary action plans were successfully implemented. Chang et al. (2006) investigated the impact of strengthening vehicle emission regulations on economic activities, focusing on the economic impact of reducing sulfur content in the diesel fuel quality standard. Graus and Worrell (2007) give an overview of the effects of SO_2 and NO_x pollution control on the energy efficiency of fossil-fired power generation in several countries, distinguishing national levels of desulfurization and denitrification. Lu et al. (2010) estimated annual SO_2 emissions in China after 2000 using a technology-based method, showing that the trend in estimated SO_2 emissions in China is consistent with trends in SO_2 concentration and the pH and frequency of acid rain in China, as well as with increasing trends in background SO_2 and sulfate concentration in East Asia. We will discuss the possible impacts of CO_2 abatement measures on SO_2 and NO_x emissions. The CO_2 abatement targets here use the NDC reduction proposals for 2030 in the Paris Agreement. Furthermore, we discuss the co-benefit effects of carbon taxes on SO_2 and NO_x emissions, respectively, in order to construct marginal abatement cost curves for SO_2 and NO_x emissions.

4.3.2 Results and Discussions

4.3.2.1 Scenario Study of CO_2, SO_2, and NO_x Emissions

CO_2, SO_2, and NO_x emissions resulting from energy consumption in 2030 are shown in Fig. 4.6. China plans to reduce the intensity of CO_2 emissions per unit of GDP in 2030 by 60–65% compared with 2005 levels. Here, we use 65% as the target abatement level. If compared with the BAU scenario in 2030, CO_2 emissions will be reduced by 32.5% in the TAR scenario. Simultaneously, SO_2 and NO_x emissions in 2030 will be reduced by 33.2% and 25.6%, respectively, in the TAR scenario compared with BAU. In Japan, CO_2 emissions in 2030 will be reduced by 34.6% in TAR compared with BAU, while SO_2 and NO_x emissions will be reduced by 24.1% and 33.8%, respectively. However, the absolute abatement values are relatively small, removing 0.13 Mt. SO_2 and 0.37 Mt. NO_x from the BAU scenario in 2030. The CO_2 reduction target for Korea in 2030 is to reduce 37% of CO_2 emissions compared with the BAU scenario. If this emission target is achieved, SO_2 emissions will be reduced by 44.3%, and NO_x will be reduced by 38.8% compared with BAU.

Efforts to reduce CO_2 emissions will result in saving energy, switching among fossil fuels, and switching from fossil fuels to renewables or energy sources with low or no emissions. This will also reduce SO_2 and NO_x emissions. The co-benefit effects of carbon reduction vary by country because of different energy consumption structures and different existing removal rates. If the existing removal rate is low, carbon abatement efforts will have greater reduction effects on SO_2 and NO_x, as in China and Korea. Otherwise, co-benefit effects are reduced, as for co-benefit reduction of SO_2 and NO_x in Japan. The existing removal rates of SO_2 and NO_x are relatively high in Japan, and carbon reduction has a limited impact on the total emissions of SO_2 and NO_x.

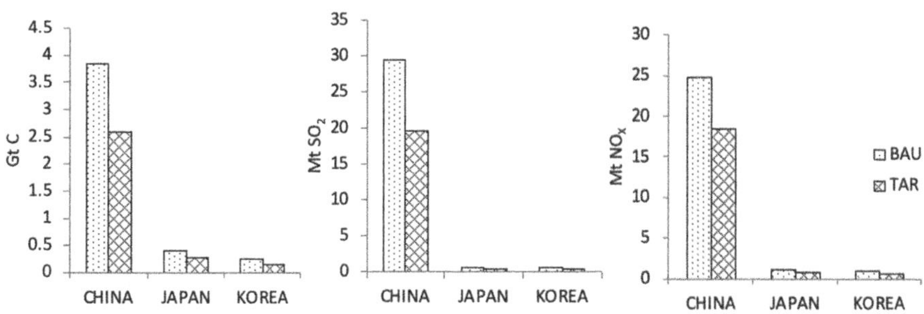

Fig. 4.6 Scenarios of CO_2, SO_2, and NO_x emissions in 2030 under business-as-usual (BAU) and target (TAR) scenarios

4.3.2.2 Marginal Abatement Cost

The marginal abatement cost reflects the cost of one additional unit of carbon emission that is abated. To plot the marginal abatement cost curve, we simulate the introduction of progressively higher carbon taxes from 0 to 300 USD (2005)/tC and record the volume of reduced carbon emissions. The reduced emissions of SO_2 and NO_x are also recorded in order to analyze the co-benefit effects of carbon taxes. The marginal abatement cost of carbon emission reduction and co-benefit effects for SO_2 and NO_x are shown in Fig. 4.7.

According to the analysis, carbon emissions in China mainly come from the consumption of coal, which accounts for 58.0% of the BAU scenario in 2030. To reduce carbon emissions, coal consumption will decrease as carbon taxes increase from 0 to 200 USD/tC in China. If carbon taxes continue to increase, then new low-carbon technology for coal consumption will be introduced. For example, more pulverized coal power plants will be built, since the carbon emission rate of pulverized coal power is 0.20 kg C/kWh, which is lower than that of existing coal power (0.28 kg C/kWh). The generating cost of pulverized coal power is 40 millions/ kWh, which is far higher than that for existing coal power (20.3 millions/kWh).

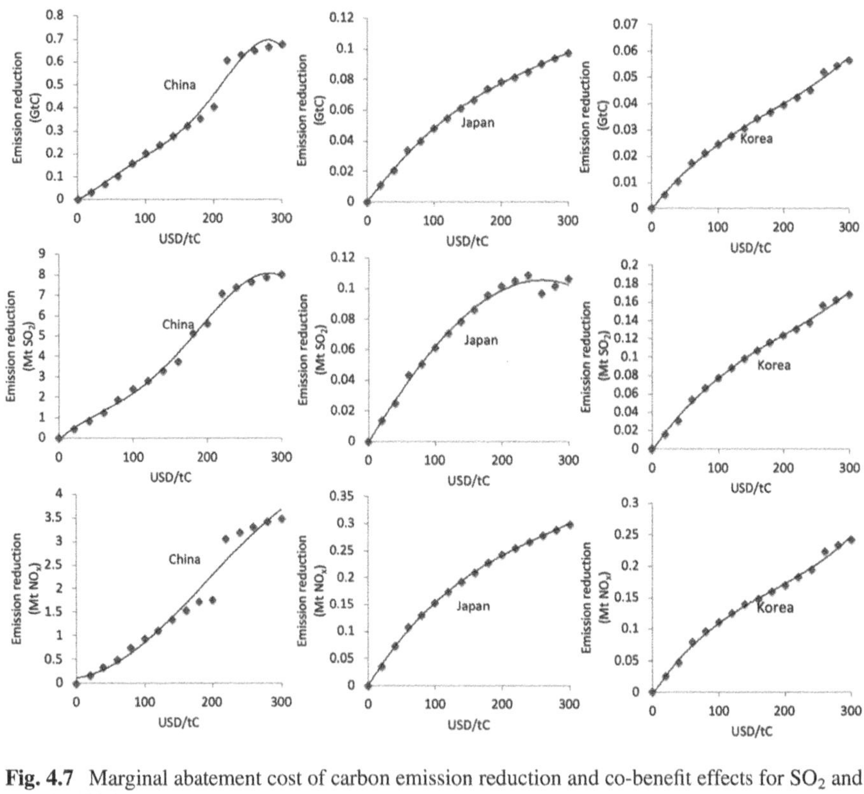

Fig. 4.7 Marginal abatement cost of carbon emission reduction and co-benefit effects for SO_2 and NO_x in 2030

Carbon abatement will become easier if carbon taxes are above 200 USD/tC, but SO_2 abatement will be less affected; SO_2 emissions may even increase to a certain extent because of an expected expansion in coal consumption. NO_x emissions in China are also mainly from coal consumption, especially hard coal. In addition, oil products are a large source of NO_x emissions. Thus, the carbon abatement effort will have a critical effect on NO_x abatement in China. Carbon taxes have co-benefit effects on SO_2 and NO_x abatement in Japan. SO_2 abatement will benefit less than that of NO_x from carbon taxes, as existing desulfurization technologies or related industry processes with low SO_2 emissions are widely used in Japan. Also, coal consumption will only account for 19.0% of total primary energy consumption in 2030 in the BAU scenario; the change in coal consumption will not be as significant as in China. For Korea, carbon taxes have a similar effect on carbon, SO_2, and NO_x abatement, which means that a similar percentage of emissions will be abated by imposing carbon taxes.

4.4 Conclusion

In this chapter, a G-CEEP model is used to provide an approach to analyze possible future energy consumption structures, co-benefit effects of carbon abatement strategies on SO_2 and NO_x emissions, and elucidate a roadmap to implement carbon emission targets, in order to identify suggestions for policymakers.

We analyzed two scenarios deriving from NDC proposals: BAU and TAR. We project primary energy consumptions; emissions of CO_2, SO_2, and NO_x; and associated GDP by 2050 with respect to the two scenarios, for China, Japan, and Korea. The results show that coal remains the most important fossil fuel in China and will account for 58.0% of energy consumption in 2030 in the BAU scenario, deceasing to 45.2% in the TAR scenario in 2030. Hydropower and renewable energy in 2030 will shift from 13.4% in BAU to 22.2% in TAR. Carbon intensity is reduced by 65% in the TAR scenario in 2030, compared to 2005 levels, as suggested in China's NDC proposal. Primary energy consumption changes little in Japan in the BAU scenario, due to relatively low economic growth. The share of coal, oil, and natural gas will account for 19.0, 54.4, and 17.3%, respectively, of total primary energy consumption by 2030 in BAU, shifting to 20.7, 41.4, and 23.3%, respectively, in TAR in 2030. Carbon intensity in 2030 also decreases from 0.09 tC per thousand USD in BAU to 0.06 tC per thousand USD in TAR. The total primary energy consumption of Korea will increase by 17.5% from 2020 to 2030 in the BAU scenario, and carbon intensity in 2030 will decrease from 0.12 tC per thousand USD in the BAU to 0.07 tC per thousand USD in the TAR. GDP losses from the BAU scenarios will be 1.8, 1.5, and 2.4 for China, Japan, and Korea, respectively, in the TAR scenario in 2030.

This chapter also provides an analysis of co-benefit effects under carbon abatement strategies. Carbon taxes play an important role in reducing carbon emissions and also emissions of SO_2 and NO_x, as the latter mainly derive from fossil fuel consumption. Co-benefit effects vary depending on energy consumption structures, switching among fossil fuels or to non-fossil fuels, and energy utilization technologies. Increased carbon taxes lead to the introduction of new energy technologies, which may have different effects on carbon, SO_2, and NO_x abatements in different countries. Existing emissions with high removal rates will be less affected by carbon taxes. In this case, SO_2 and NO_x emissions can feasibly be further reduced by relying on CO_2 abatement technology.

References

Manne AS, Richels RG (1992) Buying greenhouse insurance. The MIT Press, Cambridge, MA

Burniaux J-M, Martin J, Nicoletti G, Martins JO (1991) GREEN a multi-sector, multi-region general equilibrium model for quantifying the costs of curbing CO2 emissions: a technical manual. France

Chang HJ, Cho GL, Kim YD (2006) The economic impact of strengthening fuel quality regulation-reducing sulfur content in diesel fuel. Energ Pol 34:2572–2585

den Elzen MGJ, Hof AF, Beltran AM, Grassi G, Roelfsema M, van Ruijven B et al (2011) The Copenhagen Accord: abatement costs and carbon prices resulting from the submissions. Environ Sci Pol 14(1):28–29

Fricko O, Havlik P, Rogelj J, Klimont Z, Gusti M, Johnson N et al (2016) The marker quantification of the shared socioeconomic pathway 2: a middle-of-the-road scenario for the 21st century. Glob Environ Chang 42:251–267. https://doi.org/10.1016/j.gloenvcha.2016.06.004

Graus WHJ, Worrell E (2007) Effects of SO_2 and NO_x control on energy-efficiency power generation. Energ Pol 35:3898–3908

IEA (ed) (2010) World Energy Outlook 2010. International Energy Agency Press, Paris

IEA (ed) (2011a) Energy statistics of non-OECD countries, 2011 edition. International Energy Agency Press, Paris

IEA (ed) (2011b) Energy statistics of OECD countries, 2011 edition. International Energy Agency Press, Paris

Kahouli-Brahmi S (2008) Technological learning in energy–environment–economy modelling: a survey. Energ Pol 36(1):138–162. https://doi.org/10.1016/j.enpol.2007.09.001

Lu Z, Streets DG, Zhang Q, Wang S, Carmichael GR, Cheng YF, Wei C, Chin M, Diehl T, Tan Q (2010) Sulfur dioxide emissions in China and sulfur trends in East Asia since 2000. Atmos Chem Phys 10:6311–6331

Nurrohim A, Sakugawa H (2004) A fuel-based inventory of NO_x and SO_2 emissions from manufacturing industries in Hiroshima Prefecture, Japan. Appl Energ 78:355–369

O'Neill BC, Kriegler E, Ebi KL, Kemp-Benedict E, Riahi K, Rothman DS et al (2014) The roads ahead: Narratives for shared socioeconomic pathways describing world futures in the 21st century. Glob Environ Chang. https://doi.org/10.1016/j.gloenvcha.2015.01.004

Prywes M (1986) A nested CES approach to capital-energy substitution. Energ Econom (1):22–28

Su X, Ren H, Zhou W, Mu H, Nakagami K (2010) Study on future scenarios of low-carbon Society in East Asia Area, part 1: development of glocal century energy and environment planning model and case study. Pol Sci 17(2):85–96

Su X, Zhou W, Ren H, Nakagami K (2012a) Co-benefit analysis of carbon emission reduction measures for China, Japan and Korea. Pol Sci 19(2)

Su X, Zhou W, Nakagami K, Ren H, Mu H (2012b) Capital stock-labor-energy substitution and production efficiency study for China. Energ Econom 34(4):1208–1213. https://doi.org/10.1016/j.eneco.2011.11.002

Svoronos (1985) Duality theory and finite horizon approximations for discrete time infinite horizon convex programs. Department of Operations Research, Stanford University
UNFCCC (2020) Nationally determined contributions (NDCs) interim registry. https://www4.unfccc.int/sites/ndcstaging/Pages/Home.aspx

Part II
Urban-Rural Linkage for Low-Carbon Community

Chapter 5
Realizing a Local Low-Carbon Society Through Urban-Rural Linkage

Hongbo Ren, Weisheng Zhou, and Xuepeng Qian

Abstract The urban area is always the hot spot of academic research relating to low-carbon cities. However, because the energy consumed is mainly fossil fuels and the stock of natural resources is limited in the urban areas, even though the renewable energy resources are exhausted, the potential of local CO_2 emissions is marginal. In addition, the urban area usually has higher energy demand density than the rural area which is mainly dominated by the residential demands. On the other hand, from the viewpoint of supply side, the energy infrastructures are well maintained and can be enhanced even more in the urban area, while in the rural area, a large amount of distributed renewable energy resources (e.g., biomass, wind, solar energy) with less or no carbon emissions are not well explored. Therefore, in order to construct a low-carbon city, especially for the cities with relatively low urbanization ratio, it may be beneficial to develop a cooperative framework between the urban and rural areas.

5.1 Characteristics and Importance of Urban-Rural Linkage

In order to mitigate the impacts of climate change, development of low-carbon societies is receiving increasing attention in both developed and developing countries. In such a society, various innovations are applied to technology, institutions, and individual behavior. To outline progress toward this ideal, mid- and long-term

H. Ren
College of Energy and Mechanical Engineering, Shanghai University of Electric Power, Shanghai, China

W. Zhou (✉)
College of Policy Science, Ritsumeikan University, Ibaraki, Osaka, Japan
e-mail: zhou@sps.ritsumei.ac.jp

X. Qian
College of Asia Pacific Studies, Ritsumeikan Asia Pacific University, Beppu, Oita, Japan
e-mail: qianxp@apu.ac.jp

W. Zhou et al. (eds.), *East Asian Low-Carbon Community*,
https://doi.org/10.1007/978-981-33-4339-9_5

targets and scenarios have been promoted by several countries. Reduction of net CO_2 emissions by 2050 to 50% of current levels has been recognized as a global goal. With this in mind, countries such as the United Kingdom and Japan have announced national goals and pathways to achieve them (Fujino et al. 2008; Kevin et al. 2008; NIES 2008; Sarah et al. 2008). To allow clearer elucidation of a realistic long-term vision and contribute to international climate negotiations in the post-Kyoto period, mid-term targets for 2020 have been proposed by the United States, the United Kingdom, Germany, France, and Japan. These countries seek to reduce their CO_2 emissions by 7, 26, 40, 20, and 25%, respectively, from 1990 levels.

As the main source of CO_2 emissions, energy systems feature heavily in the mid-term and long-term pathways mentioned above. Thus, the development of a low-carbon energy system, through penetration of energy efficiency and utilization of renewable energy sources, will play a key role in achieving a low-carbon society.

While global coordination is necessary to understand the macro level, local actions cannot be neglected if policies are to be effective. In order to realize a low-carbon society at the local level, regional characteristics (e.g., natural resources, climate conditions) must be accounted for. To achieve the same low-carbon goal, different regions may require different configurations of technologies, institutions, and lifestyle changes. Many local municipalities have unveiled climate change mitigation goals and plans for the mid- to long-term. Since April 2008, 13 cities in Japan with varying local conditions have been selected as Environmental Model Cities, functioning as pilot demonstrations of actions that could be taken to achieve low-carbon development.

Although an integrated action plan has been promoted in many cities, the two main components of a municipality (urban and rural areas) are always treated separately, especially with respect to energy systems. Urban areas usually have higher energy demand density than rural areas; the latter are dominated by residential demands. From the supply side, energy infrastructure is well maintained and can be extensively enhanced in urban areas, while in rural areas renewable energies including biomass, wind, and solar energy may be enriched.

According to this analysis of the characteristics of urban and rural areas in terms of both supply and demand, feasible technical options are proposed for various types of users (residential, commercial, industrial) in both areas, as shown in Table 5.1. Most currently available high-efficiency thermal power generation and renewable technologies have been considered in the context of developing a low-carbon society through innovation of local energy systems. For residential customers, solar heaters, photovoltaic (PV) systems, micro combined heat and power (micro CHP) plants, and micro wind turbines are preferred. For industrial customers, because of high energy requirements, large-scale CHP plants and biomass energy systems are suitable. For the same type of end-user, urban and rural areas may introduce different technologies because of differences in energy stock and infrastructure as described above.

Figure 5.1 shows the concept of a local energy system based on urban-rural cooperation. Regional circulation of energy and resources (e.g., biomass) allows development a systematic partnership between urban and rural areas. Rural areas have abundant renewable energy resources, while urban areas are centers of energy

Table 5.1 Feasible technical options for different types of local energy system users

User type	Rural area	Urban area
Residential	Solar heater Photovoltaic (PV) system Micro combined heat and power (CHP) (liquefied petroleum gas (LPG), biomass) Micro wind turbine	Solar heater PV system Micro CHP (city gas) Micro wind turbine
Commercial	Micro CHP Solar heating system PV system	PV system CHP District heating and cooling Mega solar heating system
Industrial	Livestock biomass plant (CHP) Wood biomass plant (CHP) Agriculture biomass plant (CHP) Micro hydropower Mega solar heating system	CHP system Heat interconnecting network between factories Sewage sludge biomass plant (CHP) Food biomass plant (CHP)

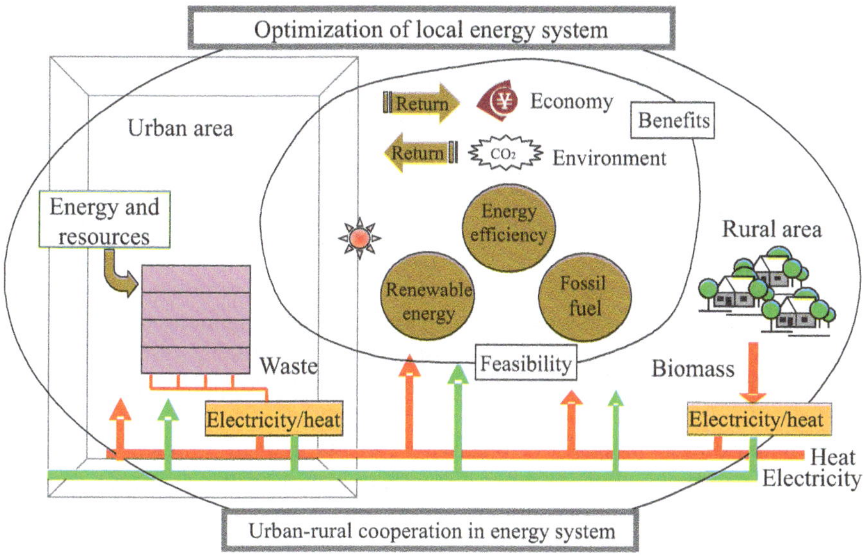

Fig. 5.1 Concept of local energy system based on urban-rural cooperation

consumption. By developing a cooperative framework, two-way flow of energy, resources, revenue, and emissions can be achieved between urban and rural areas, bridging economic and environmental gaps through rational system design.

5.2 Necessity and Potential of Distributed Energy Resources

A distributed energy resource (DER) system refers to the utilization of on-site energy sources to provide electricity and other energy to one or more buildings or facilities. Generally, DER is installed for one or more of the following applications (Huang et al. 2007; Kari and Arto 2006; Soderman and Pettersson 2006):

1. Overall load reduction: reducing overall electricity consumption by using highly efficient power generators in lieu of grid-purchased power.
2. Energy independence: using on-site power generation to meet all energy needs (usually to improve power reliability and quality).
3. Standby power: using a generator as a backup electricity source to ensure power availability during grid outages.
4. Peak shaving: using on-site generation intermittently to avoid purchasing grid power during expensive peak time. Peak shaving also refers to using on-site generation during periods of maximum electricity consumption, expressly for lowering the energy demand component of a given billing period (applies only for tariff structures with demand charge).
5. Net energy sales (net metering): generating more electricity than needed and selling the surplus to the grid.
6. Combined heat and power (CHP): using waste heat from a power generator directly (e.g., manufacturing processes, space heating, water heating) or through a thermally activated device (e.g., absorption chillers, dehumidifiers, bottoming cycles).
7. Grid support: installed by power companies to support transmission, distribution, and feeder systems for a wide variety of reasons, including meeting highest peak loads without having to overbuild infrastructure, postponing system upgrades, maintaining power quality, and maintaining uninterruptible power during planned outages.
8. Premium power: mitigates or otherwise corrects frequency variations, voltage transience, surges, dips, or other disruptions from grid power, which can trip sensitive digitally controlled motors, drives, and computer systems.

Table 5.2 shows the characteristics of various DER installation types, of which low costs and fixed maintenance are the most important. In contrast, thermal output and emissions are considered less important by various applications. Although environmental issues are increasingly important, economics is the most prioritized factor when introducing distributed energy systems. Therefore, we focus mainly on the economic aspects of distributed energy systems. Nevertheless, as awareness of environmental problems increases, environmental aspects will become more important and are sometimes transferred to economic indexes through the introduction of carbon taxes, for example.

Figure 5.2 illustrates an example of a distributed energy system. Generally, a DER system is composed of energy production devices, local energy consumption,

Table 5.2 Characteristics of distributed energy resource (DER) application types

Application	Low cost	High efficiency	Thermal output	Emission	Start-up time	Fixed maintenance	Variable maintenance
Continuous power	◐	●	○	◐	○	◐	●
CHP	◐	●	●	◐	○	◐	●
Peaking	●	◐	○	○	◐	●	◐
Green	◐	◐	◐	●	○	◐	◐
Emergency	●	○	○	○	●	●	○
Standby	●	○	○	○	◐	●	○
True premium	◐	◐	○	◐	●	◐	◐
Peaking T&D deferral	●	○	○	○	◐	●	○
Baseload T&D deferral	◐	●	◐	◐	○	◐	●
Spinning/Non spinning reserve	◐	◐	○	○	●	◐	◐
Reactive power	◐	◐	○	◐	◐	◐	◐
Voltage	◐	◐	○	◐	◐	◐	◐
Local area security	●	○	○	○	◐	●	○

Key: ● Important characteristic
◐ Moderately important or important in certain applications
○ Relatively unimportant

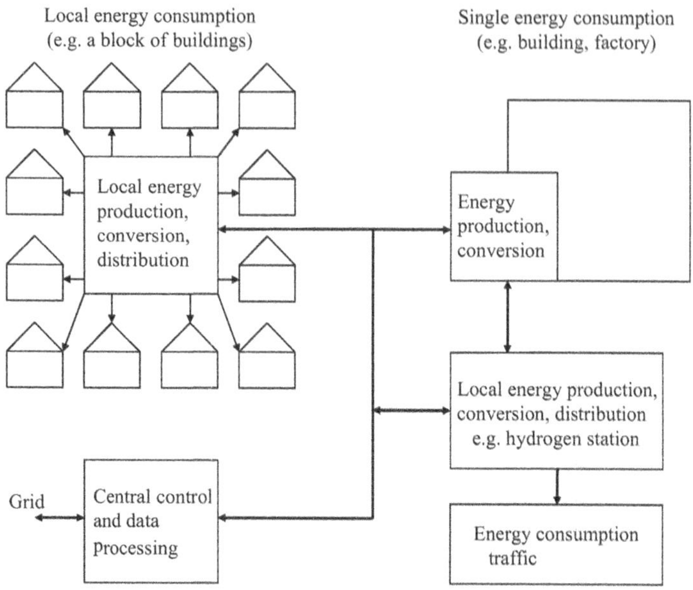

Fig. 5.2 Representation of a distributed energy system

control and processing devices, and the utility grid. Such a system can be beneficial to both consumers and the energy utility, if system integration is properly engineered. For the utility, DER can help to avoid concerns about transmission and distribution upgrades and under certain conditions can provide voltage support and stabilize the distribution network. Benefits to the customers or end-users include power quality and reliability, peak shaving, choice, and potentially lower energy costs.

Compared with a traditional central energy supply, DER can utilize a wide range of power generators, including biomass-based generators, combustion turbines, concentrating solar power and PV systems, fuel cells, wind turbines, micro turbines, diesel generator sets, hybrid systems, and electrical power storage, as well as thermal recovery technologies. Figure 5.3 gives an overview of the current status of the most common distributed generation technologies, grouped into thermal and renewable generation technologies.

CHP refers to the simultaneous production of electricity and heat or cooling at or near the point of consumption. It provides many benefits compared to separate heat and power production, including increased energy efficiency, operating cost savings, and reduced air pollution and contribution to global warming. There are additional benefits for industry, including increased reliability and power quality and higher productivity. The electric power industry and its customers can also benefit when industrial CHP capacity is used to support and optimize the overall power grid.

Conventional power plant systems create a great amount of heat while converting fuel into electricity. For an average central utility power plant, approximately two

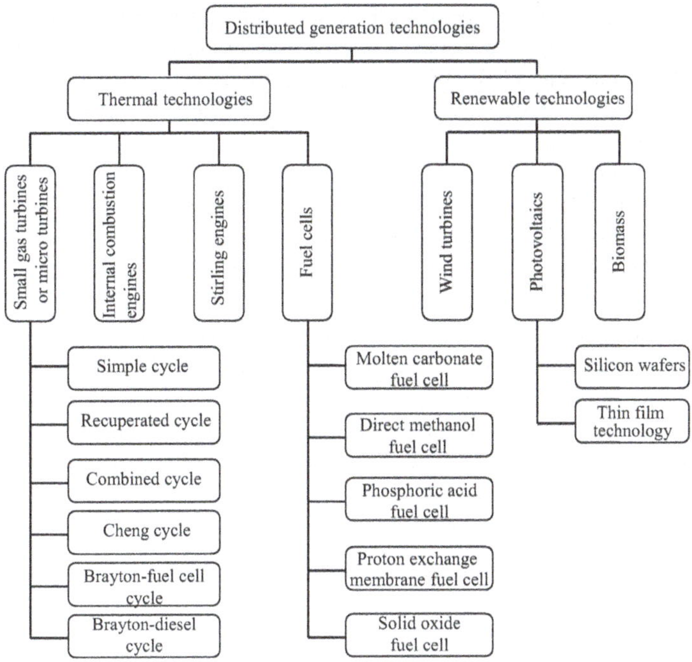

Fig. 5.3 Distributed technologies for power generation

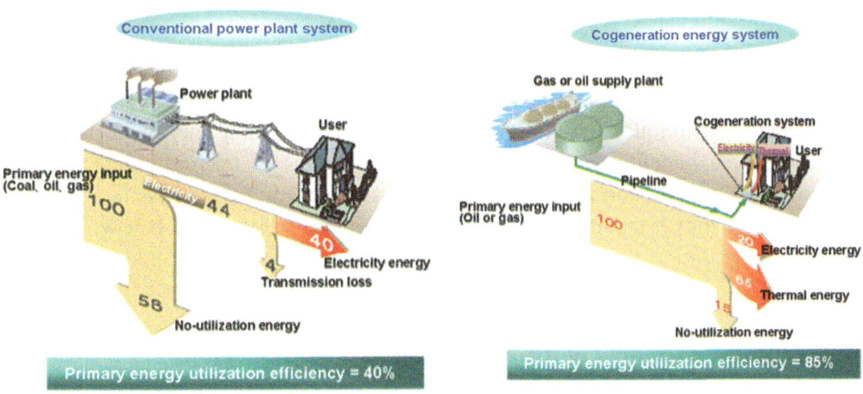

Fig. 5.4 Comparison of conventional power plant with combined heat and power (CHP) system

thirds of the energy content of the input fuel is converted to heat and subsequently wasted. As an alternative, an end-user with significant thermal and power needs can simultaneously generate both its thermal and electrical energy in a CHP system located at or near its facility. CHP can significantly increase the efficiency of energy utilization, as shown in Fig. 5.4.

Figure 5.4 shows that a typical CHP system can reduce energy requirements by close to 45% compared to separate production of heat and power. For every 100 units of input fuel, CHP converts 85 units to useful energy, of which 20 are electricity and 65 are produced as steam or hot water. Conventional separate heat and power production generates approximately 40 units of electrical energy from 100 units of input fuel.

By increasing energy efficiency, CHP also significantly reduces emissions of criteria pollutants such as NO_x and SO_2 and non-criteria greenhouse gases such as CO_2. CHP can provide environmental benefits as part of an economically attractive investment; the technology can significantly reduce emissions and compares favorably to advanced low emission central station technologies such as gas-fired combined cycle systems (US Environmental Protection Agency Combined Heat and Power Partnership 2002).

In a restructured power market, CHP and other on-site generation options can provide grid support to the local distribution utility. On-site generation can offer ancillary benefits to the grid including:

- Voltage and frequency support to enhance reliability and power quality.
- Avoidance or deferral of high costs, long lead times, and transmission and distribution upgrades.
- Bulk power risk management.
- Reduced line losses and reactive power control.
- Outage cost savings.
- Reduced central station generating reserve requirements.
- Transmission capacity release.

CHP offers enhanced reliability for customers, as well as operational and load management flexibility, the ability to arbitrate electric and gas prices, and energy management techniques including peak shaving and thermal energy storage. The value of these benefits depends on the characteristics of the individual facility, including energy use and prices, load profiles, and electric rate tariffs.

The prime mover technologies of CHP systems are summarized below.

5.3 Prime Mover Technologies for Combined Heat and Power

5.3.1 Gas Turbine

Gas turbines are an established technology available in various sizes, ranging from several hundred kilowatts to over several hundred megawatts. Gas turbines produce high-quality heat that can be used for industrial or district heating requirements. Alternatively, this heat can be recuperated to improve power generation efficiency or used to generate steam to drive a turbine in a combined-cycle plant. Gas turbine emissions can be controlled to very low levels using dry combustion techniques, water or steam injection, or exhaust treatment. Maintenance costs per unit of power output are about a third to a half of those of reciprocating engine generators. Low maintenance and high-quality waste heat render gas turbines a preferred choice for industrial or commercial CHP applications greater than 3 MW.

Gas turbines can be used in a variety of configurations: (1) simple cycle operation, in which a single gas turbine produces power only; (2) CHP operation, in which a simple cycle gas turbine operates with a heat recovery heat exchanger that recovers heat from the turbine exhaust and converts it to useful thermal energy, usually in the form of steam or hot water; and (3) combined cycle operation in which high pressure steam is generated from recovered exhaust heat and used to create additional power using a steam turbine. Some combined CHP cycle systems extract steam at an intermediate pressure for use in industrial processes.

The most efficient commercial technology for central station power-only generation is the gas turbine-steam turbine combined-cycle plant, with efficiencies approaching 60% lower heating value (LHV). Simple-cycle gas turbines for power-only generation are available with efficiencies approaching 40% LHV. Gas turbines have long been used by utilities to provide peaking capacity. However, with changes in the power industry and advancements in technology, the gas turbine is increasingly being used for base load power.

Gas turbines produce high-quality exhaust heat that can be used in CHP configurations to reach overall system efficiencies (electricity and useful thermal energy) of 70–80%. By the early 1980s, the efficiency and reliability of smaller gas turbines (1–40 MW) had progressed sufficiently to be an attractive choice for industrial and large institutional users for CHP applications.

5.3.2 Steam Turbine

Steam turbines are one of the most versatile, and oldest, prime mover technologies still in general production. They are used to drive a generator or mechanical machinery. Power has been generated using steam turbines for approximately 100 years, since reciprocating steam engines were replaced due to higher efficiencies and lower costs. The capacity of steam turbines can range from 50 kW to several hundred MW for large utility power plants. Steam turbines are widely used for CHP applications in the United States and Europe.

Unlike gas turbine and reciprocating engine CHP systems, where heat is a by-product of power generation, steam turbines normally generate electricity as a by-product of heat (steam) generation. A steam turbine is captive to a separate heat source and does not directly convert fuel to electrical energy. Energy is transferred from a boiler to the turbine through high pressure steam that in turn powers the turbine and generator. This separation of functions enables steam turbines to operate with an enormous variety of fuels, varying from clean natural gas to solid waste, including all types of coal, wood, wood waste, and agricultural by-products (e.g., sugarcane bagasse, fruit pits, and rice hulls). In CHP applications, steam at lower pressure is extracted from the steam turbine and used directly in a process or for district heating or is converted to other forms of thermal energy including hot or chilled water.

Steam turbines offer a wide array of designs and complexity to match desired applications and performance specifications. Those used for utility services may have several pressure casings and elaborate design features to maximize power plant efficiency. For industrial applications, steam turbines are generally of a simpler single casing design to increase reliability and reduce cost. CHP can be adapted to both utility and industrial steam turbine designs.

5.3.3 Micro Turbine

Micro turbines are small electricity generators that burn gaseous and liquid fuels to create high-speed rotation that turns an electrical generator. Current micro turbine technology is the result of development of small stationary and automotive gas turbines, auxiliary power equipment, and turbochargers, much of which has been pursued by the automotive industry since in the 1950s. Micro turbines entered field testing in around 1997, and commercial service began in 2000.

The size range of micro turbines currently available and in development is from 30 to 400 kW, while conventional gas turbines range from 500 kW to 350 MW. Micro turbines run at high speeds and, like larger gas turbines, can be used in power-only generation or in CHP systems. They are able to operate on a variety of fuels, including natural gas and liquid fuels such as gasoline, kerosene, and diesel fuel or distillate heating oil. In resource recovery applications, waste gases are burned that would otherwise be flared or released directly into the atmosphere.

Designed to combine the reliability of auxiliary power systems used onboard commercial aircraft with the design and manufacturing economies of turbochargers, units are targeted at CHP and prime power applications in commercial buildings and light industry. In most configurations, a high-speed turbine (100,000 rpm) drives a high-speed generator producing direct current (DC) power that is electronically inverted to 60 (or 50) Hz AC. Micro turbine systems are capable of producing power at around 25–33% efficiency by employing a recuperate that transfers exhaust heat back into the incoming air stream. Efficiencies generally decrease at elevated ambient temperatures. Systems are air-cooled, and some designs use air bearings, thereby eliminating the water and oil systems used by reciprocating engines. Low emission combustion systems that provide emissions performance comparable to larger gas turbines have been developed. The potential for reduced maintenance and high reliability and durability is currently being demonstrated in actual applications.

Micro turbines are ideally suited for distributed generation due to their flexibility in connection methods, low emissions, and their ability to provide stable and reliable power and to be stacked in parallel to serve larger loads. Types of application include:

- Peak shaving and base load power (grid parallel).
- Combined heat and power.
- Stand-alone power.
- Backup/standby power.
- Ride-through connection.
- Primary power with grid as backup.
- Resource recovery.

Target customers include financial services, data processing, telecommunications, restaurants, hotels and lodging, retail and office buildings, and other commercial sectors. Micro turbines are currently used in resource recovery operations at oil and gas production fields, wellheads, coal mines, and landfill operations, where by-product gases in essence provide free fuel. Unattended operation is important, since these locations may be remote from the grid and, even when served by the grid, may experience costly downtime when electric service is lost due to weather, fire, or interference from animals.

In CHP applications, the waste heat from micro turbines is used to produce hot water, to heat building space, to drive absorption cooling or desiccant dehumidification equipment, and to supply other thermal energy needs in a building or industrial process.

5.3.4 Reciprocating Engine

Also known as internal combustion engines, reciprocating engines are a widespread and familiar technology. A variety of stationary engine products are available for a range of power generation applications and duty cycles, including standby and emergency power, peaking service, intermediate and base load power, and CHP. Reciprocating engines are available for power generation in sizes ranging from a few kilowatts to over 5 MW.

There are two basic types of reciprocating engine: spark ignition (SI) and compression ignition (CI). Spark ignition engines used for power generation prefer natural gas as fuel, although can be run on propane, gasoline, or landfill gas. Compression ignition engines (often called diesel engines) operate on diesel fuel or heavy oil or can be set up to run in a dual-fuel configuration that burns primarily natural gas with a small amount of diesel pilot fuel.

Diesel engines have historically been the most popular type of reciprocating engine for both small and large power generation applications. However, in the United States and other industrialized nations, diesel engines are increasingly restricted to emergency standby or limited duty-cycle service because of atmospheric emission concerns. As a result, the natural gas-fueled SI engine is now a popular choice for the higher duty-cycle stationary power market.

The current generation of natural gas engines offers low capital costs, fast start-up, proven reliability when properly maintained, excellent load-following characteristics, and significant heat recovery potential. Electric efficiencies of natural gas engines range from 28% LHV for small engines (<100 kW) to over 40% LHV for very large lean-burn engines (>3 MW). Waste heat can be recovered from the hot engine exhaust and from engine cooling systems to produce either hot water or low pressure steam for CHP applications. Overall CHP system efficiencies (electricity and useful thermal energy) of 70–80% are routinely achieved with natural gas engine systems.

5.3.5 *Fuel Cell*

Fuel cells provide an entirely different approach to the production of electricity compared to traditional prime mover technologies and are currently in the early stages of development. Fuel cell stacks available and under development are silent, produce no pollutants, have no moving parts, and have potential fuel efficiencies far beyond the most advanced reciprocating engine or gas turbine power generation systems. Fuel cell systems with support ancillary pumps, blowers, and reformers have similar advantages.

As is common with new technologies, fuel cell systems have several disadvantages, such as product immaturity, overengineered system complexities, and unproven product durability and reliability. These translate into high capital costs, lack of support infrastructure, and technical risks for early adopters, which cause market resistance that reinforces these disadvantages. However, the many advantages of fuel cells over other prime movers suggest that they could well become the prime mover of choice for many applications and products in the future.

Fuel cells produce power electrochemically from hydrogen delivered to the negative pole (anode) of the cell and oxygen delivered to the positive pole (cathode). The hydrogen can come from a variety of sources, but the most economic is the reforming of natural gas or liquid fuels. There are several different liquid and solid media that support these electrochemical reactions: phosphoric acid (PAFC), molten carbonate (MCFC), solid oxide (SOFC), and proton exchange membrane (PEM) are the most common systems. Each of these media comprises a distinct fuel cell technology with its own performance characteristics and development schedule. PAFCs are the most widely deployed fuel cells currently in commercial service, with 200 kW units delivered to over 200 customers. PEM fuel cells are just entering the market, and SOFC and MCFC technologies are in field test or demonstration stages. Direct electrochemical reactions are generally more efficient than using fuel to drive a heat engine to produce electricity. Fuel cell efficiencies range from 35 to 40% for PAFC to upward of 60% for systems still in development. Fuel cells are inherently quiet and have extremely low emissions as only a small part of the fuel is combusted. Like a battery, fuel cells produce direct current that must be run through an inverter to obtain 60 Hz AC. These power electronics components can be integrated with other power quality components as part of a power quality control strategy for sensitive customers. Due to current high costs, fuel cells are best suited to environmentally sensitive areas and customers with power quality concerns.

5.3.6 Biomass Energy

Biomass is regarded as a renewable fuel and has considerable potential for use in distributed energy cogeneration systems, including those already discussed. For electricity generation, the potential energy stored in biomass is typically extracted directly by combustion or through processing. Direct combustion of biomass within a boiler can produce steam to drive a steam turbine. In this case, specific biomass materials are used in order to avoid ash accumulation, which decreases efficiency and increases costs. Alternatively, biomass can be processed through a gasifier, which converts liquids and solids into a combustible gas. This gas can then be used as fuel for a gas turbine.

5.3.7 Photovoltaic System

In 1839, French physicist Edmund Becquerel discovered that certain materials produce small electric currents when exposed to light. His early experiments were about 1–2% efficient in converting light to electricity and precipitated research into photovoltaic effects. In the 1940s, materials science evolved, and the Czochralski process of creating very pure crystalline silicon was developed. This process was used in 1954 by Bell Labs to develop a silicon PV cell that increased the efficiency of light to electricity conversion to 4%.

PV systems are commonly known as solar panels and are composed of discrete cells, connected together, that convert light radiation into electricity. The PV cells produce DC electricity, which must then be inverted for use in AC systems. Current units have efficiencies of 24% in the laboratory and 10% in practical applications, both of which are below the 30% maximum theoretical efficiency that can be attained by a PV cell.

Insolation is a term used to describe the solar energy available for conversion to electricity. Factors affecting insolation include the intensity of light and the operating temperature of PV cells. Light intensity is dependent on the local latitude and climate and generally increases the closer a site is to the equator.

Photovoltaic systems produce no emissions, are reliable, and require minimal operational maintenance. They are currently available from a number of manufacturers for both residential and commercial applications, and improvements mean that installation costs continue to reduce while efficiency increases. Applications for remote power are quite common.

5.3.8 On-Site Wind Turbine

Wind turbines convert the kinetic energy of wind into electricity. Windmills have been used for many years to harness wind energy for mechanical work such as pumping water. Wind turbines, which are in effect windmills dedicated to producing electricity, were considered the most economically viable choice within the renewable energy portfolio. Today, they retain their reputation as an environmentally sound and convenient alternative technology. Wind turbines produce electricity without requiring additional investment in infrastructure, such as new transmission lines, and are thus commonly employed for remote power applications. They are currently available from many manufacturers and improvements in installation costs and efficiency continue.

Wind turbines are packaged systems that include a rotor, generator, turbine blades, and drive or coupling device. As wind blows through the blades, the air exerts aerodynamic forces that cause the blades to turn the rotor. As the rotor turns, its speed is altered to match the operating speed of the generator. Most systems have a gearbox and generator in a single unit behind the turbine blades. Similar to PV systems, generator output is processed by an inverter to change the electricity from DC to AC before use.

References

Fujino J, Ehara T, Matsuoka Y, Masui T, Kainuma M (2008) Bask-casting analysis for 70% emission reduction in Japan by 2050. Clim Pol 8:108–124

Huang J, Jiang C, Xu R (2007) A review on distributed energy resources and MicroGrid. Renew Sustain Energy Rev 12(9):2472–2483

Kari A, Arto S (2006) Distributed energy generation and sustainable development. Renew Sustain Energy Rev 10(6):539–558

Kevin LA, Sarah LM, Alice B, Simon S, Paolo A, Paul E (2008) The Tyndall decarbonisation scenarios—part II: scenarios for a 60% CO2 reduction in the UK. Energy Policy 36 (10):3764–3773

National Institute for Environmental Studies (NIES) (2008) Japan scenarios and actions towards low-carbon societies (LCSs)

Sarah LM, Alice B, Kevin LA, Simon S, Paolo A, Paul E (2008) The Tyndall decarbonisation scenarios—part I: development of a backcasting methodology with stakeholder participation. Energy Policy 36(10):3754–3763

Soderman J, Pettersson F (2006) Structural and operational optimization of distributed energy systems. Appl Therm Eng 26:1400–1408

U.S. Environmental Protection Agency Combined Heat and Power Partnership (2002) Catalogue of CHP Technologies

Chapter 6
Development of a Tool to Optimize Urban-Rural Linkage and a Decentralized Power Supply

Hongbo Ren and Weisheng Zhou

Abstract In order to realize high economical and energy-saving potentials of local energy system, it is necessary to determine its structure rationally by selecting some kinds of equipment from many alternative ones so that they match energy demand requirements for an objective user. It is also important to determine rationally the number and capacities of each kind of equipment selected and the system's annual operating strategies corresponding to hourly variations in energy demands. Facing the abovementioned problems, a systematic optimization procedure is needed. The role of optimization is to reveal the best (under certain criteria and constraints) design and the best operational point of local energy system automatically, with no need for the designer to study and evaluate one by one the multitude of possible variations. It is a time-dependent optimization and without pre-specification, so the degree of freedom is very large, and the problem is very complex. Energy models have been widely used for solving this problem. This chapter attempts to develop a general method to find an optimal integrated system among different distributed energy combinations for a local community, minimizing the total energy cost (sometimes carbon emissions and primary energy consumptions as well) while guaranteeing reliable system operation. PV, wind, CHP, and energy storage are considered in the model.

Nomenclature

Indices

h	Hour {1,2...24}
i	Technology {the set of technologies selected}

H. Ren
College of Energy and Mechanical Engineering, Shanghai University of Electric Power, Pudong, Shanghai, China

W. Zhou (✉)
College of Policy Science, Ritsumeikan University, Ibaraki, Osaka, Japan
e-mail: zhou@sps.ritsumei.ac.jp

W. Zhou et al. (eds.), *East Asian Low-Carbon Community*,
https://doi.org/10.1007/978-981-33-4339-9_6

m Month {1,2...12}
d Day in a month
p,q,r Indicator of equipment or customer sites
u End use {electricity_only, cooling, space heating, hot water}

Customer data

Cloadm,d,h,u	Customer load in kW for end use u during hour h, day d and month h
$SR_{m,d,h}$	Irradiation data (kW/m^2)
$V_{m,d,h}$	Wind speed (m/s)

Market data

IRate	Interest rate (%)
CTax	Tax on carbon emission (Yen/kg-C)
ECInt	Carbon emission rate from marketplace generation (kg/kWh)
GPrice	Natural gas price during hour h, type of day t, and month m (Yen/m^3)
GBase	Natural gas basic service fee (Yen)
CCInt	Carbon emission rate from burning natural gas to meet heating and cooling loads (kg/kWh)
$GDCharge_m$	Natural gas flux charge for month m ($Yen/\ m^3$)
$EPrice_{m,d,h}$	Regulated tariff for electricity purchase during hour h, day d, month m (Yen/kWh)
$EDCharge_m$	Regulated demand charge (Yen/kW)
$SPrice_{m,d,h}$	Electricity selling price (Yen/kWh)

Distributed Energy Resource Technologies Information

YDCF	Annuity factor for direct-fired natural gas AC
YHSF	Annuity factor for heat storage unit
YESF	Annuity factor for electricity storage unit
YF_i	Annuity factor for DER technology i
YP	Annuity factor for pipelines
DCCap	Capacity of direct-fired natural gas absorption chiller (AC) (kW)
DCLTime	Expected lifetime of direct-fired natural gas AC (year)
DCFCost	Fixed capital cost of direct-fired natural gas AC (Yen)
DCFCost	Fixed capital cost of direct-fired natural gas AC (Yen/kW)
$FVCost_i$	Variable capital cost of technology i (Yen/kW)
$FFCost_i$	Fixed capital cost of technology i (Yen)
Eff_i	Power generation efficiency of technology i (%)
$FMTime_i$	Maximum number of hours technology i is permitted to operate during the year (h)
$FLtime_i$	Expected lifetime of technology i (year)
$Fmaxp_i$	Nameplate power rating of technology i (kW)
$Fminp_i$	Minimum load rating of technology i (kW)
$FOMf_i$	

	Fixed annual operation and maintenance costs of technology i (Yen/kW)
$FOMv_i$	Variable operation and maintenance costs of technology i (Yen/kWh)
$PVInt_i$	Rated capacity of PV cells (kW/m^2)
HSLTime	Expected lifetime of heat storage unit (year)
HSFCost	Fixed capital cost of heat storage unit (Yen)
HSVCost	Variable capital cost of heat storage unit (Yen/kWh)
ESFCost	Fixed capital cost of electricity storage unit (Yen)
ESVCost	Variable capital cost of electricity storage unit (Yen/kWh)
ESLTime	Expected lifetime of electricity storage unit (year)
ESmax	Maximum dispatch level of electricity storage (kWh)
ESmin	Minimum dispatch level of electricity storage (kWh)
U(i)	Set of end uses that can be met by technology i
l	Distance between sites (m)
c_p	Unit cost of pipeline (Yen/m)

Other Parameters

C_{total}	Annual total cost (Yen)
C_{Elec}	Electricity purchase cost (Yen)
C_{Gas}	Gas purchase cost (Yen)
C_{Inv}	Annualized investment cost (Yen)
C_{OM}	Annual operational and maintenance costs (Yen)
C_{CTax}	Carbon tax cost (Yen)
C_{Ss}	Start and stop costs (Yen)
C_{Sal}	Revenue from selling excess electricity
$C_{inv,P}$	Investment cost for pipelines (Yen)
f_C	Minimum cost objective (Yen)
f_E	Minimum carbon emissions objective (kg)
f_P	Minimum primary energy consumption objective (MJ)
CE_{Elec}	Carbon emissions from grid electricity (kg)
CE_{Gas}	Carbon emissions from gas purchase (kg)
EC_{Elec}	Primary energy consumption of grid electricity (MJ)
EC_{Gas}	Primary energy consumption of gas purchase (MJ)
$MDays_m$	Number of days in each month m
$PVp_{i,m,d,h}$	Power generation of PV technology i (kW)
IRate	Interest rate on DER investments (%)
GHRate	Natural gas heat rate (kWh/m^3)
$PVArea_i$	Area of PV cells (m^2)
α_i	The amount of heat that can be recovered from unit kW of electricity that is generated using DER technology i
β_u	The amount of heat generated from unit of natural gas purchased for end use u
$\gamma_{i,u}$	The amount of useful heat that can be allocated to end use u from unit kW of recovered heat from technology i

δ_u	The amount of heat that can be allocated to end use u from unit kW of stored heat that is released
λ_u	The amount of electricity that can be allocated to end use u from unit kW of stored electricity that is released
ε_H	The amount of heat that is not lost due to dissipation during 1 h from unit kWh of stored heat
ε_E	The amount of electricity that is not lost due to dissipation during 1 hour from unit kWh of stored electricity
ω	Unit start and stop cost (Yen)
θ	Minimum percentage of electricity purchase (%)
$Wp_{i,m,d,h}$	Power generation of wind technology i (kW)
V_{ci}	Cut in speed of wind turbine (m/s)
V_{ri}	Rated speed of wind turbine (m/s)
V_{fi}	Cutoff speed of wind turbine (m/s)
M	A very large number

Decision variables

DC	Indicator variable for installation of a direct-fired natural gas AC
HS	Indicator variable for installation of heat storage unit
ES	Indicator variable for installation of electricity storage unit
$Operate_{i,m,d,h}$	Number of units of technology i, set to run at hour h, day d, in month m
$PElec_{m,d,h,u}$	Purchased electricity from the distribution company by the customer during hour h, day d, and month m for end use u (kW)
$PGas_{m,d,h,u}$	Purchased natural gas during hour h, day d, and month m for end use u (m^3)
$EGen_{i,m,d,h,u}$	Generated power by technology i during hour h, day d, month m and for end use u to supply the customer's load (kW)
$IHeat_{i,m,d,h}$	Amount of heat from technology i that is diverted toward the heat storage unit during hour h, day d, and month m (kW)
$IGElec_{m,d,h}$	Amount of grid electricity that is diverted toward the electricity storage unit during hour h, day d, and month m (kW)
$IElec_{i,m,d,h}$	Amount of generated electricity from technology i that is diverted toward the electricity storage unit during hour h, day d, and month m (kW)
$FStart_{i,m,d,h}$	Indicator variable for the start of technology i
$FStop_{i,m,d,h}$	Indicator variable for the stop of technology i
$NInv_i$	Number of units of technology i installed by the customer
$OHeat_{m,d,h,u}$	Amount of stored heat that is released to meet the load of end use u during hour h, day d, and month m (kW)
$RHeat_{i,m,d,h,u}$	Amount of heat recovered from technology I that is used to meet end use u during hour h, type of day t, and month m (kW)
HSmax	Capacity of heat storage unit (kWh)
ESmax	Capacity of electricity storage unit (kWh)

$SHeat_{m,d,h}$	Amount of stored heat available at the start of hour h, day d, month m (kW)
$OElec_{m,d,h,u}$	Amount of stored electricity that is released to meet the load of end use u during hour h, day d, and month m (kW)
$SElec_{m,d,h}$	Amount of stored electricity available at the beginning of hour h, day d, and month m (kW)
$Flag_{O,m,d,h}$	Indicator variable for discharge of electricity from the electricity storage unit
$Flag_{I,m,d,h}$	Indicator variable for charge of electricity to the electricity storage unit
$D_{i,m,d,h}$	Indicator variable for the state of equipment operation
$Pf_{p,q}$	Indicator variable for the existence of pipeline
$H_{p,q,m,d,h}$	Heat flow between sites
$Esal_{i,m,d,h}$	Electricity sold to the gird from technology i, in hour h, day d, and month m (kWh)

6.1 General Structure of the Tool

The tool to optimize the integration of a distributed energy system is formulated as a mixed integer linear programming (MILP) model. The use of integers represents, for example, the number of pieces of equipment adopted. Figure 6.1 is a flowchart illustrating the model structure. The objective function to be minimized is the annual cost of providing energy services to a site, through either utility electricity and gas

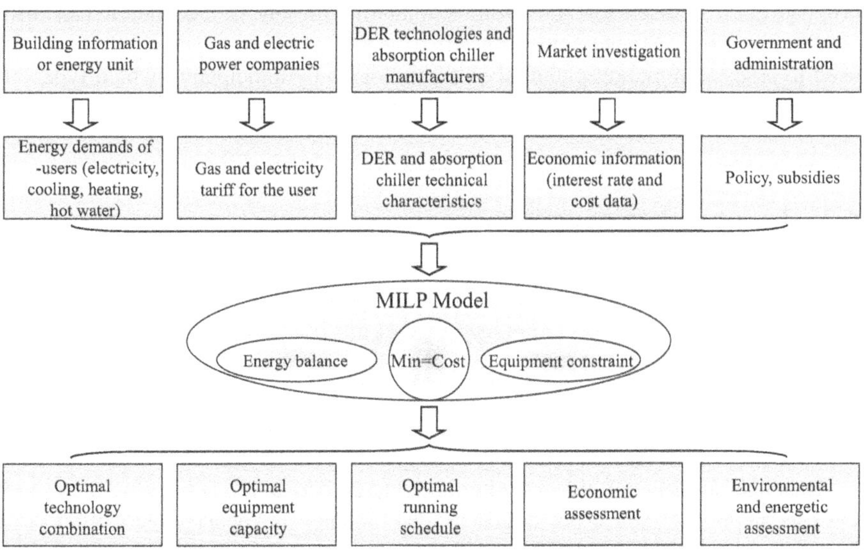

Fig. 6.1 Flow chart of model for distributed energy system optimization

purchases or distributed energy resource (DER) operation or a combination of both. This value is a summation of the costs of electricity and gas purchases, capitalized costs of DER equipment, and operating and maintenance costs. The results obtained from this process provide the optimal combination of on-site generation and heat recovery, an elementary operating schedule detailing how equipment should be used, and a summary result for each case, such as the total electricity bill and electricity generation and purchase in each hour (Gao and Ren 2011).

While constructing the model, the following issues are relevant:

1. The procedure must be flexible and generic, allowing analysis of small- and medium-sized areas in any location and climate, with different demand needs and supply possibilities.
2. Temporal matching of supply and demand is required in order to analyze diurnal and seasonal variations.
3. Demands for electricity, heating, cooling, and hot water must be considered.
4. Intermittent suppliers, alternative fuel technologies for CHP generation, heat storage and production technologies, and other uses for excess electricity must be available for consideration.
5. The procedure must aid decisions about the best mix, type, and sizing of all technologies and processes being considered and about the amount of fuel or other storage required as appropriate. Informed decisions can then be made about the suitability of potential supply systems based on subsequent cost analyses.
6. It must be possible to analyze and compare a number of potential combinations quickly and easily.

The model has three possible applications. First, it can be used to guide choices of equipment for specific sites or provide general solutions, for example, sites, and propose good choices for sites with similar circumstances. Second, it can also provide the basis for operation of installed on-site generation, and third, it can be used to assess the market potential of technologies by anticipating which types of customer might find various technologies attractive (Afzal et al. 2003).

6.2 Input and Output Data

Input data for the optimization tool can be divided into three categories: customer information, technical information, and market information.

Customer information includes the load profiles (electricity, cooling, heating, and hot water) across various time intervals, as well as local climate data (solar, wind, etc.). Load profiles can be obtained through measured data or simulated using tools such as DOE-2, Flexsim, or Building Energy Analyzer (Ren et al. 2010).

Technical information refers to thermodynamic parameters governing the use of DER technologies, such as power generation efficiency and heat recovery efficiency. Some characteristics related to the environment, such as rate of carbon emissions and natural gas from the macro grid, are also provided.

Market information includes the energy tariffs and financial data of the DER equipment, including capital, operating, and maintenance costs, together with the interest rate on the investment. Sometimes, policy-related data, such as carbon tax rates, are also provided.

The outputs to be determined by the tool while minimizing cost are:

- Technology (or combination of technologies) installed and their respective capacities
- Hourly operating schedules for installed equipment
- Total cost and carbon emissions of supplying the total energy requirement through either DER or macro grid generation or, more typically, a combination of both

6.3 Objective Function

The objective function of the model is to minimize the cost of supplying energy to a specific customer by using distributed generation to meet part or all of the electricity and heating requirements, as shown in Eq. (6.1). It is evaluated as the sum of energy costs, annual capital, and operational and maintenance costs, as well as the carbon tax cost and costs associated with starting and stopping, minus the revenue from selling excess electricity (Ren and Gao 2010).

$$\text{Min } C_{\text{total}} = C_{\text{Elec}} + C_{\text{Gas}} + C_{\text{Inv}} + C_{\text{OM}} + C_{\text{CTax}} + C_{\text{Ss}} - C_{\text{Sal}} \tag{6.1}$$

The cost of purchasing external electricity is described by Eq. (6.2) and is composed of the demand charge and total energy cost. The demand charge is calculated using the charge rate multiplied by peak electricity demand for each month. The energy cost is calculated as the cumulative amount of electricity purchased multiplied by the utility electricity rate. Here, it is assumed that the given electricity purchase tariff varies seasonally (where summer months are July through September, inclusive) and by load period (on-peak, mid-peak, and off-peak).

$$\begin{aligned} C_{\text{Elec}} = {} & \sum_{m} \text{EDCharge} \cdot \max\left(\sum_{u} \text{PElec}_{m,d,h,u} + \text{IGElec}_{m,d,h}\right) \\ & + \sum_{m}\sum_{d}\sum_{h}\left(\sum_{u} \text{PElec}_{m,d,h,u} + \text{IGElec}_{m,d,h}\right) \cdot \text{EPrice}_{m,d,h} \end{aligned} \tag{6.2}$$

The fuel cost is composed of the natural gas base service fee, natural gas flux charge, and volumetric cost, as shown in Eq. (6.3). The flux charge is calculated by multiplying the charge rate by peak natural gas demand for each month. The volumetric cost consists of DER use and non-DER use and is calculated by multiplying the cumulative gas consumption of DER equipment and direct natural gas use during each period by the fuel tariff rate.

$$
\begin{aligned}
C_{\text{Gas}} = & \sum_m \text{GBase} + \sum_m \text{GDcharge}_m \cdot \\
& \max\left(\frac{\sum_u \text{PGas}_{m,d,h,u} + \sum_i \frac{\sum \text{EGen}_{i,m,d,h,u} + \text{ESal}_{i,m,d,h} + \text{IElec}_{i,m,d,h}}{\text{Eff}_i}}{\text{GHRate}} \right) \\
& + \sum_m \sum_d \sum_h \sum_u \frac{\text{PGas}_{m,d,h,u}}{\text{GHRate}} \cdot \text{GPrice} + \\
& \sum_i \frac{\sum_m \sum_d \sum_h \left(\sum_u \text{EGen}_{i,m,d,h,u} + \text{ESal}_{i,m,d,h} + \text{IElec}_{i,m,d,h} \right)}{\text{Eff}_i \cdot \text{GHRate}} \cdot \text{GPrice}
\end{aligned}
\tag{6.3}
$$

Investment cost is calculated according to the annualized capital cost and is composed of fixed and variable parts. Annualizing capital is a means of spreading the initial cost of an option across the lifetime of that option while accounting for opportunity costs and inflation. The cost of capital is annualized as if it were being paid off as a loan at a particular discounted interest rate over the lifetime of the option. The result is a future value cost or constant annual cost of capital. The annualized investment cost, which is described in Eq. (6.4), is composed of investment in four components: DER equipment, storage tank, battery, and natural gas AC.

$$
\begin{aligned}
C_{\text{Inv}} = & \sum_i (\text{NInv}_i \cdot F\max p_i \cdot \text{FVCost}_i + \text{NInv}_i \cdot \text{FFCost}_i) \cdot \text{YF}_i \\
& + \text{YHSF} \cdot (\text{HSVCost} \cdot \text{HSmax} + \text{HS} \cdot \text{HSFCost}) \\
& + \text{YESF} \cdot (\text{ESVCost} \cdot \text{ESmax} + \text{ESFCost} \cdot \text{ES}) \\
& + \text{YDCF} \cdot (\text{DCVCost} \cdot \text{DCCap} + \text{DCFCost} \cdot \text{DC})
\end{aligned}
\tag{6.4}
$$

$$
\text{YF}_i = \frac{I\text{Rate}}{\left(1 - \frac{1}{(1+I\text{Rate})^{\text{FLTime}_i}}\right)} \forall i
\tag{6.5}
$$

$$
\text{YDCF} = \frac{I\text{Rate}}{\left(1 - \frac{1}{(1+I\text{Rate})^{\text{DCLTime}}}\right)}
\tag{6.6}
$$

$$\text{YHSF} = \frac{I\text{Rate}}{\left(1 - \frac{1}{(1+I\text{Rate})^{\text{HSLTime}}}\right)} \tag{6.7}$$

$$\text{YESF} = \frac{I\text{Rate}}{\left(1 - \frac{1}{(1+I\text{Rate})^{\text{ESLTime}}}\right)} \tag{6.8}$$

The operation and maintenance (O&M) cost is composed of fixed and variable costs, as illustrated in Eq. (6.9). The fixed O&M cost is calculated by multiplying the installed DER capacity by a unit cost coefficient. The variable O&M cost is calculated by multiplying the cumulative electrical energy for each period of DER system operation by a unit cost coefficient.

$$C_{\text{OM}} = \sum_i \sum_m \sum_d \sum_h \left(\sum_u \text{EGen}_{i,m,d,h,u} + \text{ESal}_{i,m,d,h} + \text{IElec}_{i,m,d,h} \right) \cdot \text{FOMv}_i + \sum_i \text{NInv}_i \cdot F\max p_i \cdot \text{FOMf}_i \tag{6.9}$$

The carbon tax cost is described in Eq. (6.10) as the cost for carbon emissions from utility electricity purchases, on-site electricity generation, and direct gas combustion.

$$C_{\text{Ctax}} = \sum_m \sum_d \sum_h \sum_u \text{PElec}_{m,d,h,u} \cdot \text{CTax} \cdot \text{ECInt} + \left(\sum_i \frac{\sum_m \sum_d \sum_h \left(\sum_u \text{EGen}_{i,m,d,h,u} + \text{ESal}_{i,m,d,h} + \text{IElec}_{i,m,d,h} \right)}{\text{Eff}_i} + \sum_m \sum_d \sum_h \sum_u \text{PGas}_{m,d,h,u} \right) \cdot \text{CTax} \cdot \text{GCInt} \tag{6.10}$$

The cost associated with starting and stopping is illustrated in Eq. (6.11).

$$C_{Ss} = \sum_i \sum_m \sum_d \sum_h \left(\text{FStart}_{i,m,d,h} - \text{FStop}_{i,m,d,h} \right) \cdot \omega \tag{6.11}$$

The income from selling electricity to the grid is described by Eq. (6.12).

$$C_{\text{Sal}} = \sum_i \sum_m \sum_d \sum_h \text{ESal}_{i,m,d,h} \cdot S\text{Price}_{m,d,h} \tag{6.12}$$

6.4 Constraints

A key constraint is that the energy demand during each hour must be met in one of three ways: purchase of energy from utilities, operation of a technology or set of technologies selected by the model, or a combination of these options. The energy balance is described in Eq. (6.13).

$$\mathrm{Cload}_{m,d,h,u} = \sum_{i} \mathrm{EGen}_{i,m,d,h,u} + \mathrm{PElec}_{m,d,h,u} + \beta_u \cdot \mathrm{PGas}_{m,d,h,u} + \sum_{i} \left(\gamma_{i,u} \cdot \mathrm{RHeat}_{i,m,d,h,u}\right) + \delta_u \cdot \mathrm{OHeat}_{m,d,h,u} + \lambda_u \cdot \mathrm{OElec}_{m,d,h,u} \quad \forall m, d, h, u \tag{6.13}$$

Another key performance constraint is that any installed DER technologies must operate within the minimum and maximum operating capacity limits of each unit, as illustrated in Eqs. (6.14) and (6.15).

$$\sum_{u} \mathrm{EGen}_{i,m,d,h,u} + \mathrm{ESal}_{i,m,d,h} + \mathrm{IElec}_{i,m,d,h} \leq \mathrm{Operate}_{i,m,d,h} \cdot F\max p_i \quad \forall i, m, d, h \tag{6.14}$$

$$\sum_{u} \mathrm{EGen}_{i,m,d,h,u} + \mathrm{ESal}_{i,m,d,h} + \mathrm{IElec}_{i,m,d,h} \geq \mathrm{Operate}_{i,m,d,h} \cdot F\min p_i \quad \forall i, m, d, h \tag{6.15}$$

The dispatched generators are constrained to run below the maximum operating hours of all units, as shown in Eqs. (6.16) and (6.17).

$$\sum_{m}\sum_{d}\sum_{h} \mathrm{Operate}_{i,m,d,h} \leq \mathrm{NInv}_i \cdot \mathrm{FMTime}_i \quad \forall i \tag{6.16}$$

$$\mathrm{Operate}_{i,m,d,h} \in \{0, 1, 2 \ldots \mathrm{NInv}_i\} \quad \forall i, m, d, h \tag{6.17}$$

Photovoltaics can produce electricity in proportion to the capacity of the installed system and the amount of insolation, as illustrated in Eqs. (6.18), (6.19), and (6.20).

$$\sum_{u} \mathrm{EGen}_{i,m,d,h,u} + \mathrm{ESal}_{i,m,d,h} + \mathrm{IElec}_{i,m,d,h} \leq \mathrm{NInv}_i \cdot \mathrm{PVp}_{i,m,d,h} \quad \forall i \in \mathrm{PV}, m, d, h \tag{6.18}$$

$$\mathrm{PVp}_{i,m,d,h} = \min\left\{F\max p_i, \mathrm{SR}_{m,d,h} \cdot \mathrm{PVArea}_i \cdot \mathrm{Eff}_i\right\} \quad \forall \mathrm{i} \in \mathrm{PV}, m, d, h \tag{6.19}$$

$$\text{PVArea}_i \cdot \text{PVInt}_i = F\max p_i \quad \forall i \in \text{PV} \tag{6.20}$$

The electricity produced by wind turbines is limited by wind speed, as shown in the following equations:

$$\sum_u \text{EGen}_{i,m,d,h,u} + \text{ESal}_{i,m,d,h} + \text{IElec}_{i,m,d,h} \leq \text{NInv}_i \cdot Wp_{i,m,d,h} \quad \forall i \in \text{wind}, m, d, h \tag{6.21}$$

$$Wp_{i,m,d,h} = F\max p_i \cdot \frac{V^k_{m,d,h} - V^k_{c_i}}{V^k_{R_i} - V^k_{c_i}} \quad \forall i \in \text{wind}, V_{c_i} \leq V_{m,d,h} \leq V_{R_i} \tag{6.22}$$

$$Wp_{i,m,d,h} = F\max p_i \quad \forall i \in \text{wind}, \text{V}_{R_i} \leq V_{m,d,h} \leq V_{F_i} \tag{6.23}$$

$$Wp_{i,m,d,h} = 0 \quad \forall i \in \text{wind}, V_{m,d,h} < V_{c_i} \text{U} V_{m,d,h} > V_{F_i} \tag{6.24}$$

A constraint is set by how much heat can be recovered from each type of DER technology for both immediate usage and diversion to storage, as described in Eq. (6.25).

$$\sum_u \text{RHeat}_{i,m,d,h,u} + \text{IHeat}_{i,m,d,h} \leq \alpha_i \left(\sum_u \text{EGen}_{i,m,d,h,u} + \text{ESal}_{i,m,b,h} + \text{IElec}_{i,m,d,h} \right) \quad \forall i, m, d, h \tag{6.25}$$

Additional constraints are needed to ensure the correct operation of the storage tank. Equation (6.26) shows the heat inventory balance constraint. It states that the total amount of heat stored at the beginning of an hour is equal to the non-dissipated heat stored at the beginning of the previous hour, plus recovered heat diverted toward storage during the hour, minus stored heat released to meet end-use loads during the hour. Equation (6.27) is the specific case of Eq. (6.26) for the first hour of the day. Here, the total heat stored at the beginning of the first hour of the day is equal to the non-dissipated heat stored at the beginning of the last hour of the previous day, plus the inflow and minus the outflow of heat during the hour. Equation (6.28) is similar to Eq. (6.27). Equation (6.29) prevents stored heat from being used by end-use loads such as electricity only. Equation (6.30) prevents the quantity of heat stored from exceeding the storage capacity. Equation (6.31) resets the heat stored to zero, and Eq. (6.32) indicates that stored heat is released at the end of the year. Equation (6.33) prevents the use of heat from storage during the first hour of the year. Equation (6.34) prevents the use of heat recovered from generation or storage for a particular end use unless an appropriate CHP technology is installed. Finally, Eq. 6.35 prevents the use of stored heat unless a heat storage unit has been purchased.

$$\text{SHeat}_{m,d,h+1} = \varepsilon_H \cdot \text{SHeat}_{m,d,h} + \sum_i \text{IHeat}_{i,m,d,h} - \sum_u \text{OHeat}_{m,d,h,u} \quad \forall m, d, h \quad \text{if} \quad h \neq '24' \tag{6.26}$$

$$\text{SHeat}_{m,d,1+1} = \varepsilon_H \cdot \text{SHeat}_{m,d,24} + \sum_i \text{IHeat}_{i,m,d,24} - \sum_u \text{OHeat}_{m,d,24,u} \quad \forall m, d, \quad \text{if} \quad d \neq '\text{MDays}' \tag{6.27}$$

$$\text{SHeat}_{m+1,1,1} = \varepsilon_H \cdot \text{SHeat}_{m,MDays_m,24} + \sum_i \text{IHeat}_{i,m,MDays_m,24} - \sum_u \text{OHeat}_{m,\text{MDays}_m,24,u} \quad \forall m \quad \text{if} \quad m \neq '12' \tag{6.28}$$

$$\text{OHeat}_{m,d,h,u} = 0 \quad \forall m, d, h \quad \text{if} \quad u \in \{\text{electricity\_only}\} \tag{6.29}$$

$$\text{SHeat}_{m,d,h} \leq \text{HSmax} \quad \forall m, d, h \tag{6.30}$$

$$\text{SHeat}_{1,1,1} = 0 \tag{6.31}$$

$$\text{SHeat}_{12,d,24} = 0 \tag{6.32}$$

$$\text{OHeat}_{1,1,1,u} = 0 \tag{6.33}$$

$$\delta_u \cdot \text{OHeat}_{m,d,h,u} + \sum_i \gamma_{i,u} \cdot \text{RHeat}_{i,m,d,h,u} \leq \sum_i \alpha_i \cdot \gamma_{i,u} \cdot \text{OPerate}_{i,m,d,h} \cdot F\max p_i \quad \forall m, d, h, u \tag{6.34}$$

$$\sum_u \text{OHeat}_{m,d,h,u} \leq \text{SHeat}_{m,d,h} \quad \forall m, d, h \tag{6.35}$$

Constraints are also needed to ensure appropriate operation of the battery. Equation (6.36) shows the electricity inventory balance constraint. It states that the total amount of electricity stored at the beginning of an hour is equal to the non-dissipated electricity stored at the beginning of the previous hour, plus the electricity charged during the hour, minus stored electricity discharged to meet end-use loads during the hour. Equation (6.37) is the specific case of Eq. (6.36) for the first hour of the day. Here, the total electricity stored at the beginning of the first hour of the day is equal to the non-dissipated electricity stored at the beginning of the last hour of the previous day, plus the inflow and minus the outflow of electricity during the hour. Equation (6.38) is similar to Eq. (6.37). Equation

(6.39) prevents stored electricity from being used by end-use loads such as space heating. Equations (6.40) and (6.41) prevent the quantity of stored electricity from exceeding storage capacity or falling below the minimum level; Eq. (6.42) resets stored electricity to zero. Equation (6.43) prevents the use of electricity from storage during the first hour of the year. Equation (6.44) prevents the use of stored electricity unless a battery unit has been purchased. The battery cannot simultaneously charge and discharge; this condition is constrained using Eqs. (6.45), (6.46), and (6.47).

$$\mathrm{SElec}_{m,d,h+1} = \varepsilon_E \cdot \mathrm{SElec}_{m,d,h} + \sum_i \mathrm{IElec}_{i,m,d,h} + \mathrm{IGElec}_{m,d,24} - \sum_u \mathrm{OElec}_{m,d,h,u}$$
$$\forall m, d, h \quad \text{if} \quad h \neq '24' \tag{6.36}$$

$$\mathrm{SElec}_{m,d,+1,1} = \varepsilon_E \cdot \mathrm{SElec}_{m,d,24} + \sum_i \mathrm{IElec}_{i,m,d,24} + \mathrm{IGElec}_{m,d,24} - \sum_u \mathrm{OElec}_{m,d,24,u}$$
$$\forall m, d, \quad \text{if} \quad d \neq '\mathrm{MDays}_m' \tag{6.37}$$

$$\mathrm{SElec}_{m+1,1,1} = \varepsilon_E \cdot \mathrm{SElec}_{m,MDays_m,24} + \sum_i \mathrm{IElec}_{i,m,MDays_m,24} + \mathrm{IGElec}_{m,MDays_m,24}$$
$$-\sum_u \mathrm{OElec}_{m,MDays_m,24,u} \qquad \forall m \quad \text{if} \quad m \neq '12' \tag{6.38}$$

$$\mathrm{OElec}_{m,d,h,u} = 0 \quad \forall m, d, h \quad \text{if} \quad u \in \{\text{space\_heating}, \text{water\_heating}\} \tag{6.39}$$

$$\mathrm{SElec}_{m,d,h} \leq \mathrm{ES\ max} \quad \forall m, d, h \tag{6.40}$$

$$\mathrm{SElec}_{m,d,h} \geq \mathrm{ES\ min} \quad \forall m, d, h \tag{6.41}$$

$$\mathrm{SElec}_{1,1,1} = 0 \tag{6.42}$$

$$\mathrm{OElec}_{1,1,1,u} = 0 \quad \forall u \tag{6.43}$$

$$\sum_u \mathrm{OElec}_{m,d,h,u} \leq \mathrm{SElec}_{m,d,h} \quad \forall m, d, h \tag{6.44}$$

$$\mathrm{flag}_{I,m,d,h} \cdot M \leq \sum_i \mathrm{IElec}_{i,m,d,h} \quad \forall m, d, h \tag{6.45}$$

$$\mathrm{flag}_{O,m,d,h} \cdot M \leq \sum_u \mathrm{OElec}_{m,d,h,u} \quad \forall m, d, h \tag{6.46}$$

$$\mathrm{flag}_{I,m,d,h} + \mathrm{flag}_{O,m,d,h} \leq 1 \tag{6.47}$$

Equation (6.48) prevents the use of recovered heat by end uses that cannot be satisfied by the particular DER technology. Equations (6.49) and (6.52) are boundary conditions that prevent electricity from being used directly to meet heating loads, while Eq. (6.50) prevents the direct burning of gas for electricity-only use. Equation

(6.51) prevents direct burning of natural gas to meet the cooling load if AC for this purpose has not been purchased.

$$\mathrm{RHeat}_{i,m,d,h,u} = 0 \quad \forall i, m, d, h \quad \text{if} \quad u \notin U(\mathrm{i}) \tag{6.48}$$

$$\mathrm{EGen}_{i,m,d,h,u} = 0 \quad \forall i, m, d, h \quad \text{if} \quad u \in \{\text{space\_heating water\_heating}\} \tag{6.49}$$

$$\mathrm{PGas}_{m,d,h,u} = 0 \quad \forall m, d, h \quad \text{if} \quad u \in \{\text{electricity\_only}\} \tag{6.50}$$

$$\mathrm{PGas}_{m,d,h,u} \leq \mathrm{DCCap} \cdot \mathrm{DC} \quad \forall m, d, h \quad \text{if} \quad u \in \{\text{cooling}\} \tag{6.51}$$

$$\mathrm{PElec}_{m,d,h,u} = 0 \quad \forall m, d, h \quad \text{if} \quad u \in \{\text{space\_heating water\_heating}\} \tag{6.52}$$

A constraint is set to prohibit the customer from simultaneously buying and selling energy (Eq. 6.53).

$$\begin{aligned} &\mathrm{ESal}_{i,m,d,h} = 0 \\ &\text{if} \sum_{u} \sum_{i} \mathrm{EGen}_{i,m,d,h,u} < \sum_{u} \mathrm{Cload}_{m,d,h,u} \quad \forall i, m, d, h \quad \text{if} \quad u \in \{\text{electricity\_only}\} \end{aligned} \tag{6.53}$$

According to contact with electric utilities, electricity purchases cannot be lower than a specific percentage of the total requirements (Eq. 6.54).

$$\begin{aligned} \sum_{m} \sum_{d} \sum_{h} \sum_{u} \mathrm{PElec}_{m,d,h,u} \geq \theta \cdot & \sum_{m} \sum_{d} \sum_{h} \sum_{u} \mathrm{PElec}_{m,d,h,u} \\ & + \sum_{\mathrm{i}} \mathrm{EGen}_{i,m,d,h,u} + \lambda_u \cdot \mathrm{OElec}_{m,d,h,u} \end{aligned} \tag{6.54}$$

In addition, operational experience shows that frequent starting and stopping will accelerate equipment deterioration. Here, the cost coefficient is associated with starting and stopping generation. Using the following equations, the frequency of starting and stopping is expected to be minimized.

$$D_{i,m,d,h+1} - D_{i,m,d,h} = \mathrm{FStart}_{i,m,d,h} - \mathrm{FStop}_{i,m,d,h} \quad \forall i, m, d, h \tag{6.55}$$

$$\sum_{u} \mathrm{EGen}_{i,m,d,h,u} + \mathrm{IElec}_{i,m,d,h} + \mathrm{Esal}_{i,m,d,h} - M \cdot D_{i,m,d,h} \leq 0 \quad \forall i, m, d, h \tag{6.56}$$

6.5 Model Coding and Implications

Optimization of distributed energy systems is very important. Newly emerging problems in DER are increasingly large, complex, and difficult to solve. Many professional software packages have been developed to support optimization theory, such as AMPL, GAMS, XPRESS-MP, ZINO, CPLEX, STORM, MATLAB, LINDO, and LINGO. Here, all optimization problems are solved using the LINGO software package.

LINGO is a simple tool that utilizes the power of linear and nonlinear optimization to concisely formulate and solve large problems and then analyze the solutions. It provides a completely integrated package that includes a powerful language for expressing optimization models, a full feature environment for building and editing problems, and a set of fast built-in solvers (LINDO Systems Inc 2008).

Figure 6.2 shows the LINGO programming window. The outer window, labeled LINGO, is the main frame window in which all other windows are contained. The top of the frame window contains the command menus and command toolbar. The smaller child window, labeled LINGO model, is a model window. Models are entered directly into this window, according to specific formulas.

If no formulation errors are encountered during compilation, LINGO will invoke the appropriate internal solver to begin searching for the optimal solution to the

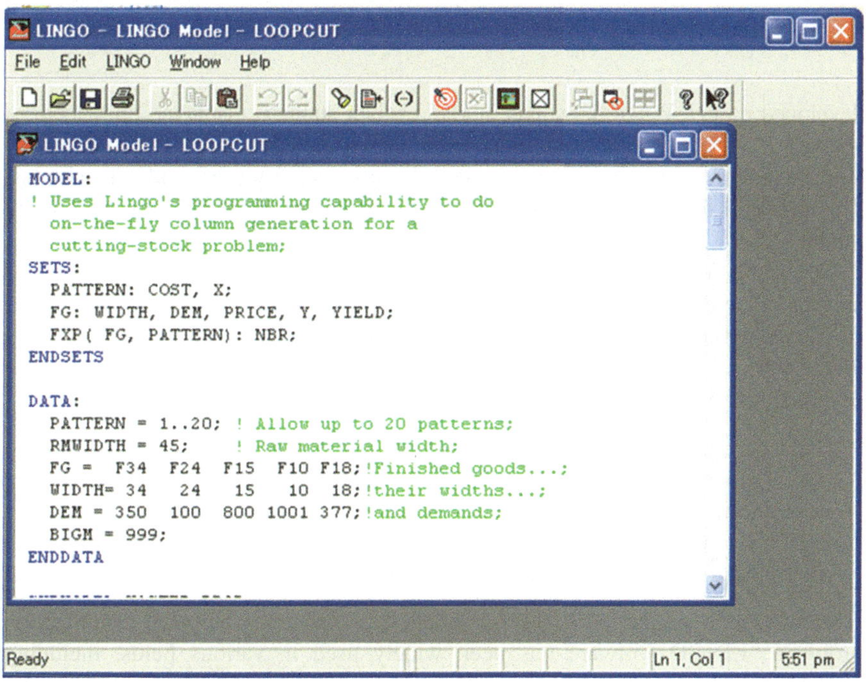

Fig. 6.2 Overall view of LINGO programming window

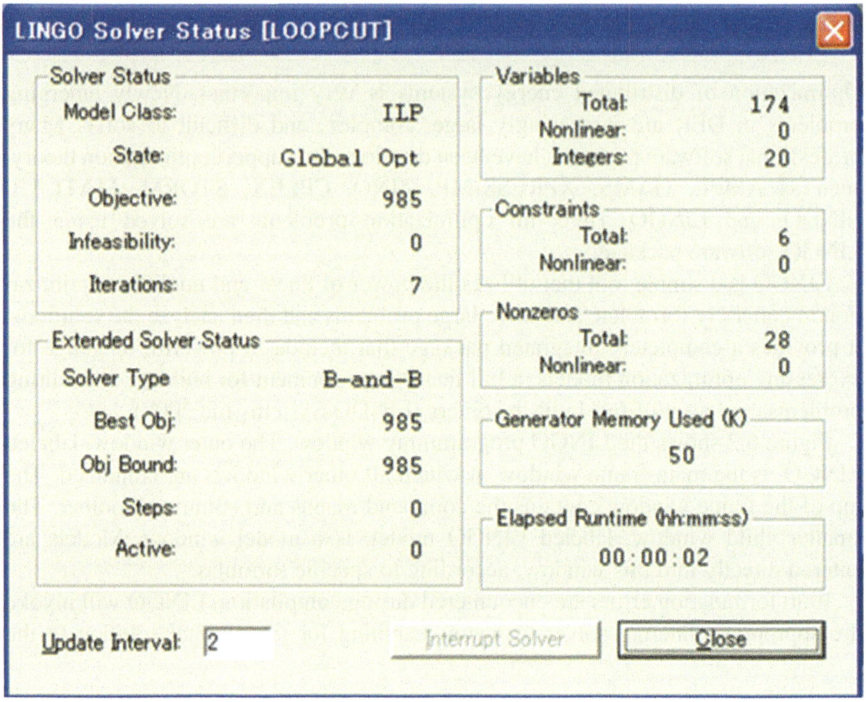

Fig. 6.3 LINGO solver status window

developed model. When the solver starts, a solver status is displayed on-screen, as shown in Fig. 6.3, which gives information about the total number of model variables, total constraints, total non-zero coefficients, memory used, and solver status. This is useful for monitoring the progress of the solver and the dimensions of the model.

When the solving process is finished, a solution report window (see Fig. 6.4) appears on-screen with details of the model solution.

After obtaining the optimal solution, the user may desire to know the sensitivity of the optimal values to changes in data input. Solution analysis in LINGO is executed based on the following concepts:

1. Reduced cost, or the amount that the objective coefficient of a variable would have to improve before it would become profitable to give that variable a positive value in the optimal solution
2. Dual price, which concerns the amount that the objective would improve as the right-hand side, or constant term, of the constraint is increased by one unit

LINGO is a simple yet powerful tool for solving and analyzing optimization problems. LINGO software has been widely used in various fields, including industry, commerce, and agriculture. In a general comparison with other optimization software, LINGO has the following advantages:

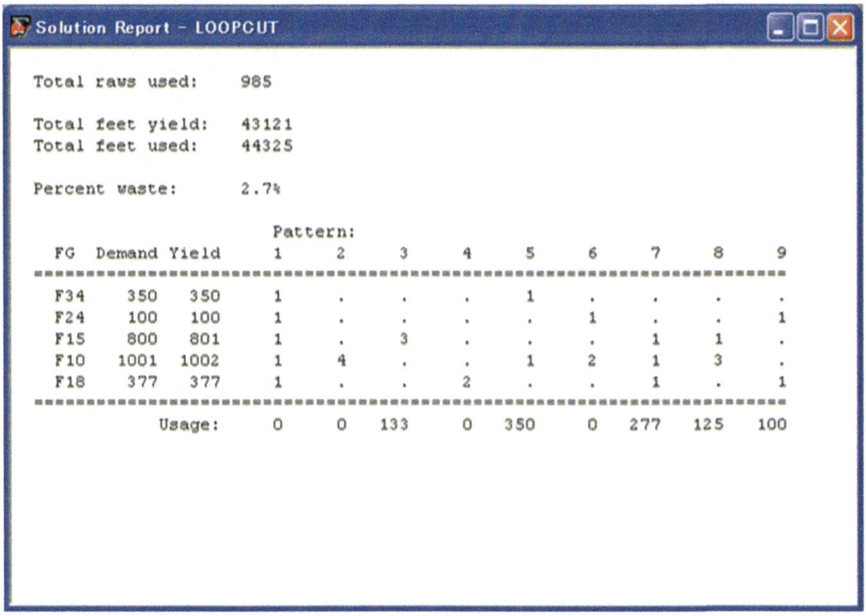

FG	Demand	Yield	1	2	3	4	5	6	7	8	9
F34	350	350	1	.	.	.	1	.	.	.	.
F24	100	100	1	.	.	.	.	1	.	.	1
F15	800	801	1	.	3	.	.	.	1	1	.
F10	1001	1002	1	4	.	.	1	2	1	3	.
F18	377	377	1	.	.	2	.	.	1	.	1
		Usage:	0	0	133	0	350	0	277	125	100

Fig. 6.4 LINGO solution report window

Easy model expression

LINGO can reduce user development time by allowing the user to formulate linear, nonlinear, and integer problems quickly in a highly readable form. The modeling language of LINGO allows the user to express models in a straightforward, intuitive manner using summations and subscripted variables, much like approaching the problem with pencil and paper. Models are easier to build, easier to understand, and, therefore, easier to maintain.

Convenient data options

LINGO makes data management easier by allowing the user to build models that pull information directly from databases and spreadsheets. Similarly, LINGO can output solution information directly into a database or spreadsheet, making it easier for the user to generate reports in their application of choice.

Powerful solvers

LINGO has a comprehensive set of fast, built-in solvers for linear, nonlinear (convex and non-convex), quadratic, quadratically constrained, and integer optimization. The user is not required to specify or load a separate solver; LINGO reads the formulation and automatically selects an appropriate solver.

Model interactively or through creation of turn-key applications

The user can build and solve models within LINGO, or LINGO can be called directly from an application written by the user. For developing models interactively, LINGO provides a complete modeling environment to build, solve, and analyze models. For turn-key solutions, LINGO comes with callable DLL and

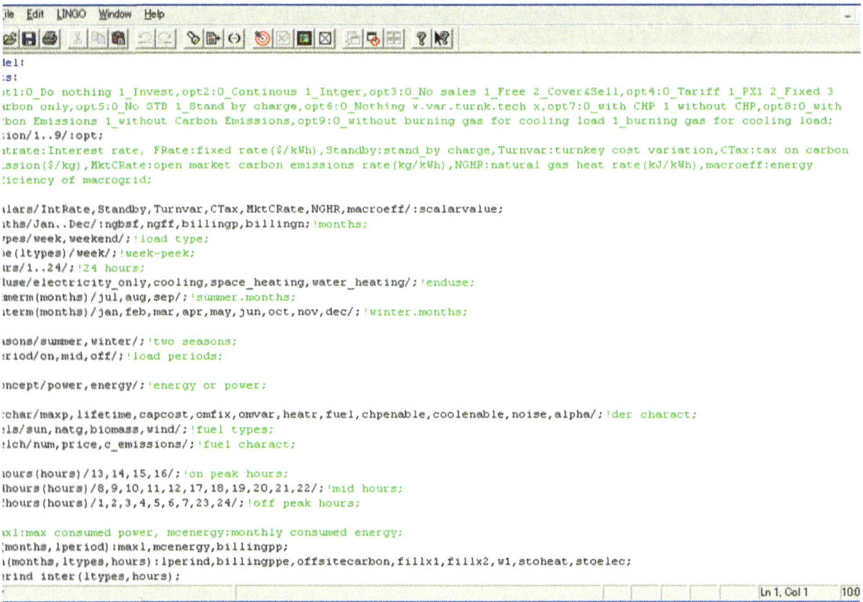

Fig. 6.5 General image of the LINGO model described in this chapter

OLE interfaces that can be called from user-written applications. LINGO can also be called directly from a Microsoft Excel macro or database application.

A LINGO model usually begins with the flag "Model:" and terminates with "End." Between these terms, the sets section, data section, calculation section, and various functions may be included. A general image of the LINGO model for the problem discussed in this chapter is shown in Fig. 6.5.

References

Afzal SS, Ryan F, Srijay G, Michael S, Jennifer LE, Chris M (2003) Distributed energy resources customer adoption modeling with combined heat and power applications, LBNL-52718

Gao W, Ren H (2011) An optimization model based decision support system for distributed energy systems planning. Int J Innovat Comput Informat Control 7(5):2651–2668

LINDO Systems Inc (2008) Lingo user's guide. LINDO Systems Inc., Chicago

Ren H, Gao W (2010) A MILP model for integrated plan and evaluation of distributed energy systems. Appl Energ 87(3):1001–1014

Ren H, Zhou W, Nakagami K, Gao W, Wu Q (2010) Multi-objective optimization for the operation of distributed energy systems considering economic and environmental aspects. Appl Energ 87 (12):3642–3651

Chapter 7
Local Low-Carbon Society Scenarios of Urban-Rural Linkage

Hongbo Ren and Weisheng Zhou

Abstract In order to realize a low-carbon society, the introduction of low-carbon energy resources in the local (both urban and rural) energy systems, which plays a main role in the total CO_2 emissions, is being paid more and more attention. In this chapter, the paths toward regional de-carbonization are indicated by conceptualizing systematic "urban-rural cooperation" that creates regional circulation of energy resources (e.g., biomass). As an illustrative example, an investigation has been conducted of assumed distributed energy systems for the Huzhou City, which is an environmental model city in China. By extending a pre-developed plan and evaluation model to include not only cost minimization but also emissions minimization, the proposed wide urban-rural energy systems are examined from both economic and environmental viewpoints. According to the simulation results, the energy system with urban-rural mutual cooperation is the best option from both economic and environmental viewpoints, and it results in a cost-effective reduction of CO_2 emissions up to 50% of current level. In addition, by endowing an economic value to the reduced CO_2 emissions through a local carbon price, the economic and environmental merits for urban and rural areas can be shared and balanced between both sides.

7.1 Research Objective and Energy Demands

In this chapter, Huzhou City, which has been declared a model environmental city by the Chinese government, is used as a case study to examine the effectiveness of local energy management based on urban-rural cooperation. According to local statistics, total electricity demand is approximately 9.6 billion kWh (4.2 billion kWh for urban

H. Ren
College of Energy and Mechanical Engineering, Shanghai University of Electric Power, Pudong, Shanghai, China

W. Zhou (✉)
College of Policy Science, Ritsumeikan University, Ibaraki, Osaka, Japan
e-mail: zhou@sps.ritsumei.ac.jp

W. Zhou et al. (eds.), *East Asian Low-Carbon Community*,
https://doi.org/10.1007/978-981-33-4339-9_7

and 5.4 billion kWh for rural areas), and thermal requirements are approximately 24.5 billion MJ (12.6 billion MJ for urban and 11.9 billion MJ for rural areas). Figures 7.1 and 7.2 illustrate monthly and hourly electricity consumption, respectively.

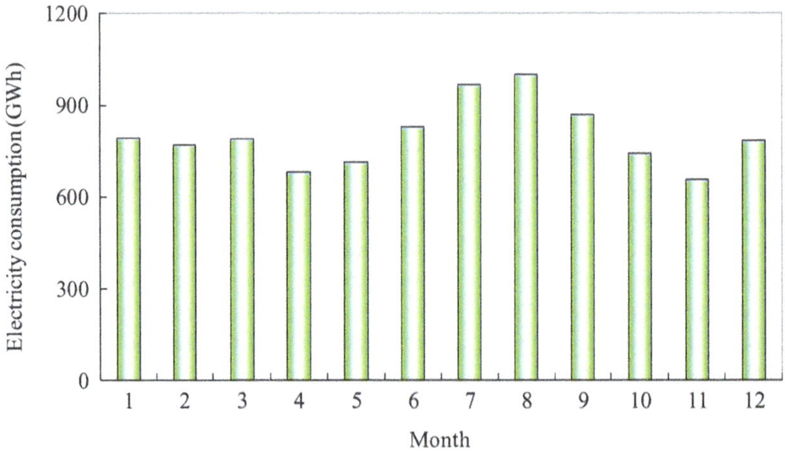

Fig. 7.1 Monthly electricity consumption in Huzhou City

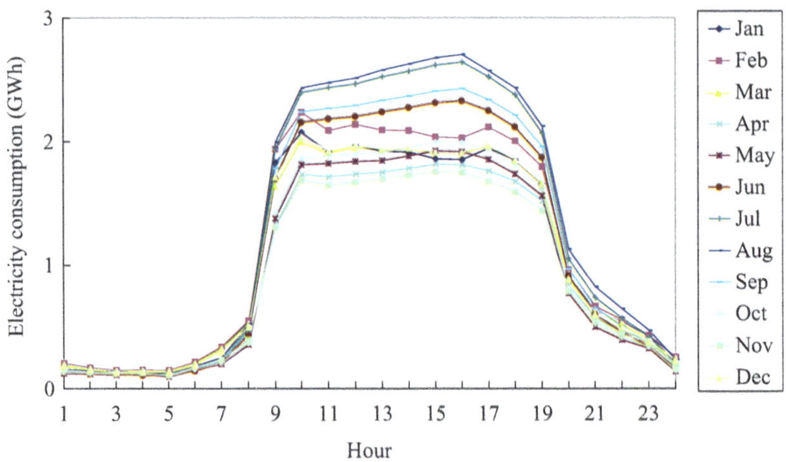

Fig. 7.2 Hourly electricity consumption in Huzhou City

7.2 Local Renewable Energy Stocks

As Huzhou City has a relatively low urbanization ratio, biomass is expected to be the main renewable energy resource in the local area. Figure 7.3 shows the annual distribution of available biomass energy stocks in urban and rural areas. Generally, agricultural biomass is the main biomass resource, followed by woody biomass, food biomass, sewage sludge, and livestock biomass. Biomass energy stocks in rural areas are about twice those in urban areas. Therefore, in order to make full use of local natural resources, rural biomass resources should be adequately considered.

Solar energy is another key local renewable energy resource that can contribute to low-carbon energy systems to a great extent. Figure 7.4 illustrates the monthly cumulative radiation and average temperature in the local area. Solar radiation is not consistent with average temperature. In July and August when the temperature is highest, monthly cumulative radiation is relatively low compared with, for example, May.

7.3 Local Energy System Scenarios

In order to understand the effect of dominance by particular areas of the city on formation of a local energy system based on urban-rural cooperation, three scenarios with different system structures are explored, plus a baseline scenario without cooperation for comparison. On the supply side, various technologies have been assumed; on the demand side, electricity and heat are the two main energy forms consumed in both urban and rural areas. The scenarios are described below (Ren et al. 2014):

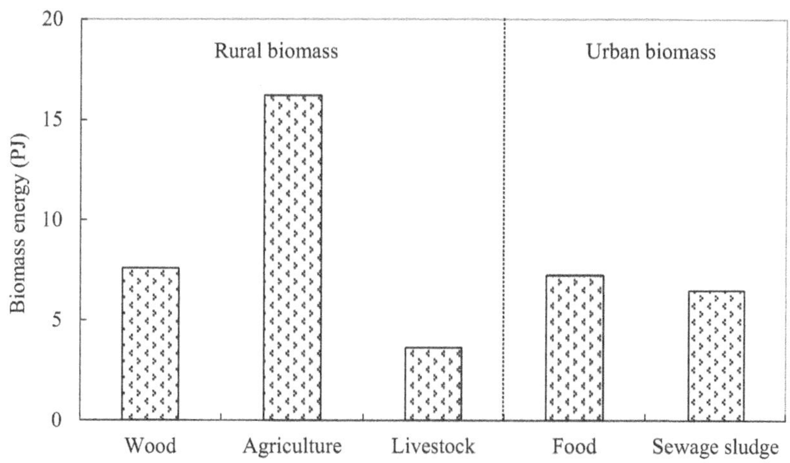

Fig. 7.3 Biomass energy reserves in urban and rural areas of Huzhou City

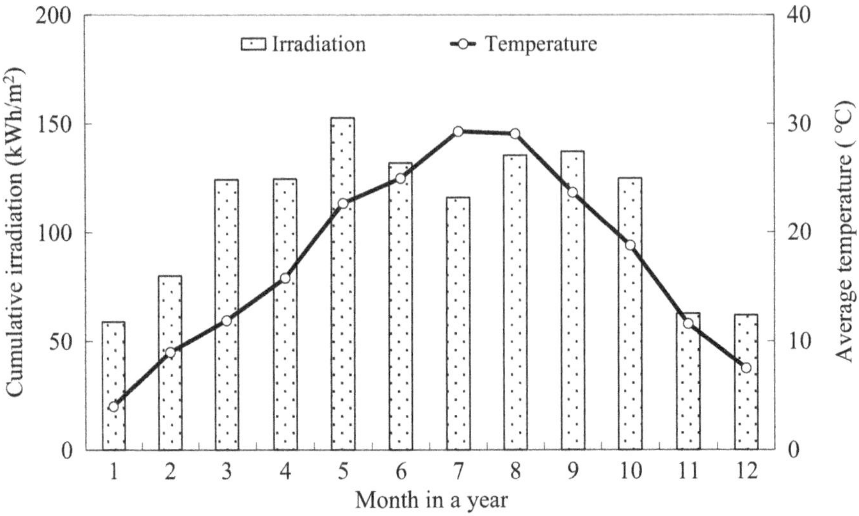

Fig. 7.4 Monthly cumulative irradiation and average temperature in Huzhou City

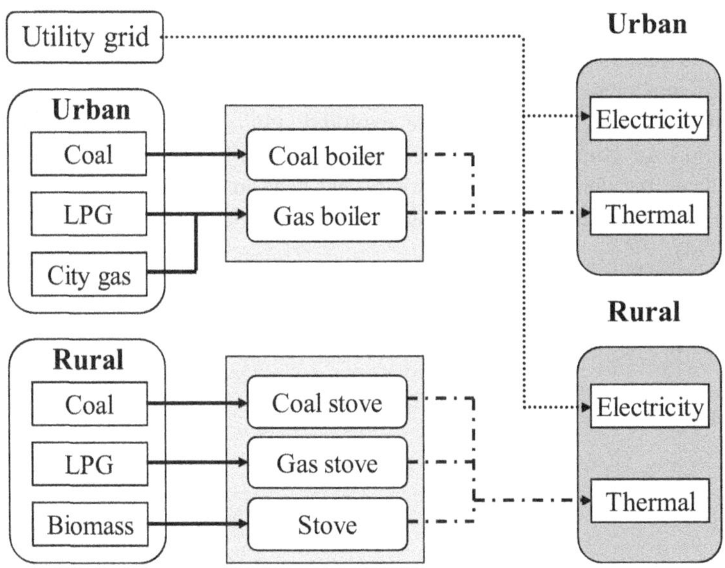

Fig. 7.5 Energy flow of Scenario 0 (no cooperation)

Scenario 0: No cooperation

In this scenario (Fig. 7.5), electricity demand in both urban and rural areas is supplied by the utility grid, generated mainly by fossil fuel combustion. Thermal requirements in urban areas are served by boilers fueled with coal, liquid

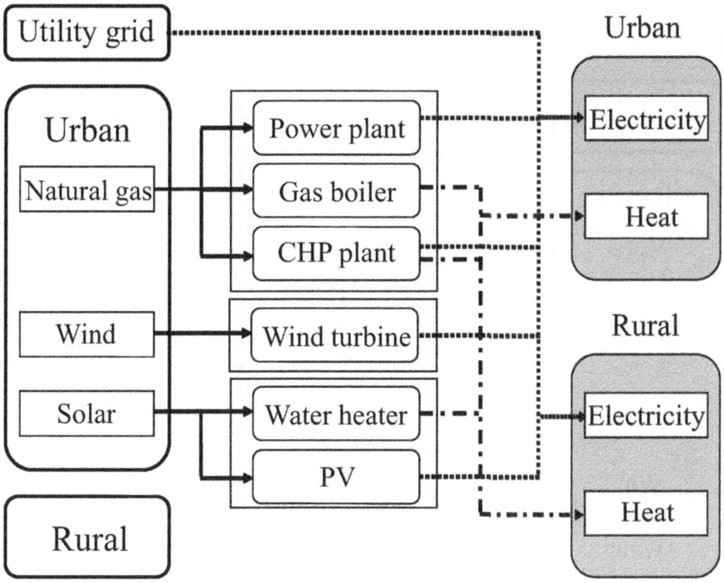

Fig. 7.6 Energy flow of Scenario 1 (urban dominant cooperation)

petroleum gas (LPG), or city gas. In contrast, conventional noncommercial biomass takes the main role in fulfilling thermal requirements in rural areas. Deficiencies are supplemented by coal or LPG-fueled stoves.

Scenario 1: Urban dominant cooperation

In this scenario (Fig. 7.6), urban areas play the main role in serving energy demands in the local area (covering both urban and rural areas). Natural gas, which is a relatively clean fossil fuel, is utilized for power generation, thermal supply, or a combination of both via combined heat and power (CHP) plants. Solar energy is also considered for both electricity and heat generation. If local power generation cannot satisfy total electricity demand, the utility grid can be used when necessary. This scenario is considered appropriate for cities with a relatively high rate of urbanization.

Scenario 2: Rural dominant cooperation

In this scenario (Fig. 7.7), rural areas are considered energy centers, through exhaustion of local natural resources. In addition to solar energy, rural biomass takes a dominant role in this scenario in order to satisfy both electrical and thermal requirements. Micro wind turbines and hydropower may also be available depending on local conditions. The utility grid and heat pumps are employed to service residual electrical and thermal demands, respectively. This scenario is anticipated to be appropriate for cities composed mainly of rural areas.

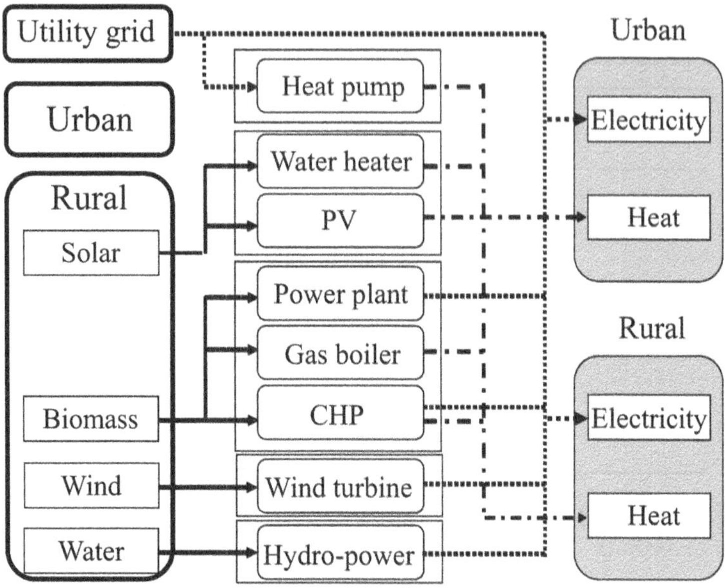

Fig. 7.7 Energy flow of Scenario 2 (rural dominant cooperation)

Scenario 3: Urban-rural mutual cooperation

In this scenario (Fig. 7.8), both urban and rural areas take an equal role in the local energy system. Natural resources in the local area are expected to be fully exploited as the main energy carrier to serve local energy requirements. Nevertheless, in order to guarantee consistent energy service, connection to local utilities remains necessary.

7.4 Energy Technology Options

As important input data for a bottom-up optimization model, technical information regarding both electricity and heat generation are assumed according to previous studies, as shown in Table 7.1 (Kojima et al. 2007; Nagata 2009; NEDO 2010; Ruan et al., 2009; Xu, 2008). Wind power and hydropower are not included in the following analysis due to local topographic conditions.

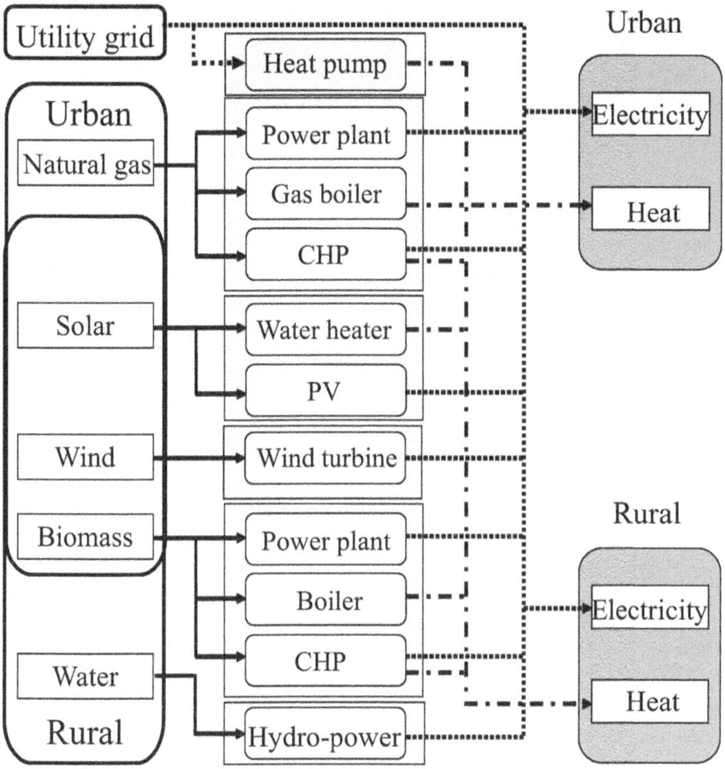

Fig. 7.8 Energy flow of Scenario 3 (urban-rural equal cooperation)

7.5 Optimal System Combination

Optimal system combination is an important part of designing a local energy system. For all three scenarios, the option with the lowest CO_2 emissions has greater total power generation capacity and smaller heat generation capacity than the option with lowest total cost (Fig. 7.9). This is because CO_2 emissions are minimized with widely used CHP technology, which leads to reduced thermal requirements from additional heat sources. In addition, when economic objectives are considered, most electricity requirements can be served by the utility grid, and thermal load is mainly supplied by solar heating supplemented by natural gas or biomass boilers. When environmental objectives are optimized, different scenarios have different optimal system combinations. For Scenario 1, a fuel cell system (PA) and natural gas power plant (NP) are the main options, with capacities of 1.9 GW and 0.7 GW, respectively. In Scenario 2, solar photovoltaic (PV) becomes dominant, followed by some biomass power generation. When urban and rural areas are coupled in a bilateral way (Scenario 3), all electrical and thermal demands can be satisfied by local generation,

Table 7.1 Cost and efficiency of technologies examined in this chapter

Fuel	Technology		Capital cost ($/kW)	Efficiency (%)	
	Index	Equipment		Electricity	Heat
Coal	CB	Boiler	89	–	70
LPG	LB	Boiler	111	–	80
	LS	Stove	22	–	40
Electricity	HP	Heat pump	5556	–	3
Natural gas	NB	Boiler	111	–	90
	NP	Power plant	1778	48	–
	DE	Diesel engine	2444	38	16
	PA	Phosphoric acid fuel cell	7778	40	40
	GE	Gas engine	2778	30	40
	GT	Gas turbine	2222	23	51
Solar	SH	Heater	222	–	30
	PV	PV	6667	14	–
Woody biomass	WF	Direct fired	5333	28	–
	WG	Gasification	7778	36	–
	WA	Boiler	1111	–	65
	WC	CHP	11,111	23	43
Agricultural biomass	AF	Direct fired	3889	18	–
	AG	Gasification	5556	25	–
	AB	Boiler	889	–	50
	AC	CHP	7444	20	40
Livestock biomass	LA	Anaerobic digestion	7556	25	–
	LG	Gasification	5556	22	–
	LB	Boiler	1667	–	55
	LC	CHP	13,333	25	45
Food biomass	FG	Gasification	7778	18	–
	FA	Anaerobic digestion	11,111	25	–
	FB	Boiler	2222	–	50
	FC	CHP	16,667	25	45
Sewage sludge	SF	Direct fired	8889	21	–
	SA	Anaerobic digestion	12,222	25	–
	SB	Boiler	1667	–	55
	SC	CHP	17,778	21	40

of which PV has the greatest capacity of approximately 2.3 GW, followed by biomass CHP systems.

Thus, although PV systems and biomass CHP plants result in reasonable environmental benefits, they are unpopular economically due to relatively high initial costs. As buffering options, solar heaters and heat pumps, as well as natural gas-fired power plants and CHP systems, can be introduced to reduce CO_2 emissions with acceptable costs over a short-term period.

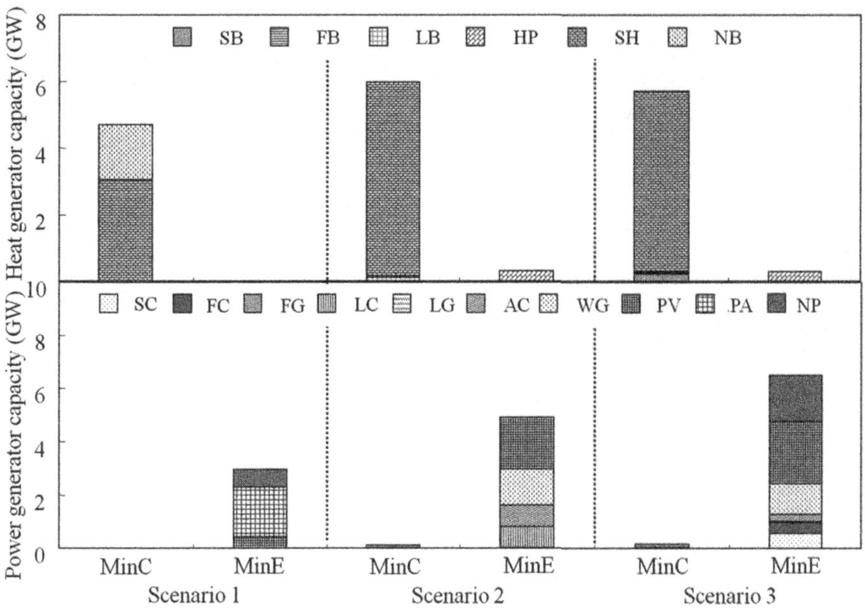

Fig. 7.9 Power and thermal generation capacity for various scenarios from economic and environmental perspectives

7.6 Economic and Environmental Performance

Simulated economic and environmental performance of different cooperative patterns in various scenarios is shown in Table 7.2. When total cost is minimized, Scenarios 2 and 3 both result in simultaneous reduction of cost and CO_2 emissions. In Scenario 1 (urban dominant cooperation), the emissions reduction ratio is relatively low and the total cost increases by 1.3% (compared with Scenario 0). When the environmental objective is minimized, annual energy costs dramatically increase, although CO_2 emissions are reduced to relatively low levels in all three scenarios. Scenario 1 has better economic and environmental performance than Scenario 2, in contrast to the situation with minimum cost. Therefore, natural gas is a good option to reduce CO_2 emissions, especially when utilized in a CHP system. In addition, from either an economic or environmental viewpoint, Scenario 3 (urban-rural mutual cooperation) is recognized as the best alternative with the lowest energy cost (economic objective) and CO_2 emissions (environmental objective).

Table 7.2 Economic and environmental performance of various scenarios

Scenario	Total cost (million USD)	Total CO_2 emissions (million ton)	Cost reduction ratio (%)	Emissions reduction ratio (%)
Total cost minimization (MinC)				
Scenario 1	1201	10.2	−1.3	7.2
Scenario 2	1044	8.4	11.9	24.0
Scenario 3	1036	7.9	12.6	28.5
CO_2 emissions minimization (MinE)				
Scenario 1	2232	4.1	−88.4	62.6
Scenario 2	3209	5.5	−170.8	50.1
Scenario 3	3564	1.6	−200.8	85.1

7.7 Influence of CO_2 Emission Constraints on System Combination

In the following analysis, while constraining annual CO_2 emissions to a 10–50% reduction, system combinations are analyzed with annual total cost as the objective function to be minimized. For Scenario 1, as the emissions reduction ratio increases, the share of energy from the utility grid gradually decreases, and natural gas becomes the dominant energy carrier (Fig. 7.10). In addition, solar heat is introduced with a constant value, which results in steady total energy supply. The emissions reduction cost increases from 0.023 USD/kg CO_2 to 0.058 USD/kg CO_2 when CO_2 emissions are reduced to half their current value.

In Scenario 2, as in Scenario 1, the power share from the utility grid gradually decreases as the emissions reduction ratio increases (Fig. 7.11). Solar energy is introduced, and the application pattern gradually changes from solar heat to solar power. Therefore, the PV system has better environmental performance than the solar heater. Rural biomass is also widely used in this scenario. Interestingly, total energy consumption increases sharply as the emissions reduction ratio increases from 10 to 20% and then decreases gradually. The system has a negative reduction cost unless CO_2 emissions are reduced to less than 35% of their current value; economic and environmental benefits can be achieved simultaneously. However, once total emissions are reduced by half, the reduction cost increases to as much as 0.08 USD/kg CO_2.

Scenario 3 is the best option in terms of cost-effectiveness in reducing CO_2 emissions because of its negative reduction cost even when annual CO_2 emissions are halved (Fig. 7.12). Solar power is not adopted in this scenario; as an alternative, biomass energy in both urban and rural areas is exhausted. Natural gas accounts for a

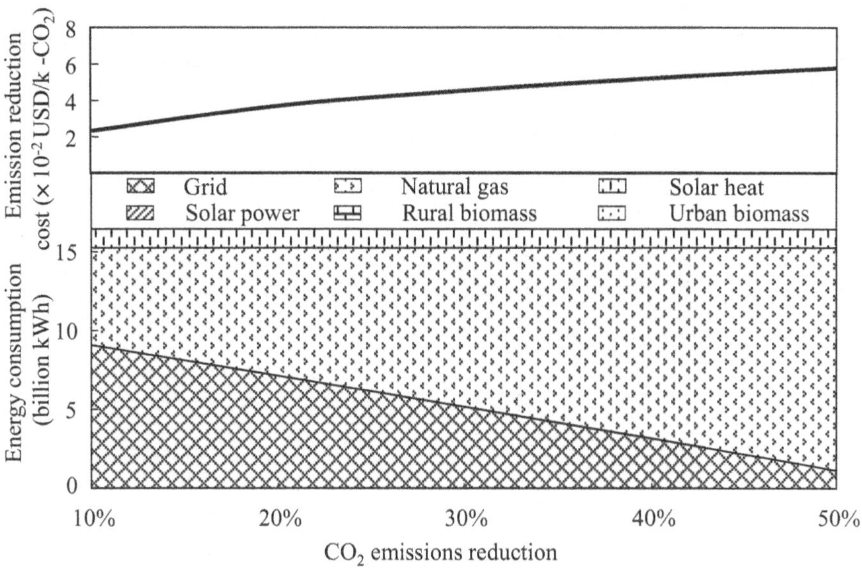

Fig. 7.10 Energy share with various CO_2 emissions reduction ratios and corresponding reduction costs (Scenario 1)

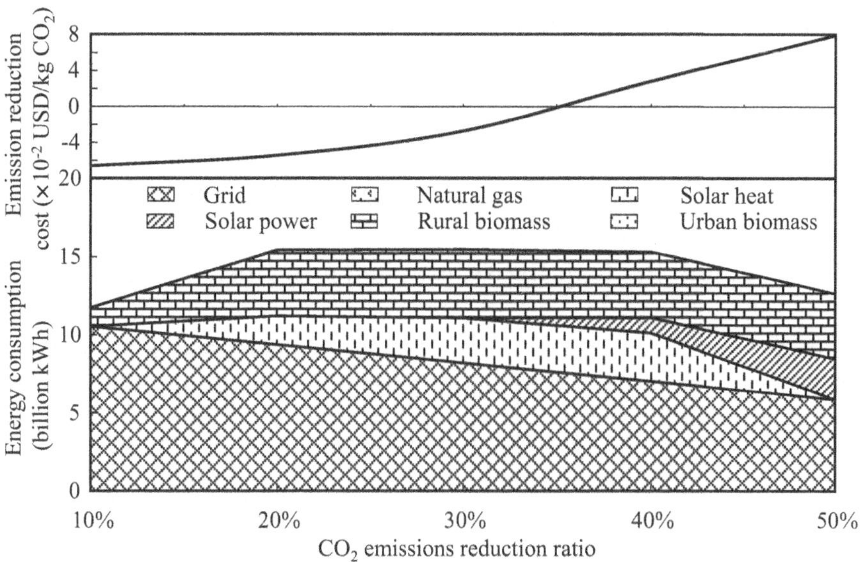

Fig. 7.11 Energy share with various CO_2 emissions reduction ratios and corresponding reduction costs (Scenario 2)

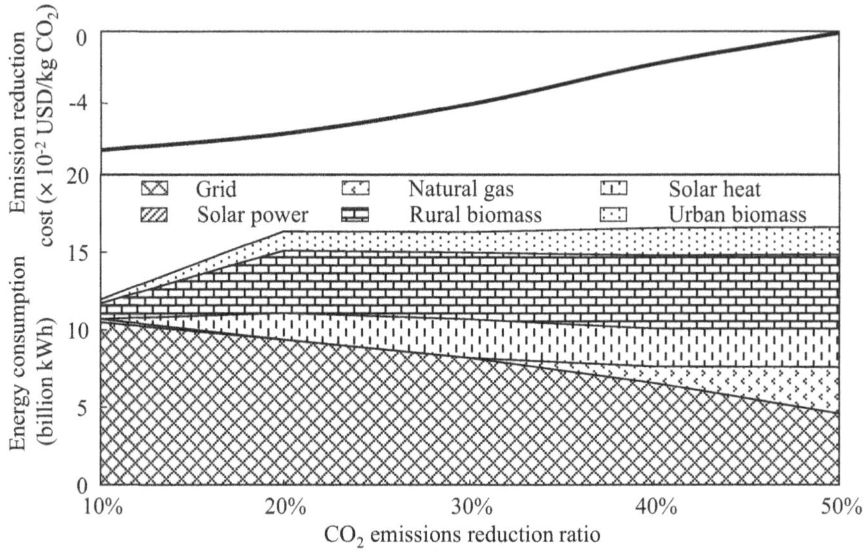

Fig. 7.12 Energy share with various CO_2 emissions reduction ratios and corresponding reduction costs (Scenario 3)

small share of total energy consumption as the reduction ratio is over 30%. Furthermore, similar to Scenario 2, a sharp increase in energy consumption appears when CO_2 emissions are reduced by 20%. This implies that CO_2 emissions and energy consumption do not have a definite linear relationship. This is because although some renewable technologies have no emissions, their efficiencies are relatively low.

7.8 Bridging the Gap Between Urban and Rural Areas

Scenario 3 is considered the best of the three options. In this section, based on the assumption illustrated in Scenario 3, the economic and environmental benefits arising from the introduction of local energy systems are reallocated and balanced between urban and rural areas. This is based on the equality of rights of urban and rural citizens to pay the same energy price and emit the same volume of CO_2. In order to realize this concept, a local carbon price should be assumed, which is employed to transfer money and emission rights between urban and rural areas (Eq. 7.1).

$$C_p = \frac{|C_u - c \cdot \mathrm{PE}_u|}{|E_u - e \cdot P_u|} = \frac{|C_r - c \cdot \mathrm{PE}_r|}{|E_r - e \cdot P_r|} \tag{7.1}$$

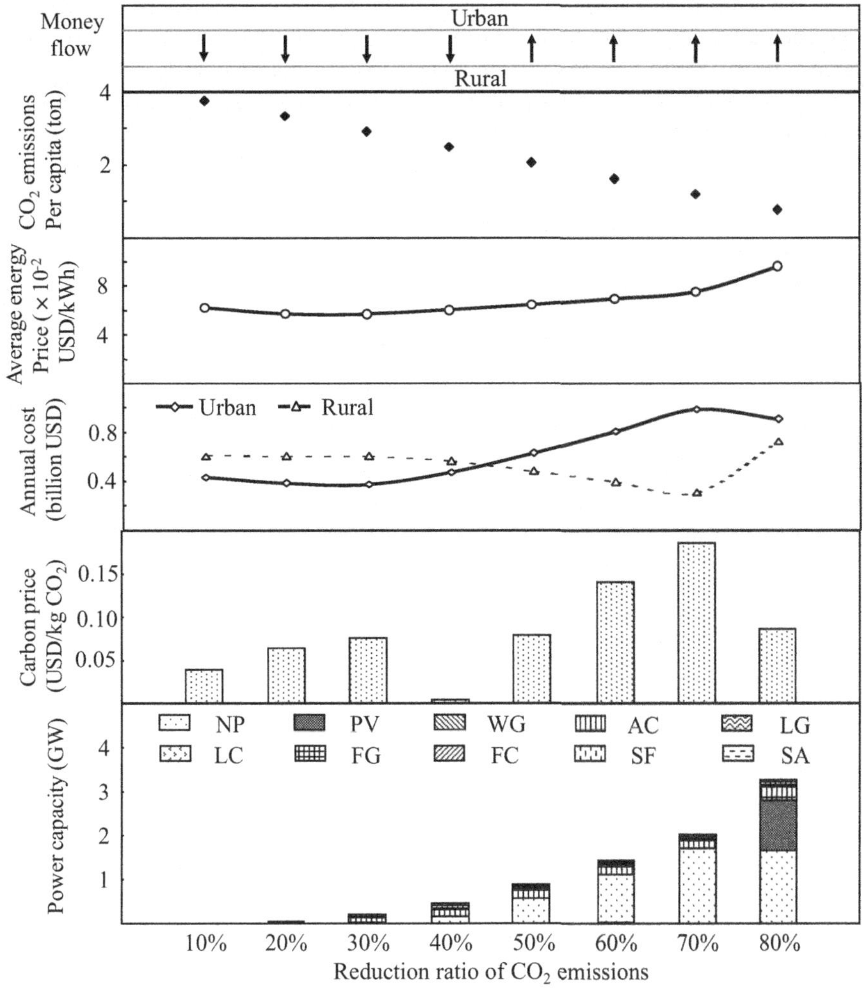

Fig. 7.13 Balance of economic and environmental performance between urban and rural areas

where C_p is the local carbon price, PE denotes annual primary energy consumption, and P is the population. Urban and rural areas are indicated by u and r, respectively, while c and e are the mean energy price and CO_2 emissions per capita, respectively.

Based on this, Fig. 7.13 shows the economic and environmental performance in both urban and rural areas under different reduction targets. Average energy price decreases with decreasing CO_2 emissions, reaching a minimum when the reduction ratio is approximately 30%, after which it gradually increases. When the reduction ratio is relatively low, rural areas have higher energy costs; conversely, the cost in urban areas is higher with larger reduction ratios.

A balanced local carbon price does not illustrate a clear relationship with the reduction ratio. When CO_2 emissions are reduced by 40%, a relatively low carbon price of 0.01 USD/kg CO_2 is sufficient to balance bilateral benefits. Local generation capacity also increases gradually. The PV system is not employed, and natural gas power plants are the dominant option until the reduction ratio approaches 80%.

References

Ren H, Wu Q, Ren J, Gao W (2014) Cost-effectiveness analysis of local energy management based on urban-rural cooperation in China. Appl Thermal Eng 64(1-2):224–232

Nagata Y (2009) The effectiveness of technology options on large-scale CO2 emission reduction toward 2050 in Japan. Research report of Central Research Institute of Electric Power Industry. Central Research Institute of Electric Power Industry, Kanagawa, pp 1–26

New Energy and Industrial Technology Development Organization (NEDO) (2010) http://www.nedo.go.jp/english/index.html

Kojima A, Takahama H, Ashizawa M (2007) Survey and analysis of domestic biomass combustion power generation system. Research report of Central Research Institute of Electric Power Industry. Central Research Institute of Electric Power Industry, Kanagawa, pp 1–37

Ruan Y, Liu Q, Zhou W, Firestone R, Gao W, Watanabe T (2009) Optimal option of distributed generation technologies for various commercial buildings. Appl Energy 86(9):1641–1653

Xu P (2008) Study on energy conservation and performance of vacuum tube solar water heater. Mater thesis, The University of Kitakyushu, 25–35

Part III
Technology Innovation for Low-Carbon Community

Chapter 8
Spatial-Temporal Distribution of Carbon Capture Technology According to Patent Data

Huan Liu, Weisheng Zhou, and Xuepeng Qian

Abstract Energy-saving technology, fuel switching, afforestation, new energy, and carbon capture use and storage (CCUS) are five technologies that can alleviate global climate change. In the 2 °C scenario (2DS), carbon capture and storage (CCS) technologies deliver 14% of cumulative CO_2 emissions reductions, with around 142 Gt CO2 captured in the period to 2060. The beyond 2 °C scenario (B2DS) deploys CCS more widely and rapidly in order to aim for the "well below 2 °C" target of the Paris Agreement. CCS is a currently available technology that can allow industrial sectors to meet ambitious emissions reduction goals. The development of carbon capture technology is vital to make CCS viable. This chapter presents an overview of carbon capture technology, in an attempt to characterize spatial and temporal distribution based on patent bibliometrics. We designed a retrieval strategy and built a database of 9847 patents. We reveal two phases in annual patent counts: slowly increasing and rapidly increasing. The year 2006 represents a turning point. Eight countries hold the most patents in terms of the top 50 patent assignee codes. This chapter describes relative technical development trends across these major countries. Based on our analysis, we speculate that an important period in carbon capture technology is approaching.

H. Liu
Economics and Social Development Institute, Hainan Construction Project Planning and Design Research Institute Co., Ltd., Haikou, China

W. Zhou (✉)
College of Policy Science, Ritsumeikan University, Ibaraki, Osaka, Japan
e-mail: zhou@sps.ritsumei.ac.jp

X. Qian
College of Asia Pacific Studies, Ritsumeikan Asia Pacific University, Beppu, Oita, Japan
e-mail: qianxp@apu.ac.jp

W. Zhou et al. (eds.), *East Asian Low-Carbon Community*,
https://doi.org/10.1007/978-981-33-4339-9_8

8.1 Basic Principles of Mitigation and Adaptation Technology

8.1.1 Climate Change and CCS

The warmest year ever recorded was 2015 (WMO 2016). Most of the observed increase in global average temperatures since the mid-twentieth century is very likely (>90% probability) due to observed increases in anthropogenic greenhouse gas (GHG) concentrations (IPCC 2008). The climate will continue to change over the coming decades as increasing amounts of heat-trapping GHGs emitted by anthropogenic activities accumulate in the atmosphere. If costs are disregarded, excessive GHG concentrations can be reduced by 15 available technologies, including carbon dioxide capture and storage (CCS) (Socolow and Pacala 2006). CCS is a currently available technology that can allow industrial sectors (e.g., fossil fuel power generation, iron, steel, and cement manufacture, natural gas processing, oil refining) to meet ambitious emission reduction goals. CCS can contribute one-sixth of the CO_2 emission reductions required by 2050 and can contribute 14% of cumulative emissions reductions between 2015 and 2050 compared to a business-as-usual approach, which would correspond to a 6 °C rise in average global temperature (IEA 2012).

CCS is not a single technology but involves integrated implementation of the following processes: separation of CO_2 from mixtures of gases (e.g., flue gases from a power station or a stream of CO_2-rich natural gas) and compression of this CO_2 to a liquid-like state; transport of the CO_2 to a suitable storage site; and injection of the CO_2 into a geologic formation where it is retained by a natural (or engineered) trapping mechanism and monitored as necessary (IEA 2013). To enhance CO_2 usage, it is integrated into CCS as CCUS in some countries, allowing CO_2 to be used in fields such as enhanced oil recovery (EOR), enhanced coal-bed methane (ECBM), CO_2 chemical utilization, and CO_2 biotransformation. There is essentially no difference between CCS and CCUS.

Many CCS technologies are commercially available and can be applied across different sectors. CO_2 capture technologies include different systems (Fig. 8.1) such as post-combustion, pre-combustion, and oxy-fuel combustion, which are available in natural gas processing, fertilizer manufacturing, and hydrogen production. CO_2 transport technologies are the most technically mature in CCS and include pipeline

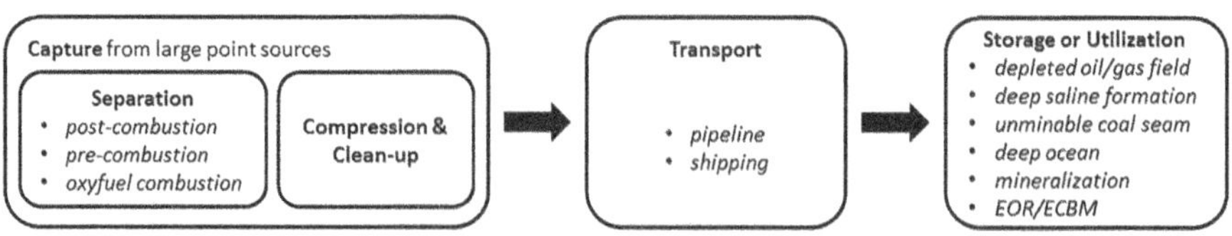

Fig. 8.1 Carbon capture and storage (CCS) chain. Source: Rubin et al. (2012), IEA (2013)

and shipping. CO_2 storage technologies in oil and gas industries are similar and are economically feasible under specific conditions.

8.1.2 Carbon Capture Innovation

In most CCS systems, the cost of capture (including compression) is the largest cost component, due to an additional high energy penalty. This can be reduced by technical development and economies of scale (Meyer et al. 2005; van Alphen et al. 2010). The development of CO_2 capture technologies is vital if CCS is to be viable (Quintella et al. 2011). In patent bibliometrics, most CCS patents refer to CO_2 capture technologies in testing extractions. Studies reveal that carbon capture patents account for a large proportion of CCS patents (e.g., Wang et al. 2010; Hong et al. 2013). Hence, suitability of CCS in industrial applications mostly depends on the costs and readiness of carbon capture (IEA 2012). Thus, we focus on carbon capture technology in this chapter.

There are many ways to measure innovation (OECD 2016) that can be divided into two categories: input-based indicators and output-based indicators. Patents are considered a significant indicator of innovation output or a tangible sign of knowledge that can give valuable insight into innovative activity in object technology (Griliches 1990). The main advantage of patents is that they are publicly available over long time periods and provide detailed technological information (Oltra et al. 2010). Patents are used to provide a comprehensive view of innovation in many domains, including low-carbon technology development. Previous information on patent counts has provided a wealth of information on innovations and inventors (e.g., Acs et al. 2002; Johnstone et al. 2010; OECD 2010; Liu et al. 2011; Leu et al. 2012; Albino et al. 2014; Park 2014; UKIPO 2014).

Several studies have investigated CCS technology or carbon capture technology using patent bibliometrics. Dechezlepretre et al. (2009), WIPO (2009), and OECD (2010) analyzed a cluster of climate change mitigation technologies and give brief overviews of CCS technology. Quintella et al. (2011), Li et al. (2013), and Wei and Man (2014) focused on specific technical routes and reagents in carbon capture, drawing on patent bibliometrics, sometimes in combination with article bibliometrics. Wang et al. (2010) drew a patent map and discuss technological features of CCS in nine countries. Hong et al. (2013) drew development paths by patent citation analysis. These studies are compared in Table 8.1.

These studies can be split into two categories based on their purpose: general reports that consider CCS as a supplement rather than essential and specialized studies that focus on technical applications. Few studies include spatial-temporal analysis across all carbon capture technologies. To the best of our knowledge, there has been no detailed consideration of developmental trends and technological distribution. Patent count differs widely across studies, from 945 to 9840, probably due to use of different databases, diverse retrieval strategies, and varying time spans. In general, searching in several databases (e.g., WIPO 2009 used six databases) or an

Table 8.1 Former studies related to carbon capture technology by patent bibliometrics

References	Description	Database set	Source
Dechezlepretre et al. (2009)	Took CCS as one of the 13 climate change mitigation technologies, provided an analysis of geographic distribution and international diffusion in an overall view. CCS is negligible, 0.35% of all fields' patents	954 CCS patents	PATSTAT
WIPO (2009)	Took CCS as one of nine alternative energy technologies, described CCS trends very briefly in terms of technology, although CCS is not, strictly speaking, an alternative energy	6858 CCS patents	EPO, WIPO, USPTO, JPO, KIPO, SIPO
OECD (2010)	Took carbon capture as one of the 10 CCMTs, presented growth rate, inventive activity, major applicants, and average patent family size with other technologies together	8069 carbon capture patents	PATSTAT
Wang et al. (2010)	Presented CCS patent key word map by Thomson Data Analyzer and Aureka, described growth and IPC distribution briefly, discussed nine countries' technological features	1171 CCS patents	DII
Quintella et al. (2011)	Gave an overview through patents applications and scientific articles together, presented five technical routes: absorption, adsorption, membranes, enzymatic, and thermodynamics	1123 carbon capture patents	EPO
Hong et al. (2013)	Put forward research ideas based on patent citation analysis approach, different from other refers in this table. Drew the CCS development map and identified the main paths by algorithm of path recognition which is based on patent citation	1498 CCS patents	USPTO
Li et al. (2013)	Gave details on seventy-nine representative patents, presented three technical routes: solvent, sorbent, and membrane, made perspectives on potential technical routes	9840 carbon capture patents	Espacenet
Wei and Man (2014)	Gave a briefly overview of four technical routes (absorption, adsorption, cryogenic, and membrane) and three reagents (absorption reagent, adsorption reagent, and membrane reagent) by patent and article analysis, discussed CCS policies in China	1344 carbon capture patents	USPTO

(Note: PATSTAT for European Patent Office Worldwide Patent Statistical Database, EPO for European Patent Office, WIPO for World Intellectual Property Office, USPTO for United States Patent and Trademark Office, JPO for Japan Patent Office, KIPO for Korean Intellectual Property Office, SIPO for State Intellectual Property Office of the People's Republic of China, DII for Derwent Innovations Index database)

integrated database (e.g., OECD 2010 used PATSTAT and Li et al. 2013 used Espacenet) provides more comprehensive results. However, Dechezlepretre et al. (2009) and Hong et al. (2013) obtained fewer patents from the integrated database PATSTAT than the Derwent Innovations Index database. This is likely due to use of an incomplete search expression.

Patent family analysis is popular. A patent family is a set of similar patents taken in various countries to protect a single invention. Therefore, analysis by patent family more accurately reflects the total number of inventions. We extend these previous studies through the creation of a more complete database using more suitable search expressions to obtain data from the Derwent Innovations Index database. We highlight developmental stages over a 40-year period and analyze characteristics by country.

8.1.3 Patent Database and Retrieval Strategy

The Derwent Innovations Index (DII) database of the Web of Science (from the I.S.I. Web of Knowledge) was used to search and analyze patent data. The DII database is a widely accepted source of patent data covering over 14.3 million basic inventions from 40 patent-issuing authorities worldwide. In patent bibliometrics, the most fundamental task is to set the range of retrieval and choose appropriate index words. Based on previous studies, search indicators include two dimensions: search topics and International Patent Classification (IPC) codes.

We collected keywords of types of carbon capture processes, combined as single topic search expressions using logical operators. For more rigorous search results, we used a thesaurus to find synonyms of IPC codes related to those keywords and combined them as single IPC code search expressions using logical operators. The IPC system is the most popular hierarchical classification of patents among countries or organizations with official patent offices and was launched by the World Intellectual Property Organization (WIPO). We tested dozens of combined search topics and IPC codes, comparing results of inputting different expressions, and then adopted the most suitable expressions. In testing, we found a dozen globally renowned vehicle companies. However, they do not operate professionally in the domain of carbon capture. We found most patents from vehicle companies were related to pollution control of automobile exhaust gases. Therefore, we excluded this technology by using the logical operator "not." This appears to be the first time that exhaust control technologies have been excluded from patent data searches using logical operators, resulting in a dataset that is more concentrated on carbon capture patents. The expressions of search topics and IPC codes used are summarized in Tables 8.2 and 8.3, respectively.

The final patent data search was conducted in April 2016. A total of 9847 patent files were found, filed over a time span of 40 years from 1976 to 2015. For each patent in our database, several data fields were extracted, including authorization year, assignee name, assignee code, and IPC code.

Table 8.2 Descriptions of search topics

	Search topics	Descriptions
1	CO_2	CO_2, CO_2
2	Carbon dioxide	Carbon dioxide
3	captur*	Capture, captured, capturing, etc.
4	recover*	Recover, recovery, recovered, recovering, etc.
5	separat*	Separate, separated, separating, separation, etc.
6	remov*	Remove, removed, removing, remover, etc.
7	absor*	Absorb, absorbed, absorbing, absorbent, absorpt, absorptive, absorption, etc.
8	adsor*	Adsorb, adsorbed, adsorbing, adsorbable, adsorp, adsorption, adsorptive, etc.
9	membran*	Membrane, membranous, etc.
10	cryogen*	Cryogen, cryogenic, etc.
11	enzym*	Enzyme, enzymes, enzymic, enzymolysis, etc.
12	combust*	Combust, combusted, combusting, combustion, combustible, etc.
13	puri*	Purify, purified, purifying, purification, etc.
14	concentrat*	Concentrate, concentrated, concentrating, concentration, etc.
15	extract*	Extract, extracted, extracting, extractive, extraction, extractable, extractant, etc.
16	compress*	Compress, compressed, compressing, compressive, compressible, compression, etc.
17	thermo*	Thermo, thermodynamics, etc.
18	car	Car.
19	auto*	Auto, automobile, automotive, etc.
20	vehicl*	Vehicle, etc.
21	engin*	Engine, etc.
22	exhaust*	Exhaust, exhausted, exhausting, etc.

8.2 Temporal Distribution of Carbon Capture Technology

Practitioners and researchers are often interested in trends within a certain field of technology. To evaluate progress in carbon capture innovations, we extracted data on authorization year from the database and identified shifts in annual growth and cumulative development.

8.2.1 Annual Development of Patent Count

Figure 8.2 presents carbon capture patents authorized annually since 1976, measured by patent count. There are two clear phases: (1) a fluctuating, slowly increasing phase and (2) a sharply increasing phase. It appears that 2006 was a turning point in technology development. In the slowly increasing phase prior to 2006, the annual

Table 8.3 Thesaurus descriptions of search IPC codes

	IPC codes	Thesaurus descriptions
1	B01D-000/00	Separation
2	B01D-053/00	Separation of gases or vapors; recovering vapors of volatile solvents from gases
3	B01D-053/02	By adsorption
4	B01D-053/04	With stationary adsorbents
5	B01D-053/14	By absorption
6	B01D-053/18	Absorbing units; liquid distributors therefore
7	B01D-053/22	By diffusion
8	B01D-053/78	With gas-liquid contact
9	B01D-053/86	Catalytic processes
10	B01D-053/94	By catalytic processes
11	C01B-003/38	Using catalysts
12	C01B-031/20	Carbon dioxide
13	F23J-015/00	Arrangements of devices for treating smoke or fumes
14	F25J-001/00	Processes or apparatus for liquefying or solidifying gases or gaseous mixtures
15	F25J-003/00	Processes or apparatus for separating the constituents of gaseous mixtures involving the use of liquefaction or solidification
16	F25J-003/02	By rectification, i.e., by continuous interchange of heat and material between a vapor stream and a liquid stream
17	F25J-003/04	For air
18	F25J-003/08	Separating gaseous impurities from gases or gaseous mixtures
19	H01M-008/04	Auxiliary arrangements or processes, e.g., for control of pressure, for circulation of fluids
20	H01M-008/06	Combination of fuel cell with means for production of reactants or for treatment of residues

patent count increases slowly, showing negative growth in some years. The average annual growth rate (AAGR) is 2.6%. The annual average is approximately 133 patents. In the sharply increasing phase after 2006, the annual patent count increases quickly; the AAGR is 23.8% from 2006 to 2013. This phase presents a linear

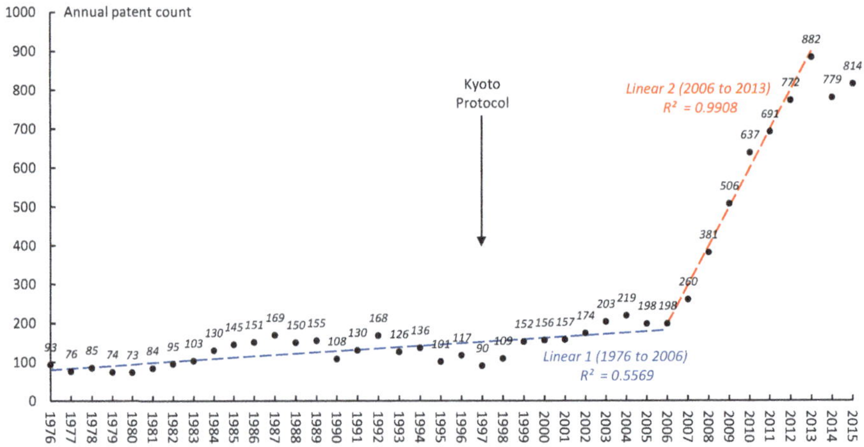

Fig. 8.2 Annual patent count of carbon capture by year authorized; linear 1 presents the trend from 1976 to 2006 and linear 2 that from 2006 to 2013. Source: Derwent Innovations Index Database (2016)

increase with high curve similarity: $R^2 = 0.9908$. The annual average count from 2006 to 2015 is approximately 636, nearly five times that of the previous phase.

The sharp increase since 2006 implies many concurrent shifts in policy and technology conditions. Firstly, there was wide agreement among 100 surveyed experts that capture facilities do not substantially differ from conventional industrial facilities (van Alphen et al. 2010). Thus in most years, carbon capture technology was applied in general sectors and grew slowly. As carbon capture became a crucial technology in CCS, innovation increased rapidly. Secondly, it usually takes several years for governments and innovators to deploy innovations. The increasing trend in carbon capture technology innovations in this century seems to reflect the significant influence of climate change policies since the signing of the Kyoto Protocol in 1997. The Kyoto Protocol affected innovations in a cluster of climate change mitigation technologies including CCS technology, probably due to the swift reaction of innovators to policy changes. The private sector received a strong signal, and many countries took action even before ratification of the protocol (Dechezlepretre et al. 2009). Thirdly, the increasing trend is also related to environmental concerns and commercial drivers (Li et al. 2013). The Sleipner project on the Norwegian continental shelf is a milestone effort in carbon mitigation, called "the mother of all CCS projects." It has separated and injected 1 Mt. CO_2 into saline formation each year since 1996 (Torp and Gale 2004). Several CCS projects are driven by commercial utilization of CO_2, such as CO_2 injection for EOR. Finally, patent counts in 2014 and 2015 do not correspond with reality, due to a time lag between application and authorization. Therefore, the last 2 years are used for comparison rather than in overall analysis of trends.

8.2.2 *Life Cycle of Carbon Capture Technology*

The concept of technology life cycles as a means to measure technological changes was proposed by Little (1981). The life cycle of a technology includes two dimensions and four stages. The two dimensions are competitive impact and integration in products or processes—a patent can be regarded as a product of innovation. The four stages are emerging, growth, maturity, and saturation. According to Little's definition, the emerging stage is a new technology with low competitive impact and low integration in products or processes. In the growth stage, there are pacing technologies with high competitive impact that have not yet been integrated in new products or processes. In the maturity stage, some pacing technologies are integrated into products or processes. In the saturation stage, a technology becomes a base technology and may be replaced by a new technology. Figure 8.3 illustrates the S-curve definition of the four stages.

Figure 8.4 presents the cumulative patent count of carbon capture technologies from 1976 to 2015. The turning period corresponding with the theoretical model in Fig. 8.3 is highlighted. Two stages of the technology life cycle are found: emerging and growth. We expect the growth stage to continue, with a sharp increase in patent counts in the future. It should be noted that these two stages of technological life cycle are defined in terms of cumulative patent count, which differs from the two phases by annual patent count shown in Fig. 8.2.

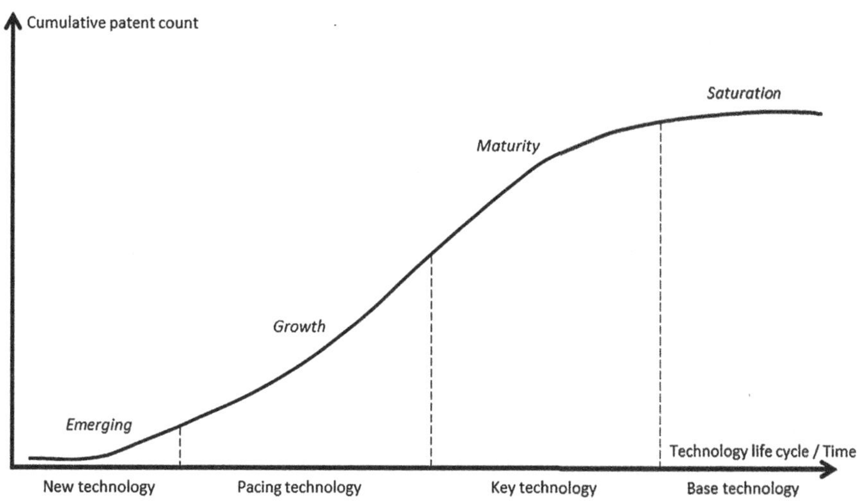

Fig. 8.3 S-curve concept of the life cycle of a technology by cumulative patent count. Source: modified from Ernst (1997)

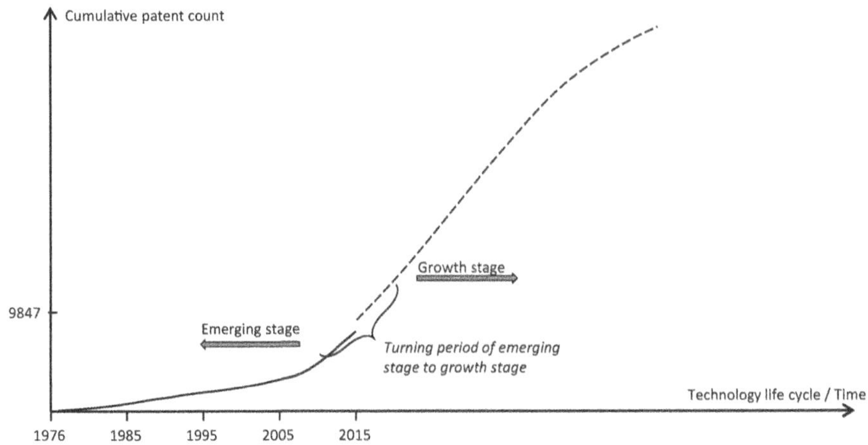

Fig. 8.4 Life cycle of carbon capture technology by cumulative patent count. Source: Derwent Innovations Index Database (2016)

8.3 Global Distribution of Carbon Capture Technology, Including Japan, China, and South Korea

We identified the spatial distribution of patents across countries in terms of assignee code and try to highlight relative development trends among high-tech countries using assignee code data. Every patentee in the Derwent Innovations Index has one assignee code. An assignee code usually contains many patentee names. Patentee names contained in one assignee code are always related in some way, such as by belonging to the same company, having combined through company merger and restructuring, or as multiple spellings of a single company name.

8.3.1 Spatial Distribution of Carbon Capture Technology

We extracted assignee code data from the carbon capture patent database. We focus on the top 50 assignee codes from a total of 4613 codes (Table 8.4). The top 50 assignee codes together own 4232 patent files, accounting for 42.97% of the 9847 patents filed. The top 50 assignees include most influential companies, institutes, and universities working on carbon capture technology globally. As six assignee codes tied in 50th place, there are 55 assignee codes in the top 50.

Figure 8.5 shows that the top 50 assignee codes and 4232 patents are concentrated in eight countries—Japan, the United States, France, China, Germany, the Netherlands, the United Kingdom, and South Korea, in descending order. East Asia, the European Union, and North America have the most patents globally, which is related to economic vitality in these areas. China is the only developing country whose

Table 8.4 Basic information of top 50 assignee codes. (Note: Totally 55 assignee codes because of six tying in 50th. Data Source: Derwent Innovations Index Database 2016)

Assignee code		Main assignee name	Patent count	Country	Classification
1	AIRL-C	Air Liquide SA	349	France	Company
2	LINM-C	Linde AG	229	Germany	Company
3	AIRP-C	Air Prod & Chem Inc	216	USA	Company
4	MITO-C	Mitsubishi Heavy Ind Co Ltd	211	Japan	Company
5	SHEL-C	Shell Int Res MIJ BV	195	Netherlands	Company
6	INSF-C	Inst Francais DU Petrole	152	France	Institute
7	ALSM-C	Alstom Technology LTD	148	France	Company
8	ESSO-C	Exxonmobil Res & Eng Co	144	USA	Company
9	UNVO-C	Uop LLC	130	USA	Company
10	PRAX-C	Praxair Technology Inc	114	USA	Company
11	BRTO-C	BOC Group	108	UK	Company
12	HITA-C	Hitachi Ltd	108	Japan	Company
13	SHAN-N	Shandong Seri Petrotech Dev Co Ltd	100	China	Company
14	TOKE-C	Toshiba KK	90	Japan	Company
15	NIIO-C	Nippon Sanso	88	Japan	Company
16	BADI-C	Basf AG	87	Germany	Company
17	SNPC-C	China Petroleum & Chem Corp	85	China	Company
18	GENE-C	General Electric Co	81	USA	Company
19	YAWA-C	Nippon Steel Corp	79	Japan	Company
20	UNIC-C	Union Carbide Corp	72	USA	Company
21	SIEI-C	Siemens AG	71	Germany	Company
22	KOER-C	South Korea Inst Energy Res	70	South Korea	Institute
23	KANT-C	Kansai Electric Power	55	Japan	Company
24	DOWC-C	Dow Chem Co	52	USA	Company
25	FUJF-C	Fuji Film Corp	51	Japan	Company
26	CALI-C	Chevron USA Inc	50	USA	Company
27	HITG-C	Babcock-Hitachi	49	Japan	Company
28	ISHI-C	Ishikawajima Harima Heavy Ind	49	Japan	Company

(continued)

Table 8.4 (continued)

Assignee code		Main assignee name	Patent count	Country	Classification
29	KEPC-C	South Korea Electric Power Corp	49	South Korea	Company
30	KOBM-C	Kobe Steel Ltd	48	Japan	Company
31	UBEI-C	Ube Ind	44	Japan	Company
32	UYZH-C	Univ Zhejiang	44	China	University
33	BEIJ-N	Beijing Yejing Technology Co Ltd	42	China	Company
34	UTIJ-C	Univ Tianjin	42	China	University
35	SEIT-C	Sumitomo Seika Chem Co Ltd	41	Japan	Company
36	CHIK-N	Chikyu Kankyo Sangyo Gijitsu Kenkyu	40	Japan	Institute
37	CHHU-N	China Huaneng Group Clean Energy Technol	39	China	Company
38	RENA-N	Res Inst Nanjing Chem Ind Group	39	China	Company
39	CHSC-N	Chinese Acad Sci Process Eng Inst	38	China	Institute
40	DUPO-C	DU Pont DE Nemours & CO E I	38	USA	Company
41	UYQI-C	UNIV Qinghua	38	China	University
42	MATU-C	Matsushita Elec Ind Co Ltd	37	Japan	Company
43	WANG-I	Wang Y	36	China	Individual
44	FLUO-C	Fluor Technologies Corp	35	USA	Company
45	ZHAN-I	Zhan-I	35	China	Individual
46	KAWI-C	Kawasaki Steel Corp	34	Japan	Company
47	MEMB-N	Membrane Technology & Res Inc	33	USA	Company
48	UYCH-N	Univ China Petroleum	33	China	University
49	UYDA-C	Univ Dalian Technology	33	China	University
50	AGEN-C	Agency of Ind Sci & Technology	32	Japan	Institute
51	BRPE-C	BP Alternative Energy Int Ltd	32	UK	Company
52	GEOR-N	George Lord Method Res & Dev Air Liquide	32	France	Company
53	JIAN-N	Jiangsu Ruifeng Technology Ind Co Ltd	32	China	Company

(continued)

Table 8.4 (continued)

Assignee code		Main assignee name	Patent count	Country	Classification
54	NIIT-C	Dokuritsu Gyosei Hojin Sangyo Gijutsu SO	32	Japan	Institute
55	UYSE-C	Univ Southeast	32	China	University

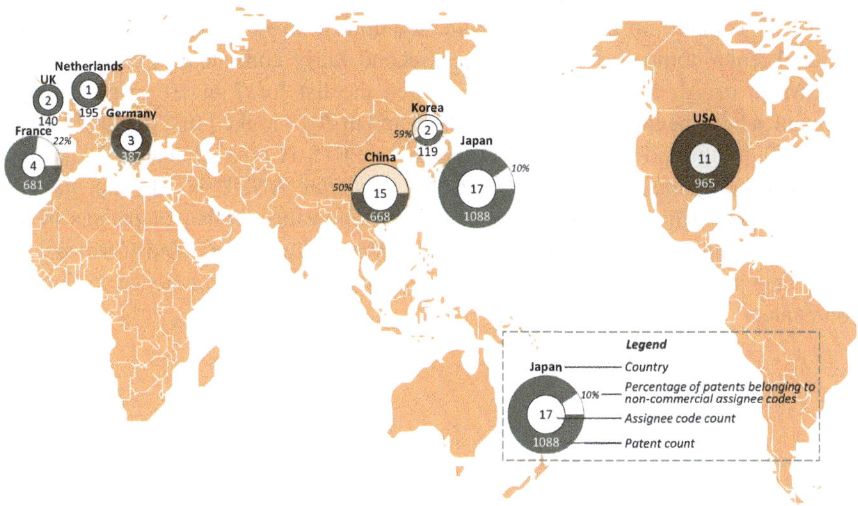

Fig. 8.5 Spatial distribution of top 50 assignee codes, number of patents, and proportion of patents belonging to non-commercial assignee codes. Source: Derwent Innovations Index Database (2016)

patent count has been increasing quickly in the last 5 years. The performance of Japan is particularly impressive as it ranks first in terms of both patent count and assignee code count. Nevertheless, inventors from the European Union have more patents on average; each inventor in the Netherlands has 195 patents, in France 170, and in Germany 129. Each inventor from the United States, the United Kingdom, Japan, South Korea, and China has 88, 70, 64, 60, and 45 patents on average, respectively. Countries in East Asia and France have more non-commercial assignees such as universities, institutes, and individuals. Non-commercial inventors have contributed 59% of patent files in South Korea, 50% in China, 22% in France, and 10% in Japan. This status highlights the fundamental role of research-based organizations that receive financial assistance from governments in these countries.

8.3.2 Changes in Rank of Top 10 Assignee Codes

To highlight relative trends across major high-tech countries, we examine the top 10 assignee codes in the last decade by country. Japan, the United States, France, China, Germany, the United Kingdom, and South Korea have more patents than other countries. Thus, companies or institutes from these countries rank in the top 10 assignee codes.

Figures 8.6 and 8.7 report the ranking of the top 10 assignee codes by year. Figure 8.7 is based on Fig. 8.6, with the addition of arrows that represent national trends. French companies rank first and second fairly consistently; either Alstom (ALSM-C) or Air Liquide (AIRL-C) topped the list for 7 in 10 years. Chinese companies have begun emerging since 2013 and in 2014 ranked first (China Sinopec SNPC-C), ranging from second to fourth in 2015 (China Huaneng CHHU-N, Shanghai Longking SHAN-N, and Anhui Huaertai ANHU-N). It appears that inventors from China will continue to file an increasing number of patents in the future. The ranks of companies from Japan, the United States, and Germany are lower than at the beginning of the decade, mainly due to the emergence of Chinese companies.

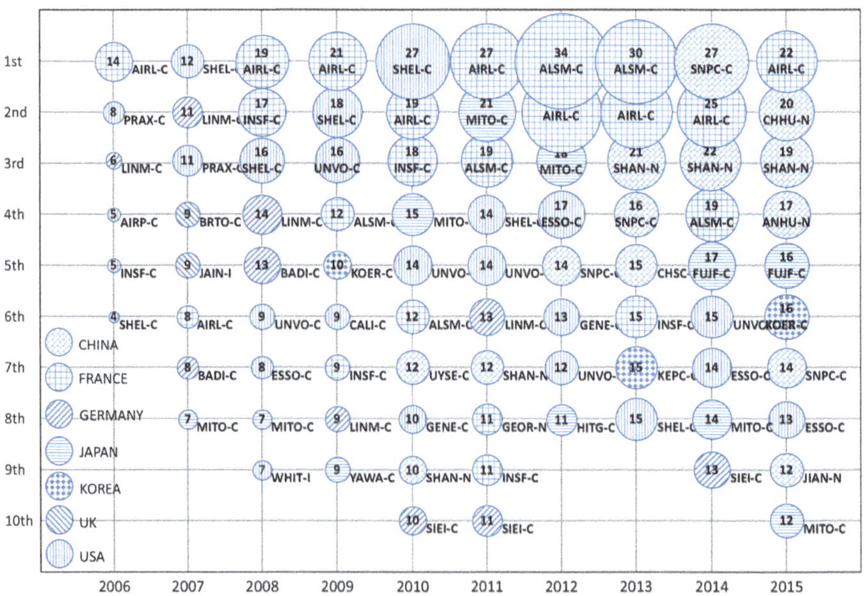

Fig. 8.6 Top 10 assignee codes across one decade. Five-letter code represents assignee code, the number in the bubble is patent count, and the bubble size represents the number of patents per assignee code

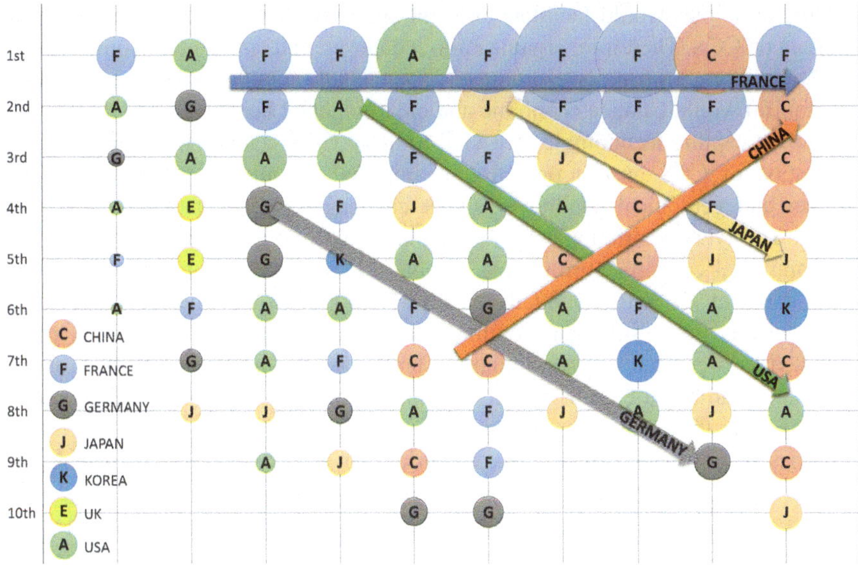

Fig. 8.7 Top 10 assignee codes across one decade, showing national trends. A bubble represents an assignee code, the capital letter is the country in which this assignee code is based, bubble size represents the patent count per assignee code, and an arrow shows relative national development trends. Due to rank ties, assignee codes in the lowest ranks are omitted in some years. Source: Derwent Innovations Index Database (2016)

8.4 Conclusion

Carbon dioxide is an inert gas that has no heating value of combustion and is an environmental concern since it is a major greenhouse gas. Varied technologies have been developed for CO_2 capture. This chapter uses six patent indicators to quantify the spatial-temporal distribution of carbon capture technology.

We speculate that a productive period for carbon capture technology development is approaching. As the count of authorized patents has continued to increase rapidly since 2006, the life cycle of carbon capture technology is or will soon be in the growth stage. It is timely for research and development in carbon capture technology to come to the fore, providing innovations to solve problems in CCS projects.

Carbon capture patents are concentrated in a few countries, such as Japan, the United States, France, China, and Germany, but CCS projects are needed worldwide. Patents are private property that belong to patentees, but the environment is a global public good. As private properties, companies, and institutes will not transfer technologies for free, carbon emissions continue to increase. Companies lack motivation to purchase technologies from innovators. Closing the gap between CCS rhetoric and technical progress is critically important to further global climate mitigation efforts. Developing strong international cooperation on CCS demonstration with global coordination, transparency, cost-sharing, and communication as guiding principles would facilitate efficient and cost-effective collaborative global learning on CCS. Therefore, organizations, countries, and companies should set up a mechanism for cooperation regarding environmentally friendly technologies. Such a mechanism would balance economic and environmental benefits.

Carbon capture technologies are vital to make CCS viable. However, commercialization of CCS project operation is impeded by limitations including CO_2 sources and geological conditions. Other challenges include lack of policy and economic drivers, restrictions due to local laws and international conventions, and environmental concerns surrounding escape of carbon from storage (Meyer et al. 2005; Ying et al. 2010; Quintella et al. 2011; IEA 2013). Therefore, advanced technology is not a sufficient condition for a CCS project. For example, many so-called low-tech but oil-rich countries should develop more CO_2 EOR projects. Increasing numbers of CCS projects should be introduced as soon as possible in countries like the United States and China, the largest developed and developing countries, respectively. Both of these countries own more carbon capture patents, are oil (gas)-producing countries, depend heavily on coal as a fuel, and have vast territories and waters that increase the likelihood of finding appropriate geologic formations for CCS.

There are some limitations to the methodology in this chapter. First, we reviewed patent counts, but this introduces some biases as patent count does not represent the whole portfolio of patent analysis. Second, there may be some biases in making cross-country comparisons using top 50 or top 10 assignee codes rather than all assignees. Third, the research methods do not allow detailed discussion of phenomena such as international technology diffusion.

References

Acs ZJ, Anselin L, Varga A (2002) Patents and innovation counts as measures of regional production of new knowledge. Res Policy 31:1069–1085. https://doi.org/10.1016/S0048-7333(01)00184-6

Albino V, Ardito L, Dangelico RM, Messeni PA (2014) Understanding the development trends of low-carbon energy technologies: a patent analysis. Appl Energ 135:836–854. https://doi.org/10.1016/j.apenergy.2014.08.012

Dechezlepretre A, Glachant M, Hascic I, Johnstone N, Meniere Y (2009) Invention and transfer of climate change mitigation technologies on a global scale: a study drawing on patent data

Ernst H (1997) The use of patent data for technological forecasting: the diffusion of CNC-technology in the machine tool industry. Small Bus Econ 9:361–381. https://doi.org/10.1023/A:1007921808138

Griliches Z (1990) Patent statistics as economic indicators: a survey. J Econom Lit 28:1661–1707. https://doi.org/10.1016/S0169-7218(10)02009-5

Hong M, Wei Z, Lucheng H (2013) Research on carbon capture and storage technology development based on patent citation. J Intell 32

IEA (2013) Carbon capture and storage technology roadmap (2013 Edition)

IEA (2012) Energy technology perspectives 2012

IPCC (2008) Climate change 2007 synthesis report. https://doi.org/10.1256/004316502320517344

Johnstone N, Hascic I, Popp D (2010) Renewable energy policies and technological innovation: evidence based on patent counts. Environ Resour Econom 45:133–155. https://doi.org/10.1007/s10640-009-9309-1

Leo MeyerMetz B, Davidson O, Coninck H, de Loos M, Meyer L (2005) IPCC special report on carbon dioxide capture and storage

Leu H, Wu C, Lin C (2012) Technology exploration and forecasting of biofuels and biohydrogen energy from patent analysis. Int J Hydr Energ 37:15719–15725. https://doi.org/10.1016/j.ijhydene.2012.04.143

Li B, Duan Y, Luebke D, Morreale B (2013) Advances in CO_2 capture technology: a patent review. Appl Energ 102:1439–1447. https://doi.org/10.1016/j.apenergy.2012.09.009

Little AD (1981) The strategic management of technology. European Management Forum, Davos

Liu J, Kuan C, Cha S, Chuang W, Gau G, Jeng J (2011) Photovoltaic technology development: a perspective from patent growth analysis. Sol Energ Mater Sol Cells 95:3130–3136. https://doi.org/10.1016/j.solmat.2011.07.002

OECD (2010) Climate policy and technological innovation and transfer: an overview of trends and recent empirical results

OECD (2016) Main science and technology indicators

Oltra V, Kemp R, De Vries FP (2010) Patents as a measure for eco-innovation. Int J Environ Technol Manag 13:130. https://doi.org/10.1504/IJETM.2010.034303

Park J (2014) The evolution of waste into a resource: examining innovation in technologies reusing coal combustion by-products using patent data. Res Pol 43:1816–1826. https://doi.org/10.1016/j.respol.2014.06.002

Quintella C, Hatimondi SA, Musse APS, Miyazaki SF, Cerqueira GS, De Araujo Moreira A (2011) CO_2 capture technologies: an overview with technology assessment based on patents and articles. Energy Procedia 4:2050–2057. https://doi.org/10.1016/j.egypro.2011.02.087

Rubin E, Mantripragada H, Marks A, Versteeg P, Kitchin J (2012) The outlook for improved carbon capture technology. Prog Energ Combust Sci 38:630–671. https://doi.org/10.1016/j.pecs.2012.03.003

Socolow R, Pacala S (2006) A plan to keep carbon in check. Sci Am 295:50–57. https://doi.org/10.1038/scientificamerican0906-50

Torp TA, Gale J (2004) Demonstrating storage of CO_2 in geological reservoirs: the Sleipner and SACS projects. Energy 29:1361–1369. https://doi.org/10.1016/j.energy.2004.03.104

UKIPO (2014) Eight Great Technologies energy storage—a patent view

van Alphen K, Hekkert M, Turkenburg W (2010) Accelerating the deployment of carbon capture and storage technologies by strengthening the innovation system. Int J Greenh Gas Control 4:396–409. https://doi.org/10.1016/j.ijggc.2009.09.019

Wang X, Zeng J, Qu J (2010) Patent analysis on the International Carbon Capture and Storage Technologies. Sci Focus 5(4). https://doi.org/10.15978/j.cnki.1673-5668.2010.04.003

Wei Z, Man Z (2014) CO_2 capture situation analysis based on patents and articles. Sci Technol Manag Res

WIPO (2009) Patent-based technology analysis report–alternative energy technology

WMO (2016) WMO statement on the status of the global climate in 2015

Ying F, Lei Z, Xiaobing Z (2010) An Analysis on Carbon Capture and Storage Technology, Regulations and Its Emission Reduction Potential. Adv Clim Chang Res 6:362–369

Chapter 9
Low-Carbon Technology Integration

Baoju Jia, Faming Sun, and Weisheng Zhou

Abstract Energy scarcity and global warming are two critical global issues in this century that must be appropriately solved in a simultaneous manner to realize sustainable development of the world. The objective of this chapter is to present some potential low-carbon technologies for power generation for East Asian Low-Carbon Community (LCC) to help the LCC to find its sustainable energy solutions. Current situation, basic principles, potential analysis, and technology integration of the technologies are generally introduced in detail. Results show that the wind power generation, PV, and OTEC have great potential contribution of clean electricity toward sustainable development of East Asian Low-Carbon Community. In addition, the solar thermal power generation also shows a good choice for China in clean electricity production. Thus, the governments in East Asian Low-Carbon Community should stimulate their development widely and effectively by carrying out effective energy policies to realize a sustainable decarbonized society.

9.1 Introduction

Energy scarcity and global warming are two critical global issues that must be appropriately solved to achieve sustainable global development. Adoption of low-carbon technologies for electricity generation is a promising way to solve these issues simultaneously, since clean power can be generated with low or no CO_2 emissions.

B. Jia
Ritsumeikan Research Center for Sustainability Science, Ritsumeikan University, Ibaraki, Osaka, Japan

F. Sun
Research Institute of Global 3E, Kyoto, Japan

W. Zhou (✉)
College of Policy Science, Ritsumeikan University, Ibaraki, Osaka, Japan
e-mail: zhou@sps.ritsumei.ac.jp

W. Zhou et al. (eds.), *East Asian Low-Carbon Community*,
https://doi.org/10.1007/978-981-33-4339-9_9

In this chapter, some potential low-carbon technologies for power generation are introduced in detail, which are considered as potential ones for East Asian Low-Carbon Community to realize a sustainable decarbonized society.

9.2 Low-Carbon Technologies for Power Generation

The current status, basic principles, analysis of potential, and technology integration of some potential low-carbon technologies are described below.

9.2.1 Solar Cells

Solar cells, also called photovoltaic (PV) cells, can convert sunlight directly into electricity. Their energy conversion efficiency is given in the following equation:

$$\eta = \frac{\mathrm{FF} \times I_{\mathrm{sc}} \times V_{\mathrm{oc}}}{P_{\mathrm{in}}} \times 100\% \tag{9.1}$$

where P_{in} is the energy density of sunlight, FF is the fill factor of solar cells, V_{oc} is the open circuit voltage, and I_{sc} is the short circuit current. Then, electricity generation is given in the following equation:

$$P_{\mathrm{out}} = P_{\mathrm{in}} \times \eta \tag{9.2}$$

Solar cell efficiency tables (version 51) (Green et al. 2018) show that the world record for solar cell efficiency (46.0%) was achieved using multi-junction concentrator solar cells. The potential of PV in China, Japan, and South Korea using these cells is given in Table 9.1. The required area of PV per household is approximately 1.9–3.3 m^2 in China, 8.0–9.7 m^2 in Japan, and 5.0–5.7 m^2 in South Korea.

The potential supply of PV power generation is defined by the following equation:

$$\begin{aligned} &\text{PV power generation potential supply} \\ &= \frac{\text{Potential of the electricity production}}{\text{Total electricity consumption in 2017}} \times 100\% \end{aligned} \tag{9.3}$$

The potential supply of PV power generation, in terms of 2017 electricity consumption levels, is 89–148% in China, 22–27% in Japan, and 11–13% in South Korea (Table 9.1).

Table 9.1 Analysis of photovoltaic (PV) potential in China, Japan, and South Korea

Items	Country		
	China	Japan	South Korea
Solar radiation (kWh/m^2/year)	1050–1750	1182–1427	1270–1408
Potential of the electricity production (kWh/m^2/year)	483–805	544–656	584–648
Electricity consumption in 2014 (kWh/year/household)	1591	5275	3290
Needed area the PV per household (m^2/household)	1.9–3.3	8.0–9.7	5.0–5.7
Area of the country (km^2)	9,596,960	377,972	100,210
Potential of the electricity generation (TWh/year)	4635–7726	206–248	59–65
Total electricity consumption in 2017 (TWh)	5219	927	512
PV generation potential supply (%)	89–148	22–27	11–13

Source: World Energy Resources (2016), World Energy Council (2014), The World Factbook (2017), Global Energy Statistical Yearbook (2017), Zhao et al. 2011, Jung and Frank, 2014

9.2.2 *Wind Power Generation*

Wind power generation uses air flow through a wind turbine to mechanically power a generator to generate electricity. Its energy conversion efficiency is given in the following equation:

$$\eta = \eta_{\mathrm{t}} \times \eta_{\mathrm{m}} \times \eta_{\mathrm{e}} \tag{9.4}$$

where η_{t} is turbine efficiency, η_{m} is mechanical efficiency, and η_{e} is electrical efficiency. Then, electricity generation is given in the following equation:

$$P_{\mathrm{out}} = P_{\mathrm{in}} \times \eta \tag{9.5}$$

where $P_{\mathrm{in}} = 0.5\rho A V^3$ is the total wind power flowing into the wind inlet of the wind turbine, ρ is air density, A is the wind blade swept area, and V is wind velocity.

The maximum efficiency of a wind turbine, known as the Betz limit, is 59.26% (Blackwood 2016). Based on this efficiency, the potential of wind power generation in China, Japan, and South Korea is given in Table 9.2. The potential supply of wind power generation is defined by the following equation:

$$\begin{aligned}&\text{Wind power generation potential supply}\\&= \frac{\text{Potential of the electricity production}}{\text{Total electricity consumption in 2017}} \times 100\%\end{aligned} \tag{9.6}$$

The potential supply of wind power generation based on total electricity consumption in 2017 is 495% in China, 209% in Japan, and 130% in South Korea.

Table 9.2 Analysis of wind power generation potential in China, Japan, and South Korea

Items	Country		
	China	Japan	South Korea
Global distribution of wind power (TWh)	43,600	3270	1120
Potential of the electricity production (TWh)	25,837	1938	664
Total electricity consumption in 2017 (TWh)	5219	927	512
Wind power generation potential supply (%)	495	209	130

Source: Lu et al. 2009, Global Energy Statistical Yearbook (2017)

9.2.3 Solar Thermal Power Generation

Solar thermal power generation is a technology for harnessing solar thermal energy to generate electricity. Its energy conversion efficiency is given in the following equation:

$$\eta = \eta_{sc} \cdot \eta_{pgc} \tag{9.7}$$

where η_{sc} [−] is the efficiency of the solar collector, which is given in the following equation:

$$\eta_{sc} = \eta_0 - a_1 \cdot \frac{(t_{sc} - t_0)}{G} - a_2 \cdot \left(\frac{t_{sc} - t_0}{G}\right)^2 \tag{9.8}$$

where η_0 [−] is zero-loss efficiency, a_1 [W/(m^2·K)] is the linear heat loss coefficient, a_2 [W/(m^2·K^2)] is the quadratic heat loss coefficient, G [W/m^2] is solar irradiance, t_{sc} [°C] is mean solar collector temperature, and t_0 [°C] is ambient temperature.

The efficiency of the power generation cycle in KCS-11 (Sun et al. 2012a, 2013a, b) is given by η_{pgc}, which is a more efficient technology for converting low to mid temperature heat sources into electricity:

$$\eta_{pgc} = 0.64\Delta P^3 - 4.90\Delta P^2 + 14.67\Delta P + 0.22 \tag{9.9}$$

where ΔP [MPa] is the pressure difference of the evaporator and condenser and has an effective range of 2.5 MPa.

Then, electricity generation is given in the following equation:

$$P_{out} = P_{in} \times \eta \tag{9.10}$$

where P_{in} is the solar thermal energy input for electricity generation and the potential supply of solar thermal power generation is defined by the following equation:

Table 9.3 The coefficients of typical solar collectors

Collector type	η_0	a_1	a_2
Unglazed	0.85	20	–
Glazed with nonselective absorber	0.75	6.5	0.030
Glazed with selective absorber	0.78	4.2	0.015
Vacuum single tube (flat absorber)	0.75	1.5	0.008
Vacuum tube Sydney	0.65	1.5	0.005
Flat plate	0.75	3.5	0.015
CPC solar collector (flat absorber)	0.70	2.6	7.5

Source: Matuska (2017), Tripanagnostopoulos et al. (2000)

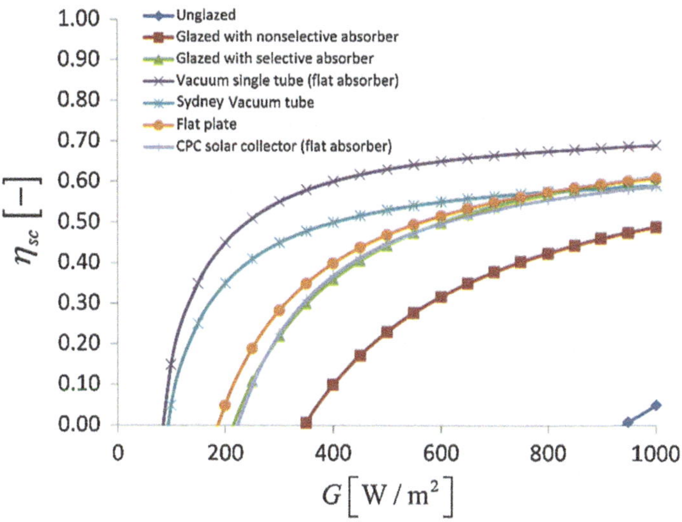

Fig. 9.1 Relationship between solar irradiance and solar collector efficiency when $t_0 = 20$ °C and $t_{sc} = 60$ °C

$$\text{Solar thermal power generation potential supply} = \frac{\text{Potential of the electricity production}}{\text{Total electricity consumption in 2017}} \times 100\% \tag{9.11}$$

Coefficients of typical solar collectors are given in Table 9.3 (Matuska 2017; Tripanagnostopoulos et al. 2000) for performance comparison, and their efficiencies are shown in the following figures. These figures show the relationship between solar irradiance and solar collector efficiency when $t_0 = 20$ °C and $t_{sc} = 60$ °C (Fig. 9.1), $t_{sc} = 70$ °C (Fig. 9.2), $t_{sc} = 80$ °C (Fig. 9.3), and $t_{sc} = 90$ °C (Fig. 9.4). A vacuum single tube with flat absorber shows the maximum efficiency during the whole solar irradiance period under all temperature conditions and therefore is chosen to harness solar thermal energy for the KCS-11 to generate electricity.

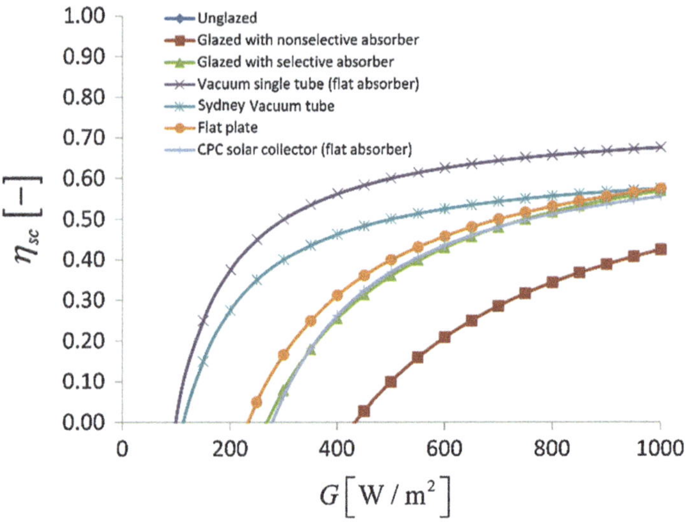

Fig. 9.2 Relationship between solar irradiance and solar collector efficiency when $t_0 = 20$ °C and $t_{sc} = 70$ °C

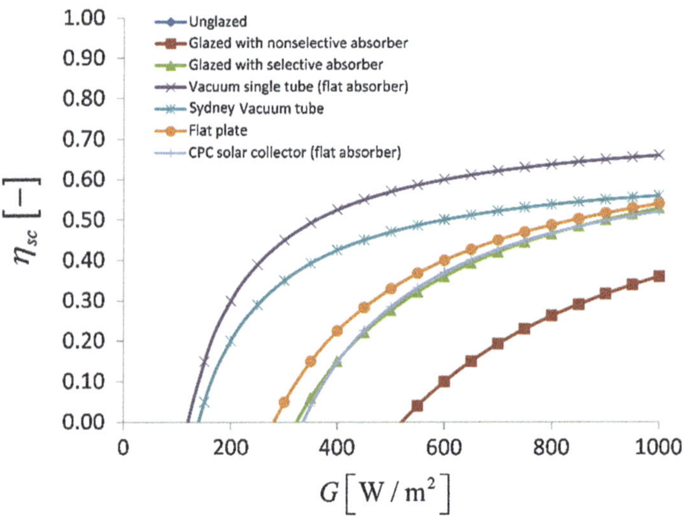

Fig. 9.3 Relationship between solar irradiance and solar collector efficiency when $t_0 = 20$ °C and $t_{sc} = 80$ °C

Figure 9.5 shows the relationship between mean solar collector temperature and the energy conversion efficiency of solar thermal power generation in KCS-11 using a vacuum single tube with flat absorber. Energy conversion efficiency increases with increasing mean solar collector temperature when solar irradiance is greater than 300 W/m^2 and decreases when solar irradiance is less than 300 W/m^2. Energy

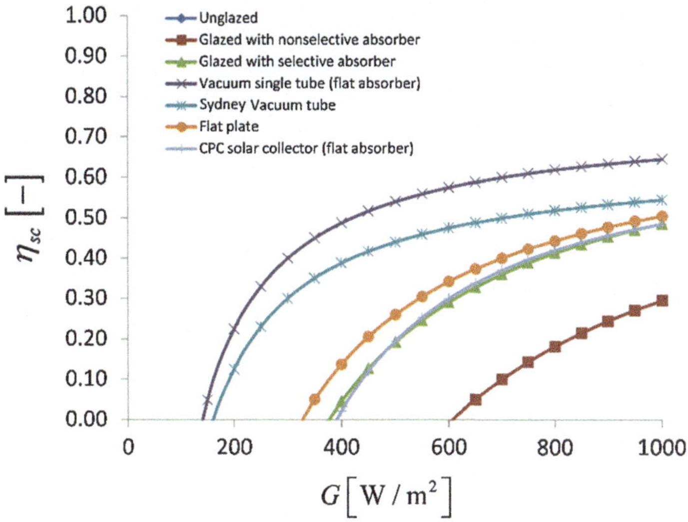

Fig. 9.4 Relationship between solar irradiance and solar collector efficiency when $t_0 = 20$ °C and $t_{sc} = 90$ °C

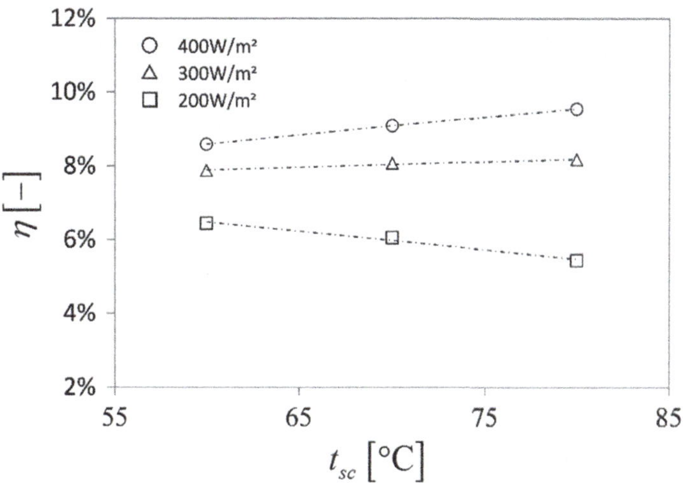

Fig. 9.5 Relationship between mean solar collector temperature and energy conversion efficiency

conversion efficiency remains constant as mean solar collector temperature changes when solar irradiance is 300 W/m^2. Thus, maximum energy conversion efficiency in KCS-11 is obtained with mean solar collector temperature $t_{sc} = 60$ °C when average annual solar irradiance (W/m^2) ranges from 0 to 300 (Fig. 9.6).

Solar thermal power generation potential in China, Japan, and South Korea is given in Table 9.4. The potential supply of solar thermal power generation, based on

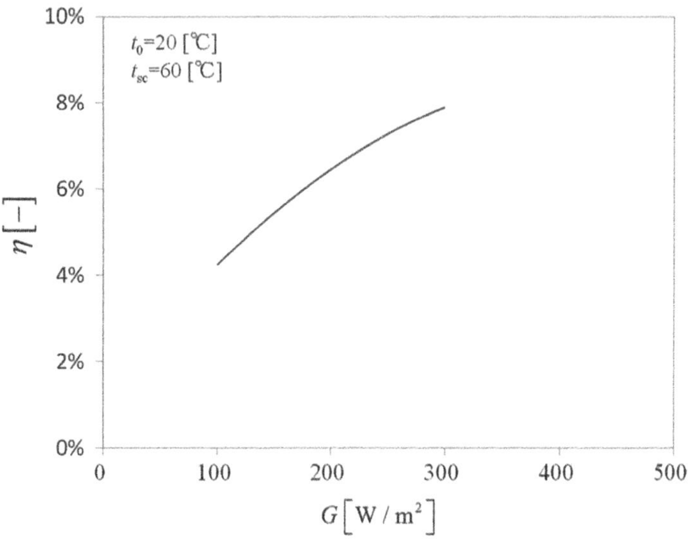

Fig. 9.6 Relationship between average annual solar irradiance and energy conversion efficiency when $t_0 = 20$ °C and $t_{sc} = 60$ °C

Table 9.4 Analysis of solar thermal power generation potential in China, Japan, and South Korea

	Country		
Items	China	Japan	South Korea
Solar radiation (kWh/m^2/year) (W/m^2)	1050~1750 (120~200)	1182~1427 (135~163)	1270~1408 (145~161)
Area of the country (km^2)	9,596,960	377,972	100,210
Potential of the electricity generation (TWh/year)	479~1084	23~31	7~8
Total electricity consumption in 2017 (TWh)	5219	927	512
Solar thermal power generation potential supply (%)	9.2~20.8%	2.5~3.3%	1.3~1.6%

Source: World Energy Resources (2016), Zhao et al. (2011), Jung and Frank (2014), The World Factbook (2017), Global Energy Statistical Yearbook (2017)

total electricity consumption in 2017, is 9.2–20.8% in China, 2.5–3.3% in Japan, and 1.3–1.6% in South Korea.

9.2.4 Ocean Thermal Energy Conversion

Ocean thermal energy conversion (OTEC) generates electricity using the temperature difference between water on the surface of the sea and at great depths. OTEC has been considered a promising solution, especially since the Fukushima nuclear accident in Japan, so an analysis of regional potential is appropriate. Here, OTEC potential in the East Asia Ocean using Rankine cycle technology is analyzed and discussed.

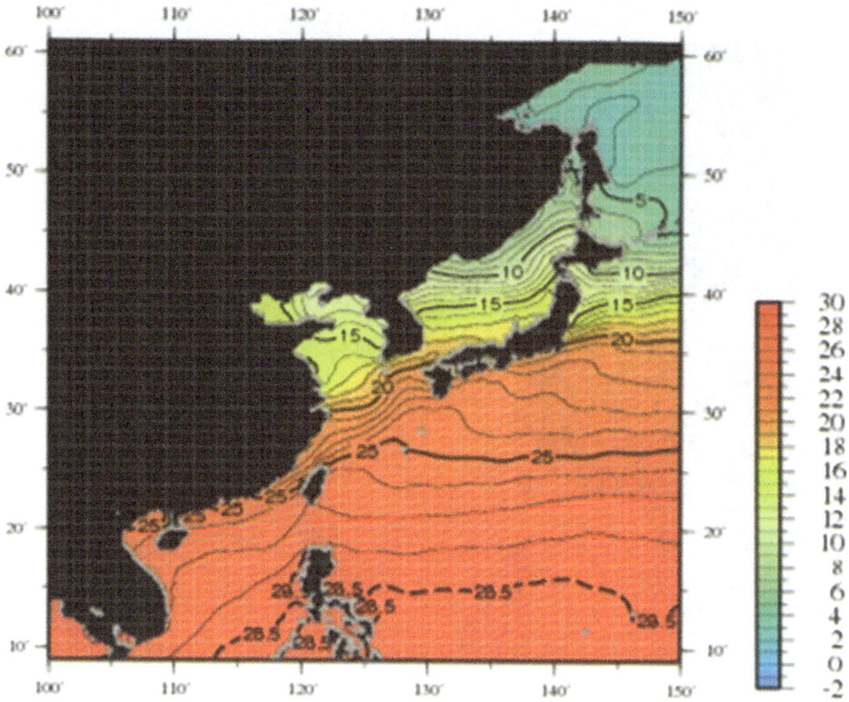

Fig. 9.7 Annual average temperature (°C) at surface of East Asia Ocean in 2009 (World Ocean Data 2009)

Firstly, the oceanic surface area (A_{OTEC}) supplied in the East Asia Ocean (100° E–150° E, 10 °N–60° N) was calculated using temperature data from the ocean surface (Fig. 9.7) and at a depth of 1000 m (Fig. 9.8). The temperature difference required for normal OTEC operation is greater than 20 °C.

Based on theory and optimized designs of OTEC using a Rankine cycle with ammonia as the working fluid (Sun et al. 2012b), the power needed to run the plant is assumed to be 30% of the gross electrical power in design conditions (Nihous 2005). It follows that the maximum net power output of the East Asia Ocean can be approximated as:

$$\dot{W}_{\mathrm{net,\,max}} = n \cdot \left\{ \dot{m}_{\mathrm{ws}} c_{\mathrm{P}} \cdot \left(\begin{array}{c} \left(1 - \dfrac{\phi}{\vartheta}\right) \cdot \left(t_{\mathrm{ws}} - \dfrac{t_{\mathrm{ws}} - \gamma \cdot (1 - e^{\alpha})}{e^{\alpha}}\right) \\ -0.3\left(1 - \dfrac{\phi_i}{\vartheta_i}\right) \cdot \left(t_{\mathrm{wsi}} - \dfrac{t_{\mathrm{wsi}} - \gamma_i \cdot (1 - e^{\alpha})}{e^{\alpha}}\right) \end{array} \right) \right\} \tag{9.12}$$

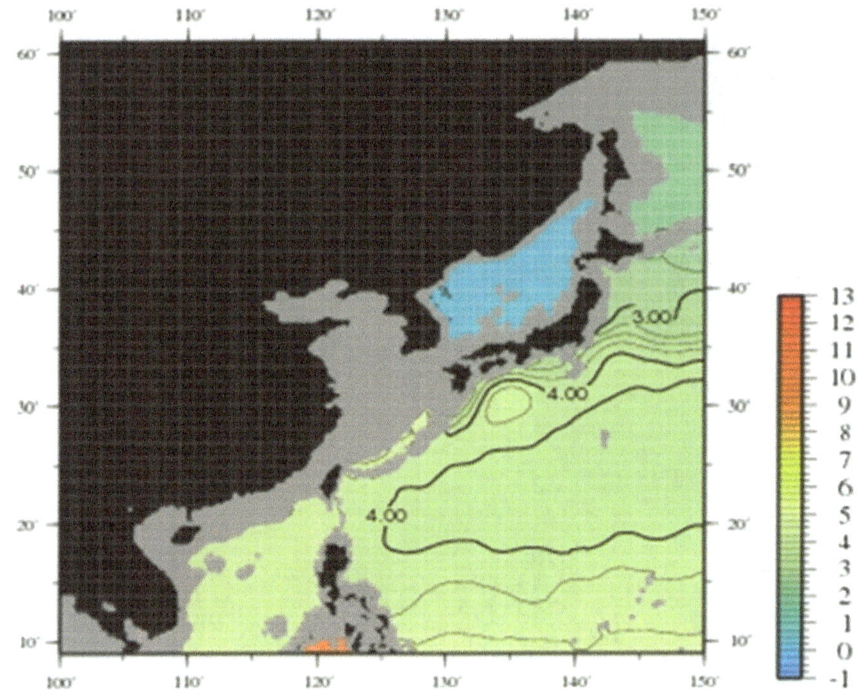

Fig. 9.8 Annual average temperature (°C) at 10^3 m depth in East Asia Ocean in 2009 (World Ocean Data 2009)

Table 9.5 Constant coefficients of ammonia

	Ammonia		
i	a_i	b_i	c_i
1	−0.0115	4.3728e-04	8.8558e-04
2	−0.0165	−5.4304e-04	6.9750e-04
3	4.6050	0.6841	2.8641e-04
4	–	−4.7073	0.7147
5	–	1267.2557	0.2310
6	–	–	0.5811

Source: Sun et al. (2012b)

where $\phi = (\xi + T_0) \cdot (a_1\gamma^2 + a_2\gamma + a_3\xi + a_4)$, $\phi_i = (\xi_i + T_0) \cdot (a_1{\gamma_i}^2 + a_2\gamma_i + a_3\xi_i + a_4), \alpha = \psi/(\dot{m}_{\mathrm{ws}}c_{\mathrm{P}})$, $\vartheta = b_1\xi^2 + b_2\gamma^2 + b_3\gamma + b_4\xi + b_5$, $\vartheta_i = b_1{\xi_i}^2 + b_2{\gamma_i}^2 + b_3\gamma_i + b_4\xi_i + b_5$, $\xi = \mathrm{c}_1 t_{\mathrm{cs}}^2 + \mathrm{c}_2 t_{\mathrm{ws}} \cdot t_{\mathrm{cs}} + \mathrm{c}_3 t_{\mathrm{ws}}^2 + \mathrm{c}_4 t_{\mathrm{cs}} + \mathrm{c}_5 t_{\mathrm{ws}} + \mathrm{c}_6$, $\xi_i = \mathrm{c}_1 t_{\mathrm{csi}}^2 + \mathrm{c}_2 t_{\mathrm{wsi}} \cdot t_{\mathrm{csi}} + \mathrm{c}_3 t_{\mathrm{wsi}}^2 + \mathrm{c}_4 t_{\mathrm{csi}} + \mathrm{c}_5 t_{\mathrm{wsi}} + \mathrm{c}_6$, $\gamma = -\mathrm{c}_1 t_{\mathrm{cs}}^2 - \mathrm{c}_2 t_{\mathrm{ws}} \cdot t_{\mathrm{cs}} - \mathrm{c}_3 t_{\mathrm{ws}}^2 + (1 - \mathrm{c}_4) t_{\mathrm{cs}} + (1 - \mathrm{c}_5) t_{\mathrm{ws}} - \mathrm{c}_6$, $\gamma_i = -\mathrm{c}_1 t_{\mathrm{csi}}^2 - \mathrm{c}_2 t_{\mathrm{wsi}} \cdot t_{\mathrm{csi}} - \mathrm{c}_3 t_{\mathrm{wsi}}^2 + (1 - \mathrm{c}_4) t_{\mathrm{csi}} + (1 - \mathrm{c}_5) t_{\mathrm{wsi}} - \mathrm{c}_6$.

In addition, a_{ii} (ii = 1,2,3,4), b_{ii} (ii = 1,2,. . .5), and c_{ii} (ii = 1,2,. . .6) are constant coefficients of ammonia, as shown in Table 9.5; n represents the potential number of

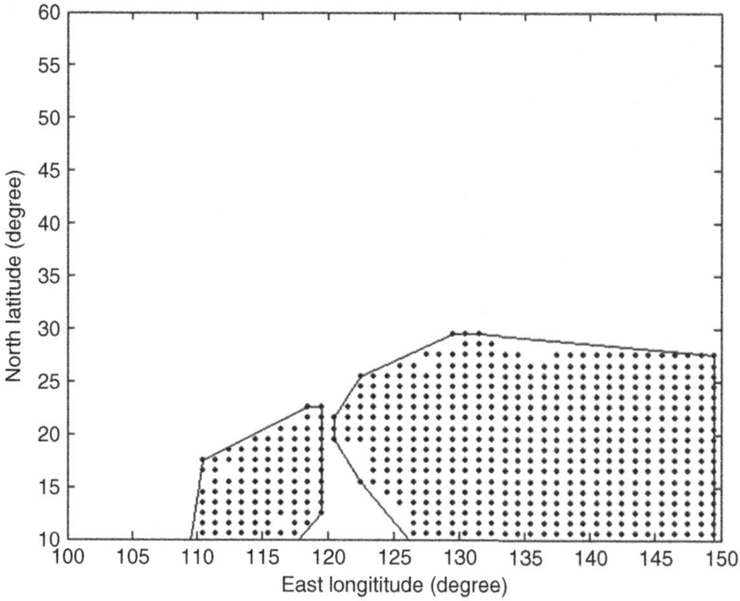

Fig. 9.9 Area required for ocean thermal energy conversion (A_{OTEC}) in the East Asia Ocean

OTEC plants, which is decided by $n = (A_{OTEC} \cdot h_m \cdot \rho)/(\dot{m}_{ws} \cdot \tau_m)$; $\dot{m}_{ws}$ is the mass flow rate of warm seawater, c_P shows the specific heat at constant pressure, T_0 represents the absolute temperature of 0 °C; h_m reflects the thickness of the warm seawater layer; ρ is average seawater density; τ_m is the utilization time of the mixed layer; and ψ shows the performance of the heat exchanger. Finally, t_{csi} and t_{wsi} are the initial temperature in OTEC of cold and warm seawater, respectively; and t_{cs} and t_{ws} are the OTEC operating temperatures.

It can be assumed that OTEC operations exert little influence on the temperature of warm seawater ($t_{ws} = t_{wsi}$), since energy is supplemented from solar radiation and surface ocean currents. However, the temperature of cold seawater is easily affected by bulk OTEC operations. Through reference (1D) analysis of OTEC (Nihous 2005), t_{cs} can be given as follows:

$$t_{cs} = \frac{-\kappa\frac{\phi}{\vartheta}\cdot\frac{h_m t_{cs}}{\tau_m}}{w-\kappa\frac{\phi}{\vartheta}\cdot\frac{h_m}{\tau_m}} + \left(\frac{(t_{ws}+t_{cs}\cdot\kappa\cdot\phi+\vartheta)}{1+\kappa\cdot\phi/\vartheta} - \frac{-\kappa\frac{\phi}{\vartheta}\cdot\frac{h_m t_{cs}}{\tau_m}}{w-\kappa\frac{\phi}{\vartheta}\cdot\frac{h_m}{\tau_m}}\right) \cdot e^{\left(w-\kappa\frac{\phi}{\vartheta}\frac{h_m}{\tau_m}\right)\left(\frac{z_{mix}-z_{cs}}{K}\right)} \quad (9.13)$$

where w reflects the upwelling rate; K is the vertical eddy diffusion coefficient; z_{cs} and z_{mix} are vertical water column coordinates of cold seawater withdrawal and mixed effluent discharge, respectively; $z_{mix} = h_m - \ln([t_{wsi} + t_{csi} \cdot (\kappa \cdot \varphi/\vartheta)]/(1 + \kappa \cdot \varphi/\vartheta)/t_{wsi}) \cdot K/w$; and κ is the additional seawater mixing coefficient that has a positive impact on mariculture.

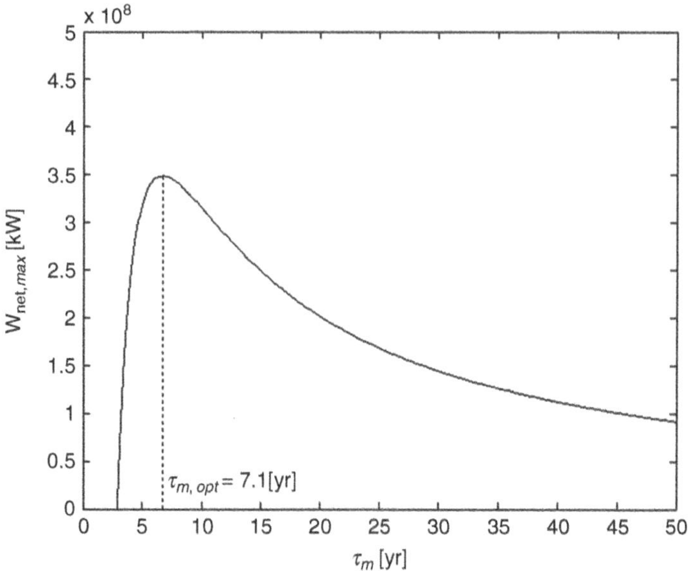

Fig. 9.10 A_{OTEC} in the East Asia Ocean

Table 9.6 Analysis of OTEC power generation potential in China, Japan, and South Korea

Items	Country		
	China	Japan	South Korea
Potential of the electricity production (TWh)	3066		
Total electricity consumption in 2017 (TWh)	5219	927	512
OTEC power generation potential supply (%)	46.1%		

Source: Global Energy Statistical Yearbook (2017)

Thus, the Rankine cycle OTEC plant is designed with the following parameters: $T_0 = 273.15$ K, $\psi = 2.5$ MW/K, $\dot{m}_{\mathrm{ws}} = 6.0 \times 10^3$ kg/s, $c_{\mathrm{p}} = 3.9$ kJ/(kg·K), $t_{\mathrm{wsi}} = 25$ °C, $t_{\mathrm{csi}} = 5$ °C, $\rho = 1025$ kg/m^3, $h_{\mathrm{m}} = 75$ m, $w = 4$ m/year, $K = 2300$ m^2/year, $z_{\mathrm{cs}} = 925$ m, and $\kappa = 0.4$ (unitless). Based on this, A_{OTEC} can be calculated to be approximately 7.5×10^{12} m^2, as shown in Fig. 9.9. The OTEC discharge outlet is located at a depth of 145 m, which has a positive impact on mariculture, and $z_{\mathrm{mix}} = 220$ m.

In this way, up to 0.35 TW or 3066 TWh of steady-state OTEC power may be available in the East Asia Ocean (Fig. 9.10). Approximately 429,170 of the described 2.5 MW OTEC plants could be installed. The $\tau_{\mathrm{m,opt}}$ and $t_{\mathrm{cs,opt}}$ are approximately 7.1 years and 9.3 °C, respectively. It is recommended that OTEC plants are built on ships, which would provide greater flexibility for power generation across the ocean.

The potential of ocean thermal energy conversion for power generation in China, Japan, and South Korea is given in Table 9.6. The potential supply of power generated by OTEC is defined by the following equation:

$$\text{OTEC power generation potential supply} = \frac{\text{Potential of the electricity production}}{\text{Total electricity consumption in 2017}} \times 100\% \quad (9.14)$$

The potential supply of OTEC power generation in East Asia (China, Japan, and South Korea) is 46.1% of total electricity consumption levels in 2017.

9.3 Conclusion

In this chapter, four kinds of potential low-carbon technologies for power generation are introduced in detail, which are considered as potential ones for East Asian Low-Carbon Community. In which, the wind power generation shows the highest potential supply, which is China 495%, Japan 209%, and South Korea 130%. The next one is PV, which is China 89–148%, Japan 22–27%, and South Korea 11–13%. The third one is OTEC, which is 46.1% for East Asia (China, Japan, and South Korea), and the solar thermal power generation potential supply is China 9.2–20.8%, Japan 2.5–3.3%, and South Korea 1.3–1.6%.

From this study we know that the wind power generation, PV, and OTEC have great potential contribution of clean electricity toward sustainable development of East Asian Low-Carbon Community. In addition, the solar thermal power generation also shows a good choice for China in clean electricity production. Thus, the governments in East Asian Low-Carbon Community should stimulate their development widely and effectively by carrying out effective energy policies to realize a sustainable decarbonized society.

In addition, geothermal power generation, wave and tidal power, nuclear power, waste heat, biogeneration, hydro (electric) power, ocean energy from concentration gradients, snow and ice heat, fuel cells, and biomass and biogas also have potential for East Asian Low-Carbon Community although they are not introduced here.

References

Blackwood M (2016) Maximum efficiency of a wind turbine. Undergrad J Math Model 6(2):2

Global Energy Statistical Yearbook (2017). https://yearbook.enerdata.net/

Green MA, Hishikawa Y, Dunlop ED, Levi DH, Hohl-Ebinger J, Anita WY, Baillie H (2018) Solar cell efficiency tables (version 51). Prog Photovolt Res Appl 26:3–12

Jung S, Frank R (Supervisor) (2014) Solar energy potential in China, South Korea and Japan. MA thesis, University of Vienna, AC Number: AC12179397

Lu X, McElroy MB, Kiviluom J (2009) Global potential for wind-generated electricity. PNAS 106 (27):10933–10938

Matuska T (2017) Solar collectors. AES-L2 report

Nihous GC (2005) An order-of magnitude estimate of ocean thermal energy conversion resources. J Energy Resour Technol 127:328–333
Sun FM, Ikegami Y, Jia BJ (2012a) A study on Kalina solar system with an auxiliary superheater. Renew Energy 41:210–219
Sun FM, Ikegami Y, Jia BJ, Arima H (2012b) Optimization design and exergy analysis of organic rankine cycle in ocean thermal energy conversion. Appl Ocean Res 35:38–46
Sun FM, Ikegami Y, Arima H, Zhou WS (2013a) Performance analysis of the low temperature solar-boosted power generation system: part I. comparison between Kalina solar system and Rankine solar system. J Solar Energy Eng (ASME) 135:011006.1–011006.11
Sun FM, Ikegami Y, Arima H, Zhou WS (2013b) Performance analysis of the low temperature solar-boosted power generation system: part II. Thermodynamic characteristics of the Kalina solar system. J Solar Energy Eng (ASME) 135:011007.1–011007.8
The World Factbook (2017). https://www.cia.gov/library/publications/the-world-factbook/
Tripanagnostopoulos Y, Yianoulis P, Papaefthimiou S (2000) CPC solar collectors with flat bifacial absorbers. Sol Energy 69(3):191–203
World Energy Council (2014) Average electricity consumption per electrified household. https://www.worldenergy.org/
World Energy Resources (2016) Solar World Energy Council. https://www.worldenergy.org/
World Ocean Data (2009). http://www.nodc.noaa.gov/OC5/indprod.html
Zhao R, Shi G, Chen H, Ren A, Finlow D (2011) Present status and prospects of photovoltaic market in China. Energy Policy 39:2204–2207

Chapter 10
Economic Assessment of Japan's Nuclear Power Policy

Masato Yamazaki and Weisheng Zhou

Abstract This chapter assesses the importance of nuclear power from the viewpoints of gross domestic product (GDP) and carbon dioxide emissions in Japan. The assessment is conducted using a computable general equilibrium (CGE) model. The CGE model is constructed based on the GTAP-Power database, which is an electricity-detailed extension of the GTAP 9 database, a global input–output table. The simulation results indicate that nuclear power generation is not significantly important for Japan in terms of GDP, because fossil fuel power compensates for shortages in the electricity market. In terms of CO_2 emissions, nuclear power is an important power source for Japan. An increase in fossil fuel power generation causes a non-negligible amount of CO_2 emissions.

10.1 Introduction

Since the Fukushima Daiichi nuclear accident in 2011, power source composition has been debated in Japan. Advocates of nuclear power focus on its cost advantages, claiming that the replacement of nuclear power with relatively expensive electricity generated using fossil fuels or renewables would place a burden on the Japanese economy. Japan is largely energy-independent, although it is affected by the politics of major oil-producing countries. Further, through increased CO_2 emissions, fossil fuel power generation would accelerate climate change. Thus, advocates assert that nuclear power is an important energy source politically, environmentally, and economically. Those who are opposed to nuclear power desire the phasing out of nuclear power plants and the development of renewable energy sources. Public

M. Yamazaki
Disaster Mitigation Research Center, Nagoya University, Nagoya, Aichi, Japan
e-mail: yamazaki.masato@nagoya-u.jp

W. Zhou (✉)
College of Policy Science, Ritsumeikan University, Ibaraki, Osaka, Japan
e-mail: zhou@sps.ritsumei.ac.jp

W. Zhou et al. (eds.), *East Asian Low-Carbon Community*,
https://doi.org/10.1007/978-981-33-4339-9_10

anxiety about nuclear power is behind this opposition. Debate has failed to narrow the differences between the two viewpoints.

Comprehensive and quantitative economic and environmental assessment is required to make the debate more constructive. This chapter assesses the importance of nuclear power from the viewpoints of gross domestic product (GDP) and carbon dioxide emissions in Japan. The assessment is conducted using a computable general equilibrium (CGE) model. CGE models are widely used to analyze the economic impacts of trade, energy, and environmental policies. Through the CGE model, scenarios with and without nuclear power are compared.

10.2 Consequences of the Fukushima Daiichi Nuclear Accident

On March 11, 2011, nuclear reactors located in the Fukushima Daiichi nuclear power station were damaged by a tsunami caused by the Great East Japan Earthquake. Nuclear meltdown occurred at reactors No. 1, No. 2, and No. 3, causing leakage of radioactive materials. Within a few days, hydrogen explosions occurred in reactors No. 1, No. 3, and No. 4.

In addition to leaking radioactivity, East Japan suffered a severe electricity shortage in the aftermath of the accident. In order to close the gap between electricity supply and demand, the Tokyo Electric Power Company (TEPCO) implemented rolling blackouts across many cities in suburban Tokyo for 3–6 h daily from March 14 to March 28 in 2011. In summer of the same year, the government ordered high-volume users of TEPCO and the Tohoku Electric Power Corporation to reduce their electricity usage by 15% from the previous year during peak weekday hours. This order was effective from July 1, 2011, to September 9, 2011. In September 2012, the Nuclear Regulation Authority (NRA) was launched in Japan. In April 2013, the NRA proposed new safety standards, requiring electric power companies to make many improvements to nuclear plants. Further, even if the electric power companies pass an inspection by the NRA, local governments of the area in which a nuclear power station is located is required to give permission to allow idle reactors to resume activity. Nationwide and local public anxiety about existing nuclear power reactors has made it difficult for local governments to grant approval. Figure 10.1 shows total power generation and power source composition from 2007 to 2017 in Japan. After the accident in 2011, the contribution of nuclear power generation to Japanese electricity generation is negligible.

Fig. 10.1 Total power generation and power source composition in Japan

10.3 Simulation Methods

10.3.1 Overview of a CGE Model

A CGE model is a comprehensive economic simulation model that is widely used to provide economic impact assessments of public policy in areas such as trade, energy, and the environment. The general equilibrium model, which is the theoretical basis of a CGE model, describes an overall economy as an interaction among economic agents such as households and firms in markets. In a standard CGE model, the behavior of economic agents is assumed to be rational. Given the prices of goods and services, households maximize their utility subject to budgetary constraints. Firms minimize unit production costs with production technology constraints, given wage rates, capital rental prices, and the prices of intermediate goods. A household budget comprises labor income, capital income, and income transfer. General equilibrium theory asserts that demand and supply in all markets considered in the model is balanced through a price adjustment mechanism.

CGE models are referred to as computable because the mathematical formation of the general equilibrium model as a nonlinear simultaneous equation is solvable numerically. A CGE model itself is a nonlinear simultaneous equation whose parameters are calibrated using an input–output table that describes interindustrial trade flow and industry–household trade flow on various regional scales. The current study uses a global input–output table, the Global Trade Analysis Project (GTAP) 9 database. Hosoe et al. (2015) explain the structure of CGE models in detail.

10.3.2 GTAP-Power Database

This chapter provides an economic and environmental assessment of nuclear power using a multicountry multisector CGE model. The CGE model is constructed based on the GTAP-Power database, which is an electricity-detailed extension of the GTAP 9 database, a global input–output table. The GTAP-Power database includes all the data of the GTAP 9 database. Electricity is disaggregated into base load and peak load power sources. Base load power sources consist of nuclear, coal-fired, oil-fired, hydro, wind, and other power sources. Peak load power sources consist of gas-fired, oil-fired, hydro, and solar power sources. Details are described in Peters (2016). Yamazaki and Takeda (2017) construct a CGE model based on the GTAP 8.1 database and analyze nuclear phaseout in Japan. The benchmark year of GTAP 8.1 is 2007.

10.3.3 Structure of the CGE Model

10.3.3.1 Production Function

In the CGE model, the production function is described by a nested constant elasticity of substitution (CES) function. In the following, the output of the CES function is called a CES composite. The Leontief function and the Cobb–Douglas function are special types of CES function, and Leontief composites and Cobb–Douglas composites are used as special types of CES composite. The nested structure of the CES function is depicted in Fig. 10.2. The sectoral output is a Leontief composite of the input of intermediate goods and the input of an energy-value-added composite. The energy-value-added composite is a CES composite consisting of the input of an energy composite and the value added. The value added is generated by the input of labor and capital services. The energy composite is a CES composite of the input of a fossil fuel composite together with electricity and electricity distribution services. The fossil fuel composite is a CES composite of the input of crude oil and petroleum products. In Fig. 10.2, the input of intermediate goods and the energy-value-added composite are linked by a horizontal line. This means that production requires a fixed proportion of input in order to increase output. Inputs joined by a curved line in the figure are substitutable. For example, the energy composite and value added are assumed to be substitutable in the model. The degree of substitutability is determined to be 0.5. This value is the elasticity of substitution between the energy composite and value added.

This model assumes that each production sector can choose its power source mix at the time of electricity consumption. The electricity user can change the power mix elastically, taking into account electricity prices. Given the electricity prices generated using various power sources, the user minimizes the electricity cost per unit. The electricity input is a CES composite of electricity generated by different power

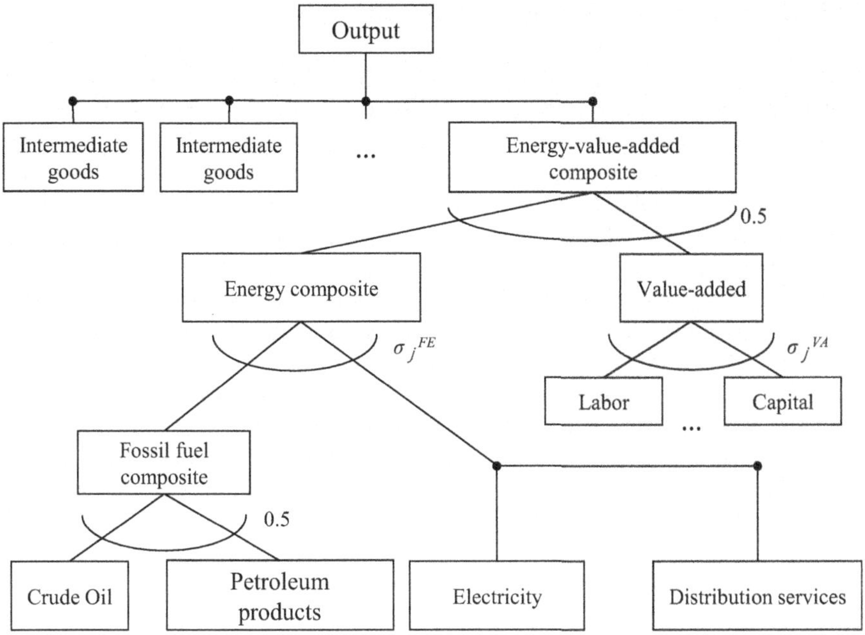

Fig. 10.2 Structure of the production function

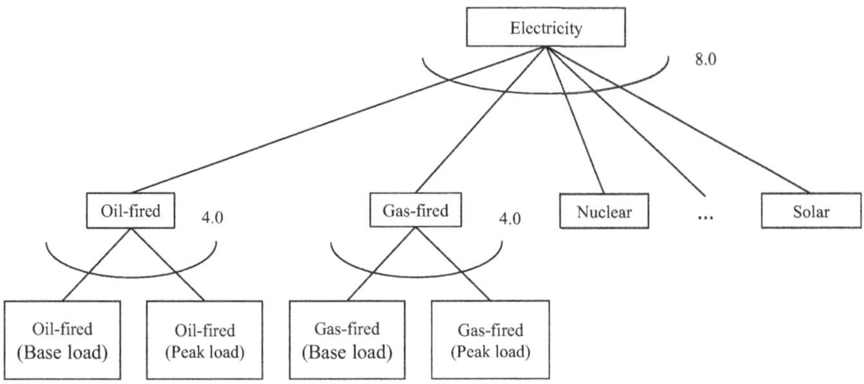

Fig. 10.3 Structure of electricity input

sources, as depicted in Fig. 10.3. Thus, the model considers various power sources; specifically, base load and peak load power sources are distinguished. The peak load power source comprises oil-fired, gas-fired, hydrogen, and solar power generation. The base load power source comprises nuclear, coal-fired, gas-fired, oil-fired, wind, and hydro power sources.

10.3.3.2 Structure of Electricity Generation

Electricity used by the production sector and households is a CES composite of various power sources. As depicted in Fig. 10.4, base load power generation is modeled as a Leontief composite of the input of intermediate goods, an energy–labor composite, and region- and source-specific capital stock. These factors are assumed to be non-substitutable. Thus, the output level of the base load power source is limited by the amount of specific capital stock, which means that the elements that comprise the base load power source have an upper limit on their supply capacity. However, as depicted in Fig. 10.5, the peak load power source can use noncapital goods as substitutes for specific capital stock, meaning that the elements that comprise the peak load power source do not have an exogenous upper limit to their supply capacity. The distinctive feature of the peak load power supply, excluding solar power, is that the cost of electricity generation rises as supply increases. The solar power source is assumed to use ordinary capital stock. Thus, as the electricity price increases, or when the feed-in tariff is introduced, investment in solar power is based on the market mechanism.

10.3.3.3 Structure of Utility Function

The utility function is also described using a nested CES function. The nested structure of the CES function is depicted in Fig. 10.6. In the uppermost nest, the utility of a representative household in each region is modeled as a CES composite of the consumption of a composite of consumption goods and leisure time. This composite consumption is a CES composite of energy composite consumption and

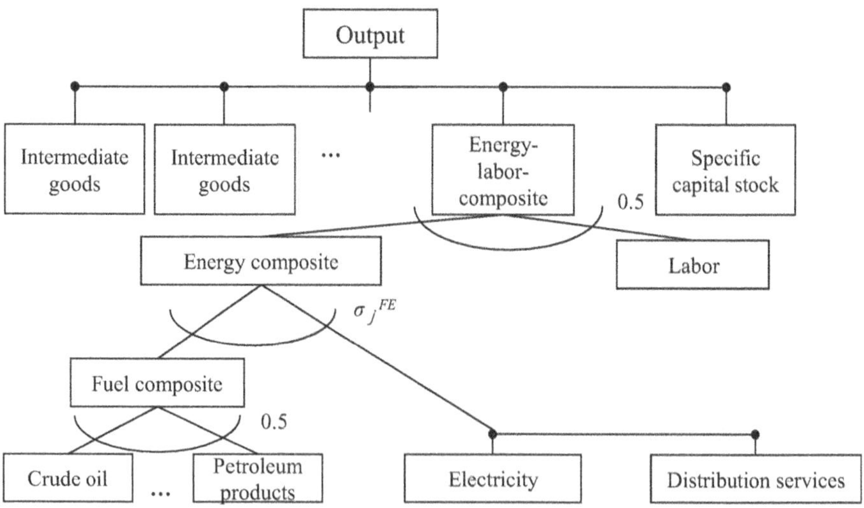

Fig. 10.4 Structure of electricity generation

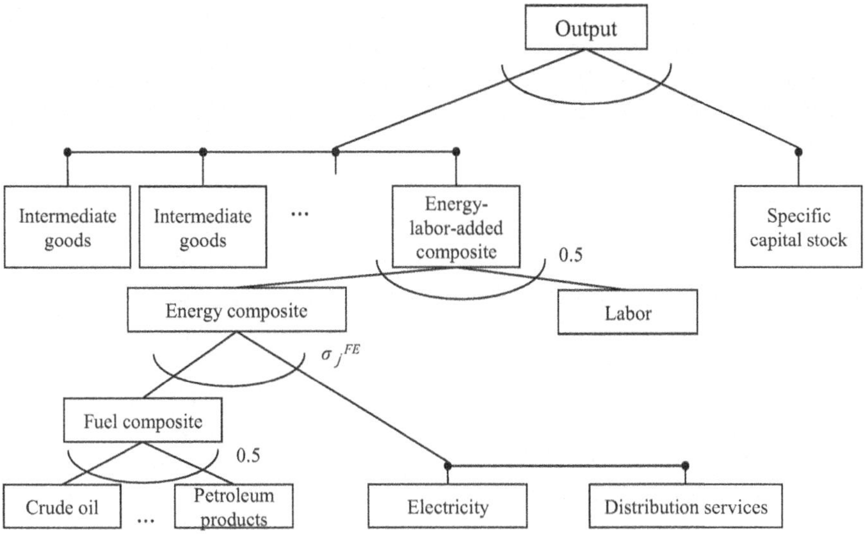

Fig. 10.5 Structure of peak load power generation

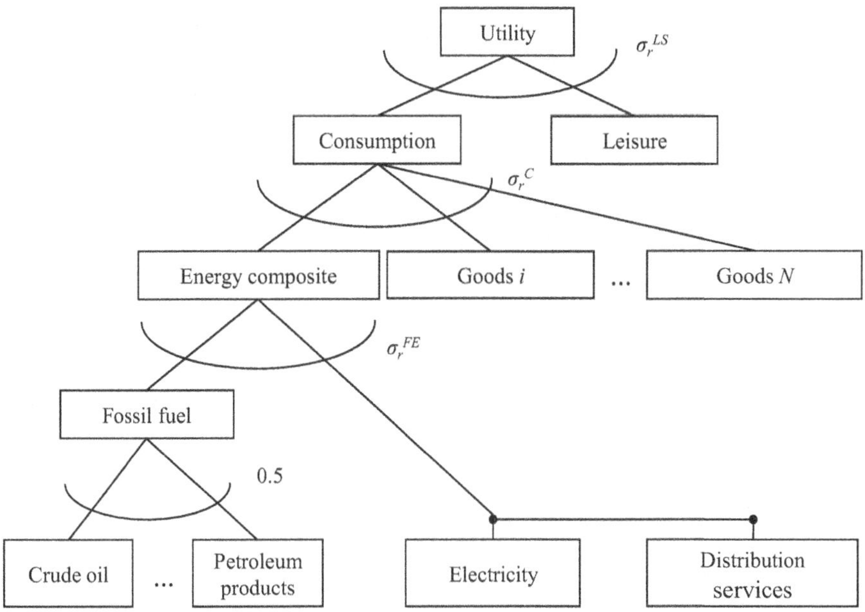

Fig. 10.6 Structure of utility function

various final consumptions of goods. The energy composite is a CES composite of the consumption of a fossil fuel composite and electricity that is accompanied by electricity distribution services. The fossil fuel composite is a CES composite comprising the consumption of crude oil, gas, coal, and petroleum products. The representative household maximizes its utility subject to time and budget constraints, given the prices of goods and services, wage rates, the rental prices of capital stock, and the unit prices of leisure time available to household members, the opportunity cost of which is wage rates.

10.3.3.4 Feed-in Tariff

The economic and environmental impacts of eliminating nuclear power may depend heavily on the potential of Japan's renewable energy supply and the capacity of the fossil fuel electricity supply. The Japanese government promotes renewable energy through a solar photovoltaic (PV) feed-in tariff (FIT) to address climate and energy issues. Under the FIT system, introduced in November 2009, an electric power company must purchase electricity generated from solar PV at a fixed price and pass on the cost to their customers. The FIT system was applied only to surplus electricity; electricity generated by solar PV must be consumed on-site first. Purchase rates at this time were 48 JPY/kWh for residential use and 24 JPY/kWh for nonresidential use. The purchase duration was 10 years. Following the Fukushima Daiichi nuclear accident, the Japanese government decided to include wind, geothermal, biomass, and micro hydro resources in the FIT targets.

Here, the FIT is modeled as a subsidy paid for electricity generation through the use of renewables and as a tax on electricity sales. The budget for renewable energy subsidies is financed by taxes on electricity sales. The tax rate of electricity sales is determined endogenously in order to balance the subsidy budget and tax revenue.

10.3.3.5 Simulation Scenarios

The benchmark year of the GTAP-Power database is 2011. In the database, nuclear power generation accounts for 7.9% of total electricity generation in Japan. Before the accident, for example, in 2010, this figure was 32%. In CGE simulation, the economic agents are assumed to adapt rationally to the economic situation described in the GTAP-Power database. The assumption of rational adaptation of economic agents is useful for calibrating the parameters of CES functions. However, this assumption could underestimate the importance of nuclear power in 2011, thus the benchmark year affects the results of the simulation.

Two scenarios are assessed. By adjusting the amount of specific capital of the nuclear power source, the power generation level can be controlled. The first scenario is one in which nuclear power generation increases to account for 20% of total power generation in Japan. This scenario is called the 20% nuclear power scenario. The Japanese government proposed a 20% nuclear power share in the *Basic Energy Plan* released after the Fukushima nuclear accident. The second scenario reproduces actual nuclear power use in Japan after the accident. In the second scenario, nuclear power accounts for almost zero percent of total power generation in Japan. The second scenario is called the zero nuclear power scenario. Simulations of these two scenarios are compared, and the importance of nuclear power generation in Japan is assessed. In both scenarios, FIT is introduced as a 20% ad valorem subsidy for renewables.

10.4 Simulation Results

10.4.1 Power Source Composition

Figures 10.7, 10.8, and 10.9 show the power source composition for each simulation scenario in Japan. Figure 10.7 is the power source composition of the benchmark year (2011). Even though nuclear power operations decrease following the Fukushima Daiichi accident, some reactors were still in use. Figure 10.8 shows the simulated power source composition in the 20% nuclear power scenario in 2011. Here, the share of nuclear power rises exogenously from 7.9 to 20%. Compared with Fig. 10.7, the shares of oil-fired and gas-fired power generation decrease notably. Figure 10.9 depicts the simulated power source composition in 2011 in the zero

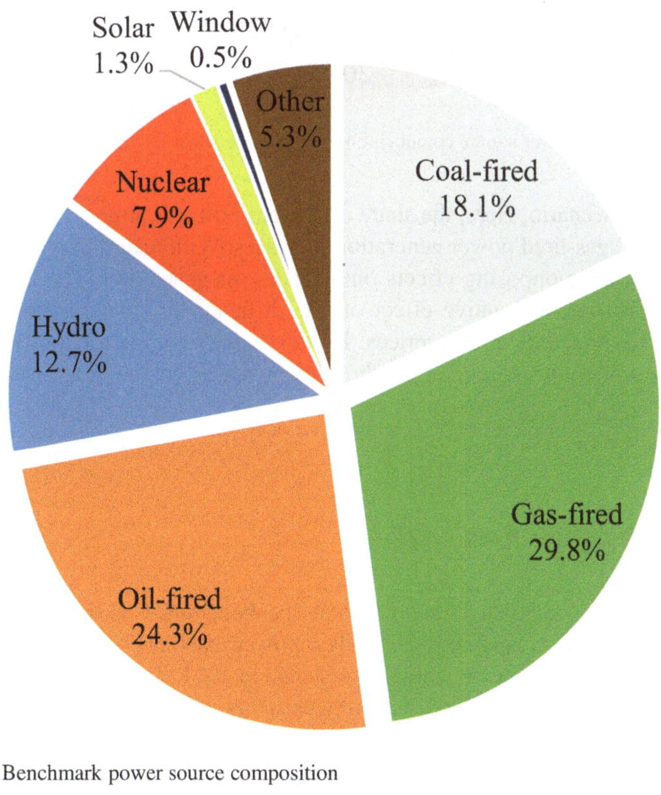

Fig. 10.7 Benchmark power source composition

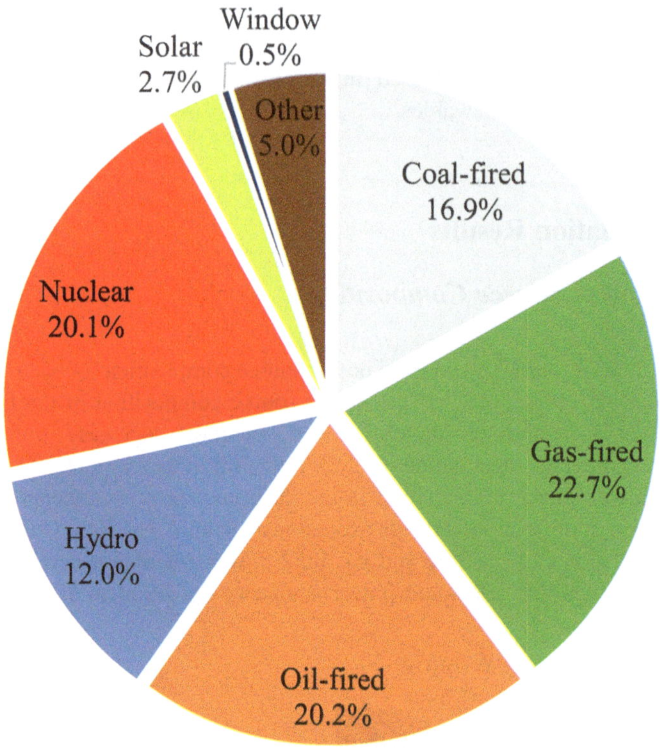

Fig. 10.8 Simulated power source composition in the 20% nuclear power scenario

nuclear power scenario. Here, the share of nuclear power is almost zero. Compared with Fig. 10.7, gas-fired power generation increases. With regard to renewables, the two scenarios have opposing effects on solar power generation. The 20% nuclear power scenario has a negative effect on the diffusion of solar power generation because of restrained electricity prices. However, the zero nuclear power scenario encourages solar power generation through increases in the price of generated electricity.

10.4.2 *Effects on Real GDP*

Figure 10.10 shows Japan's real GDP in the benchmark year and in the two simulation scenarios. Suspending all nuclear power reactors does not have a significantly negative impact on the real GDP of Japan in 2011. This is because fossil fuel power generation compensates for shortages in the electricity market. When the share of nuclear power increases to 20%, the real GDP of 2011 increases by 0.18% compared with the zero nuclear power scenario.

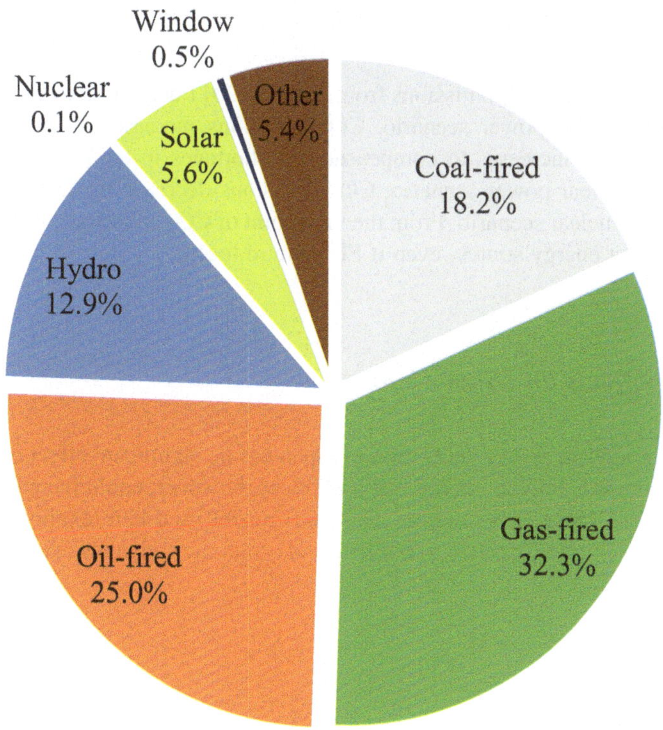

Fig. 10.9 Simulated power source composition in the zero nuclear power scenario

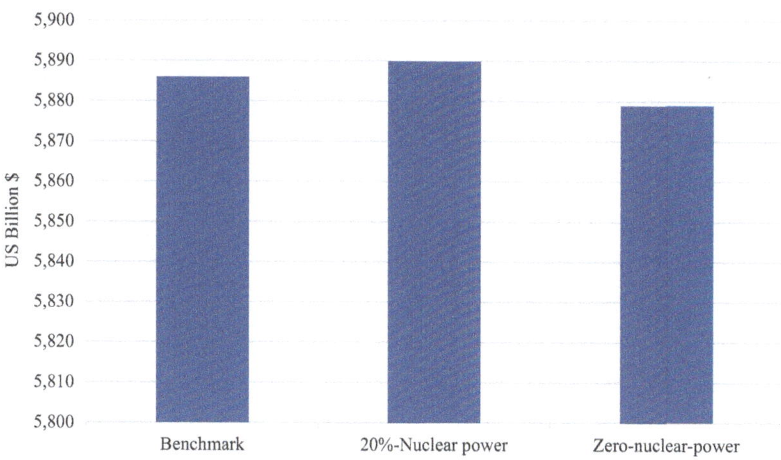

Fig. 10.10 Simulated real gross domestic product (GDP) in Japan in 2011 in different scenarios

10.4.3 Effects on CO_2 Emissions

Figure 10.11 depicts CO_2 emissions from Japan in 2011 according to each scenario. In the zero nuclear power scenario, CO_2 emissions increase because fossil fuel power generation increases to compensate for shortages in the electricity market. In the 20% nuclear power scenario, CO_2 emissions decrease by 7.68% compared with the zero nuclear scenario. From the viewpoint of CO_2 emissions, nuclear power is an important energy source, even if FIT is introduced.

10.4.4 Effects on Other Countries

The level of nuclear power generation in Japan has no significant effect on the GDP of other countries. International fossil fuel prices, however, could increase negative impacts on the Japanese economy. The CGE model used here explicitly includes oil-producing countries, although political and economic strategies of such countries are not considered.

10.5 Conclusion

This chapter assesses the importance of nuclear power generation in Japan from the perspective of real GDP and CO_2 emissions. The assessment is conducted using a multicountry multisector CGE model. The model is based on the GTAP-Power database, which is an electricity-detailed extension of the GTAP 9 database. The

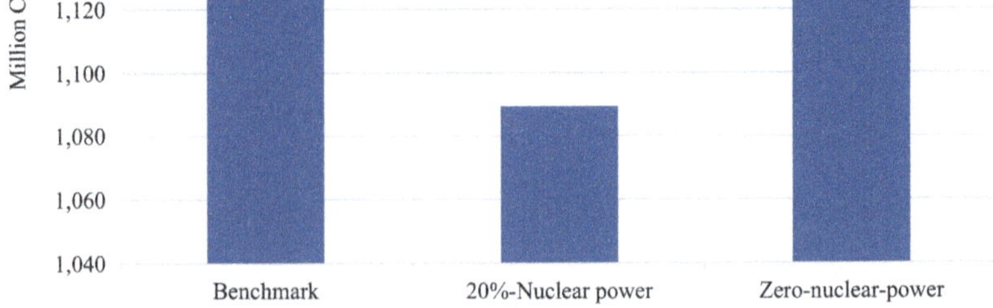

Fig. 10.11 Simulated CO_2 emissions from Japan in 2011 in different scenarios

simulation results indicate that nuclear power generation is not significantly important for Japan in terms of GDP, because fossil fuel power compensates for shortages in the electricity market. However, this is based on the assumption that Japan can buy fossil fuels on the international market without difficulty. In this regard, political risks associated with oil-producing countries should be considered. In terms of CO_2 emissions, nuclear power is an important power source for Japan. An increase in fossil fuel power generation causes a non-negligible amount of CO_2 emissions. The simulation results indicate that it is difficult to diffuse enough renewable energy to offset increases in CO_2 emissions in the zero nuclear power scenario, even if a FIT is introduced. However, the risk of natural and man-made accidents and the challenge of radioactive waste disposal cannot be ignored. The long-term view indicates that nuclear power should be replaced with renewable power. Further, the risk management system for nuclear power should build on the experience of the Fukushima Daiichi nuclear accident. Such a system should be shared among East Asian countries with nuclear power plants.

References

Hosoe N, Gasawa K, Hashimoto H (2015) Textbook of computable general equilibrium modeling: programming and simulations. Palgrave Macmillan, London

Peters JC (2016) The GTAP-power Data Base: disaggregating the electricity sector in the GTAP Data Base. J Glob Econ Anal 1(1):209–250

Yamazaki M, Takeda S (2017) A computable general equilibrium assessment of Japan's nuclear energy policy and implications for renewable energy. Environ Econ Policy Stud 19(3):537–554

Chapter 11
Construction of an East Asia Nuclear Security System

Weisheng Zhou, Kenzo Ibano, and Xuepeng Qian

Abstract After a huge impact of the Fukushima nuclear power plant accident, discussions on phasing out nuclear power have been growing in Japan. Meanwhile, influences of nuclear accidents at neighbors will not be mitigated by decommissioning domestic reactors. We examined current status of nuclear power plants in Japan, China, and South Korea. Then, using the nuclear accident database, we analyzed causes of incidence by four factors: country, reactor type, cause, and operational period. Reviewing nuclear regulation policies in Japan, China, and South Korea by comparing with the United States and European Union, we anticipate necessity of a construction of a Japan-China-Korea nuclear security framework.

11.1 Introduction

According to the United Nations Intergovernmental Panel on Climate Change (IPCC), in order to limit average global temperature increase to 2 °C (preferably 1.5 °C, set by the Paris Agreement) above pre-industrial levels, use of fossil fuels must be drastically reduced, and carbon-reducing power sources must account for at least 80% of power generation by 2050. Fossil fuels currently account for 70% of

---

This chapter is an improved version of the psaper, "Zhou, W., Xu S., Ibaano, K., Qian, X., Nakagami K. 2014. "Toward Establishing Nuclear Power Safety and Security System in East Asia: Statistical Analysis of Nuclear Power Plant Accidents around the World," Policy Science, 22–1:1–10."

W. Zhou (✉)
College of Policy Science, Ritsumeikan University, Ibaraki, Osaka, Japan
e-mail: zhou@sps.ritsumei.ac.jp

K. Ibano
Graduate School of Engineering, Osaka University, Suita, Osaka, Japan
e-mail: kibano@eei.eng.osaka-u.ac.jp

X. Qian
College of Asia Pacific Studies, Ritsumeikan Asia Pacific University, Beppu, Oita, Japan
e-mail: qianxp@apu.ac.jp

W. Zhou et al. (eds.), *East Asian Low-Carbon Community*,
https://doi.org/10.1007/978-981-33-4339-9_11

electricity generation. As a measure to combat global climate change, nuclear power is expected to be an indispensable tool to prevent the release of CO_2, a greenhouse gas, during power generation. However, there are many obstacles to relying on nuclear power, including high costs, long-term management of radioactive waste, and the dangers of nuclear accidents in particular.

The accident at the Fukushima Daiichi nuclear power plant, caused by the Great East Japan Earthquake, occurred in March 2011 (IAEA 2011) and had a huge global impact. Shocking images were released of hydrogen explosions in power plant buildings. Nuclear power policies in many countries were immediately modified. The change in Japan, in particular, was remarkable. All nuclear reactors were temporarily stopped in order to undergo inspections. In February 2020, some have resumed operation. There are currently seven operating reactors: Genkai Units 3 and 4, Sendai Units 1 and 2, Ohi Units 3 and 4, and Takahama Unit 4. In addition, Takahama Unit 3 and Ikata Unit 3 are currently undergoing inspection but are scheduled to restart operations.

Prior to the Fukushima Daiichi nuclear power plant accident, concerns about nuclear energy use were uncommon in Japan. Some people preferred electric power production by nuclear plants. Compared to thermal power plants, nuclear plants can reduce environmental pollution and operate with a small amount of fuel. Expectations for nuclear power stalled temporarily after the accident at the Chernobyl power plant in 1986. However, in the 1990s, a construction boom of nuclear power plants began, in support of economic development in developing countries such as China, resulting in a reduction in criticisms of nuclear power. Even after the Fukushima Daiichi nuclear power plant accident, primary energy demands in developing countries remain high, and it is unlikely that nuclear power phaseout will be achieved globally.

The impact of nuclear accidents is not limited to single countries; influences on peripheral countries should be considered. Hence, even if Japan successfully phases out nuclear power, risks of impacts arising from nuclear accidents in South Korea and China cannot be excluded. There are many nuclear power plants in East Asia (China National Nuclear Corporation 2014; Japan Atomic Energy Agency 2014; Korea Electric Power Corporation 2014). In order to protect people from nuclear accidents, it is essential to enhance the safety of nuclear power plants both domestically and internationally. In this chapter, we analyze trends in nuclear accidents to provide insight to the design of an East Asia nuclear security system.

11.2 Previous Analyses of Accidental Events at Nuclear Power Plants

In *Japan-US Comparative Analysis of Nuclear Power Plant Operating Rates and Trouble Incidence Rates* (Kainou 2009), accidents at nuclear power plants in Japan and the United States from 1999 to 2008 are analyzed. The "trouble" here covers all nuclear accidental events from level 0 of the International Nuclear Event Scale (INES). Operating rates of and accidents at nuclear power plants in Japan and the United States are classified by type and age, and the relationship between operating rates and frequency of accidents is statistically analyzed. Kainou divided the causes of accidents into two categories: force majeure (natural disasters such as earthquakes and tornadoes, as well as man-made external factors such as terrorist attacks and plane crashes) and manageable (problems caused by reactor design, human operational error, and mechanical failures that should have been discovered during inspections). The conclusion is that Japan and the United States do not differ greatly in terms of nuclear power plant utilization rates.

In Japan and the United States, only two types of reactor, pressurized and boiling water, were in operation, so the type of machine was given little consideration. In other countries, many other types of nuclear power plants are also in operation, including Russian light water reactors and CANDU heavy water reactors, which require more complex analysis. Causes of accidents were grouped into internal and external causes. However, as became clear in the Fukushima Daiichi nuclear power plant accident, if a natural disaster such as an earthquake occurs outside the site, it should be possible to prevent problems occurring if appropriate countermeasures are taken in advance. In the case of intrinsic causes, although problems such as machine failure can be prevented by inspection, it is unreasonable to classify these as identical to human-induced causes such as operational errors.

Then, accidents can be classified by five causes: design, manufacture, maintenance, operation, and under investigation. This allows analysis from the viewpoint of inspection. The frequency of troubles caused by manufacturing and maintenance tends to be higher than those resulting from other causes. Troubles caused by design can include cracks that occur during service due to failure to consider fatigue, mechanical and electrical resonance in the design, and stress corrosion cracking, a phenomenon that was unknown at the time of design. Troubles caused by manufacturing include deterioration of material properties due to defective joints or excessive heat input and cracking during use due to impurities adhering to the product during manufacturing. Thus, troubles caused in the design and manufacturing stages may be due to quality assurance issues such as procurement management or by events related to phenomena that were not understood during the design and manufacturing stages. Troubles caused by maintenance are due to errors in inspection and maintenance work performed at the power plant during service. Examples include equipment failure due to errors in restoration following disassembly and inspection, leaks that occur due to a foreign object biting into the valve, and wire breakage due to loose screws. Troubles caused by operation are due to errors in operation and monitoring at the power plant during service. For example, reactor power can be reduced due to errors in switching the operation of equipment, and reactors can be automatically shut down due to errors in the monitoring of instruments and operation information displays.

This analysis categorizes all occurrences of trouble as human-caused. While some unanswered questions remain regarding troubles under investigation, troubles caused by design, manufacturing, maintenance, and operations are all considered avoidable. In this context, design and manufacturing issues are related to the fundamentals of nuclear technology; problems are gradually solved as technological innovations are made. Nevertheless, current technology has its limits, and not all technological problems can be solved.

Both studies described here limited their scope primarily to two countries. The absence of a global analysis of accidents means that differences in types of nuclear power plant have not been taken into account. Therefore, this chapter analyzes accidental events in private nuclear power plants worldwide according to four factors: country, reactor type, cause, and operational period.

11.3 Database of Accidents

The International Nuclear Event Scale (INES) (IAEA 2013) is a measure used to evaluate nuclear accidents, developed by the International Atomic Energy Agency (IAEA) and the Organization for Economic Cooperation and Development Nuclear Energy Agency (OECD/NEA). The INES level increases depending on the severity of an accident, as expressed in Fig. 11:1. Following INES criteria, a nuclear event up to level 3 is called an incident and that of level 4 or higher is an accident. Before INES was formally adopted in 1992, there was no international scale with respect to nuclear accidents, and data collection was inadequate. In addition, nuclear accidents tended to be concealed from the international stage. Thus, data on nuclear accidents before 1992 are incomplete. Some nuclear incidents with INES level 3 and higher that occurred before 1992 are summarized in Table 11.1. Only nine accidents are shown here; however, there are records of several dozen accidental events before 1992. INES level was not evaluated for these events; thus the data lack integrity. In this chapter, we use INES-evaluated data from 1992 onward. A database of private nuclear power plant accidental events is provided in the appendix. All incidents with

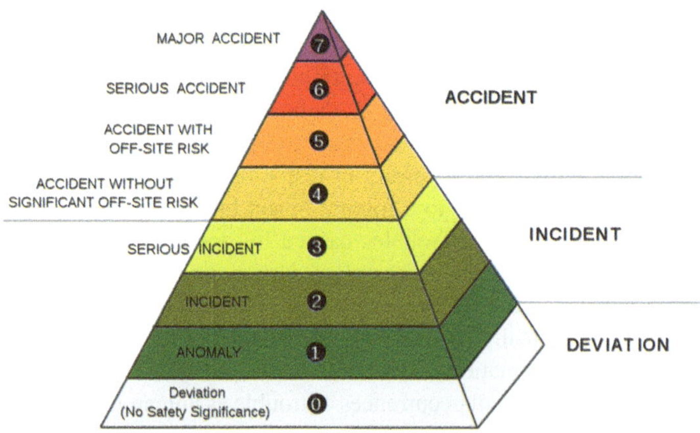

Fig. 11.1 International Nuclear Event Scale (INES)

Table 11.1 Major accidental events in nuclear power plants before 1992

Event time	Operation years	Reactor name	Reactor type	INES level
January 21, 1969	3	Lucens	Heavy water-moderated CO_2 gas-cooled	4
October 17, 1969	0	Saint Laurent	Graphite-moderated gas-cooled	4
December 7, 1975	1	Greifswald	VVER	3
July 5, 1976	4	Jaslovské Bohunice	Heavy water-moderated CO_2 gas-cooled	4
February 22, 1977	5	Jaslovské Bohunice	Heavy water-moderated CO_2 gas-cooled	4
March 28, 1979	0	Three Mile Island	PWR	5
March 13, 1980	9	Saint Laurent	Graphite-moderated gas-cooled	4
April 26, 1986	3	Chernobyl	RBMK	7
October 19, 1989	3	Vandellòs	Graphite-moderated gas-cooled	3

INES level 2 or higher that occurred in private nuclear power plants are listed. Data were acquired from IAEA PRIS (IAEA 2014) and the LAKA (2020). In the following sections, accidental events are summarized by country, reactor type, cause, and operational period.

11.4 Analysis of Accidental Events at Nuclear Power Plants

11.4.1 Analysis of Accidental Events by Country

Based on the database given in the appendix, nuclear accidental events are summarized by country in Table 11.2. Data from countries with ten or more nuclear reactors are included. Here, the accidental event rate is defined as the average number of accidental events during 1 year of reactor operation. There is a remarkably low accidental event rate in the United States, despite the presence of approximately 100 reactors. The United States ranks fifth in terms of nuclear dependence, third for number of events or accidents, and ninth for incidence rate. In contrast, France shows the highest number of accidental events, and the accidental event rate is also high. It is worthwhile to consider differences in nuclear regulation in these two countries.

The American Nuclear Regulatory Commission (NRC) is responsible for the low accidental event rate. Members of the NRC are appointed directly by the President of the United States; thus, they have a high degree of independence. The NRC actively seeks data and opinions from concerned citizens, operators, and stakeholders of

Table 11.2 Number of nuclear power plants and accidents in major nuclear power-producing countries

Country	Number of reactors[a]	Dependencies on nuclear power	Number of events after 1992	Accidental event rate (‰)
France	58 (59)	74.8	21	0.29
Ukraine	15 (17)	46.2	7	1.32
Sweden	10 (12)	38.1	6	2.08
Republic of Korea	23 (23)	30.4	3	0.37
Japan	33 (60)	6.2	5	0.09
The United States	100 (106)	19	6	0.03
The United Kingdom	16 (33)	18.1	3	0.17
Russia	33 (34)	17.8	4	0.18
Canada	19 (23)	15.3	3	0.31
India	21 (21)	3.6	5	0.8

[a]Numbers in brackets indicate number of reactors including ones shut down after 1992

nuclear power plants. Claims are assessed, fairly evaluated, and used to inform decisions. The reasons leading to decisions taken are clearly stated and documented. The NRC specializes in the supervision of operational nuclear power plants and examination of new power plants. It is not responsible for extending nuclear power use in the United States nor developing nuclear power policies. With its authority, the NRC can seize control of nuclear power generation businesses in order to increase the safety of nuclear power plants. This regulatory role of the NRC was reconstructed after the Three Mile accident and works well. The regulatory and operational sides are well separated, while maintaining communication.

In contrast, France has an accidental event rate almost ten times that of the United States and has the highest nuclear power dependence worldwide. In the French regulatory system, there are two safety monitoring agencies. The Nuclear Safety Council (ASN) is independent from government agencies, and the Radiation Protection Nuclear Safety Institute (IRSN) provides external assistance to the ASN. Nuclear regulation strategies in France are represented in remarks made by the director of ASN, "even with the ultimate vigilance, nuclear accidents can never be eliminated. This is the basis of all of our actions." Since 1992, all nuclear incidents in France have been INES level 2 or lower, meaning that their impacts were contained within the power plant site and did not affect the wider population. France assumes the occurrence of accidental events as a given and focuses on minimizing their impacts.

11.4.2 Analysis of Accidental Events by Reactor Type

In Fig. 11.2, the accidental event number and accidental event rate by reactor type are summarized for currently operational power plants, including pressurized water reactor (PWR), boiling water reactor (BWR), Russian-type pressurized water reactor (VVER), heavy water reactor (CANDU), graphite-moderated boiling light water pressure tube type reactor (RBMK), and advanced gas-cooled furnace, graphite-moderated carbon dioxide-cooled reactors (AGR). PWR shows more than twice the number of events of the reactor type with the second highest event frequency (VVER). However, this is because PWR is the most popular reactor type (438 units); the accidental event rate is low, as indicated by a polygonal line. RBMK shows a remarkably high accidental event rate. Design problems have been identified in the RBMK, which is consistent with its high accidental event rate. Fortunately, most RBMK reactors are currently decommissioned, and few remain in operation. VVER shows the second highest accidental event rate, including serious accidents, yet 30 reactors remain in operation. In particular, three VVER reactors have experienced two or more events; safety performance of these reactors is questionable. Concerns about VVER design—lack of a containment vessel and insufficient emergency core cooling system performance in the event of a coolant loss accident—are highlighted by the IAEA.

11.4.3 Analysis of Accidental Events by Cause

Although there are safety concerns related to reactor type, if an accidental event is caused by obvious external factors, such as human error and natural disasters, it is difficult to determine reactor type as the primary cause. Although identifying the unique cause of an accident is challenging, we categorize accidental events into three

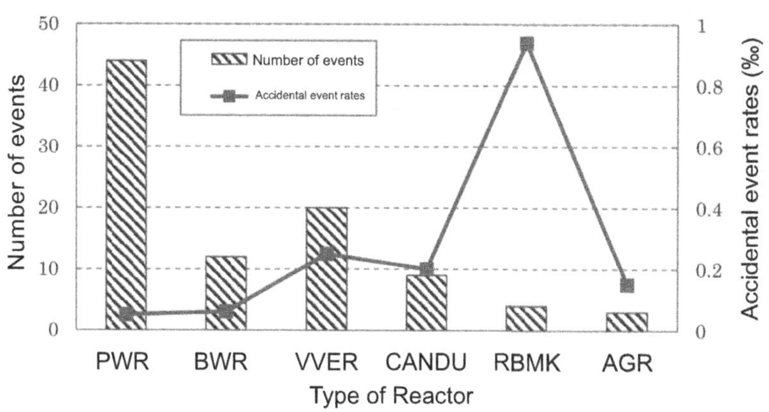

Fig. 11.2 Number of accidental events and rates for different reactor types

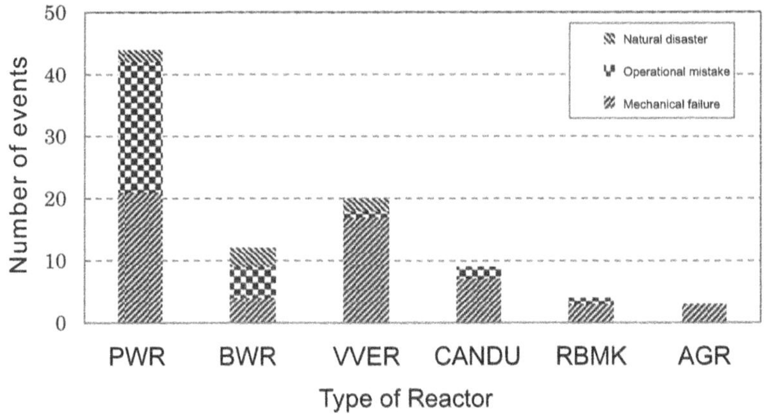

Fig. 11.3 Number of accidental events and their primary cause for different reactor types

groups according to their primary causes: natural disasters (earthquakes and tsunami), operational mistakes, and mechanical failure (in situations where there is no external force, such as natural disasters or operational mistakes) Based on the database in the appendix, reactor types and the cause of accidents are summarized in Fig. 11.3. Accidental events at VVER are mainly caused by intrinsic mechanical failures, demonstrating that VVER clearly has safety problems associated with its design.

11.4.4 Analysis of Accidental Events by Operational Period

We summarize the relationship between operational period at the time of an accidental event and the number of accidents in Fig. 11.4. Significant increases in accidental events are seen for reactors at 16–20 years of operation. These events correspond to reactors made in the 1970s and 1980s, which were part of the global nuclear construction rush. The large number of accidental events is explained by the large number of reactors in this age group. As shown in Fig. 11.4, the operating life accidental event rate uniformly remains below 2% for 40 years of operational life. However, calculated accident event rates are uniformly below 1 per-mil (‰) for reactors with less than 40 years of operation.

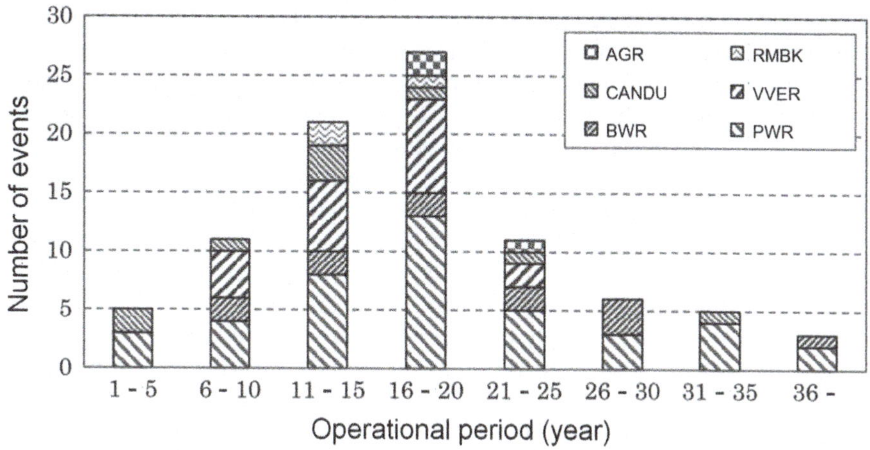

Fig. 11.4 Number of accidental events of each reactor type according to operational period at the time of event

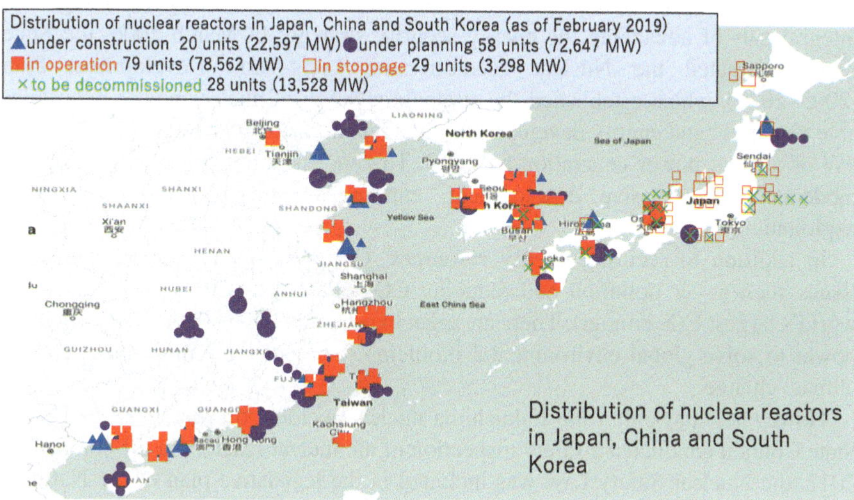

Fig. 11.5 Distribution of nuclear power plants in Japan, China, and Korea

11.5 Importance of an East Asia Nuclear Security System

11.5.1 The State of Nuclear Power in China

Figure 11.5 shows the distribution of nuclear power plants in China, Japan, and Korea. Nuclear power in China has gone through three stages of development and is now entering a period of active and rapid development. In the early 1970s, China began independent research and development, design, and construction of commercial nuclear power plants based on experimental studies. In 1991, the 300,000 kW Qinshan Phase I nuclear power plant, developed in China, was connected to the grid for power generation, ending the history of zero nuclear power in mainland China and making it the seventh country in the world to independently design and build a nuclear power plant.

In the 1990s, electricity supply in China was relatively plentiful. Nuclear energy was positioned as a "supplement to our country's energy," and development policies were moderate. At the end of 2004, China had built a total of six units, including two Qinshan units, with an installed capacity of 4.7 million kilowatts, and three nuclear power bases in Guangdong, Zhejiang, and Jiangsu provinces.

Since 2004, electricity supply has gradually become a bottleneck limiting economic and social development in China; an important role for nuclear power in China's sustainable energy development has gradually emerged. Today, China is a net importer of coal resources and has become more dependent on foreign countries for crude oil. Due to instability, such as wars with oil suppliers, risks to oil imports in terms of volume and price have increased, and energy security in China will become increasingly threatened. As a source of renewable energy, nuclear can completely avoid security issues associated with imports of oil and gas energy, making it an integral part of advancing the energy security strategy. In March 2006, the State Council created the Nuclear Medium and Long-Term Development Plan (2005–2020), which established the strategic position of nuclear power in sustainable economic and energy development in China, announcing a target of 40 million kW of nuclear power generation by 2020. The nuclear industry has changed from moderate to aggressive development, entering a new phase of large-scale exploitation.

In addition to securing energy resources, China must address environmental issues such as air pollution and reducing CO_2 emissions (China is currently the world's largest CO_2 emitter). There are great expectations for the potential of nuclear power to solve global environmental problems such as overcoming pollution and climate change.

However, 5 days after the Fukushima nuclear accident on March 11, 2011, the State Council conducted a safety inspection of all nuclear reactors in the country. In 2012, the Nuclear Safety Law was included in the legislative plan of the National People's Congress, which came into effect in January 2018. China has not changed its original policy position of nuclear power expansion in the aftermath of the Fukushima nuclear accident.

At the end of December 2019, a total of 48 research and first-critical reactors were in operation in China (Fig. 11.5). If all reactors currently in construction (10) and planning (42) phases are completed, a maximum of 100 reactors could be in commercial operation by 2030. As shown in Fig. 11.5, China ranks third in terms of the number of operating reactors out of 31 countries with nuclear power, behind the United States (96 reactors) and France (58 reactors) (IAEA-PRIS 2019). The share of nuclear power in annual power generation in China is expected to increase from 1.8% in 2010 to 4.1% in 2018 and 5% in 2020 (cumulative power generation of 1 trillion kWh). Abandoning nuclear power at this stage would be difficult, as it is a

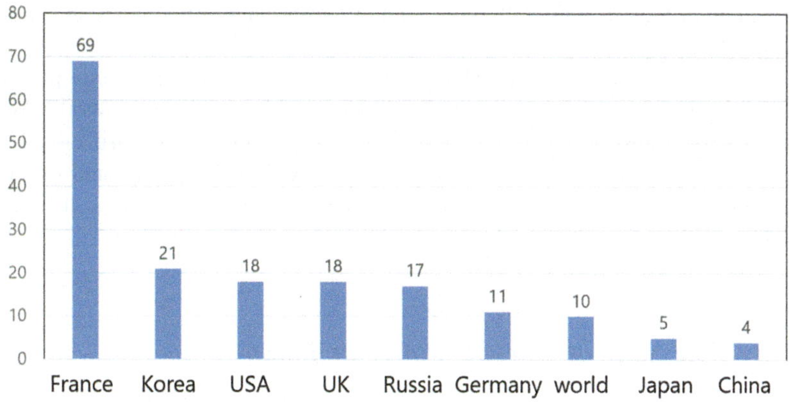

Fig. 11.6 Ratio of nuclear power to total electricity supply globally (percentage). (Source: IEA Key World Energy Statistics 2019)

means of simultaneously securing energy and addressing environmental issues. As shown in Fig. 11.6, the share of nuclear power in China remains small compared to other countries but will increase in future.

To date, there have been no nuclear accidental events of INES level 2 or higher in China. However, nuclear security remains an urgent and important issue because of the Fukushima nuclear accident in Japan, the danger of a high-level accident due to aging nuclear power plants, and the South Korean policy of denuclearization, as well as the rapid progress in nuclear power plant construction plans in China. Nevertheless, while ostensibly maintaining their pronuclear stance, even major nuclear power companies have begun to shift to renewable energy. For more information on new energy in China, see Chap. 19.

11.5.2 *The State of Nuclear Power in South Korea*

As South Korea lacks fossil fuel resources, governments prior to the current government of Moon Jae-In have all promoted nuclear power as a national policy, and the share of nuclear power in South Korea has been steadily increasing. Korea joined the International Atomic Energy Agency in 1957 and has been developing nuclear energy since then. There were only two research reactors operating in the 1960s, but construction and operation of commercial nuclear power plants have continued since 1978. Two reactors were chosen for decommissioning in 2017. There were 23 reactors in operation at the end of June 2019, with a total capacity of 22,529 MW, ranking sixth largest globally. South Korea has the highest density of nuclear power plants per unit of land area (Fig. 11.5). Most commercial reactors started operation in the 1980s and will have been operating for more than 30 years by 2020, placing them near the end of their designed lifetime (30 or 40 years) and exacerbating the issue of nuclear safety.

Currently, the government of Moon Jae-In has set out a major energy policy of denuclearization, but a total of six nuclear reactors with a capacity of 8400 MW are currently under construction. The policy is, in effect, gradual denuclearization. If it is assumed that plants currently under construction will all be operational by 2024, then there will be nuclear power plants in operation in South Korea at least up to 2083.

11.5.3 The State of Nuclear Power in Japan

Since the enactment of the Atomic Energy Basic Law in December 1955, Japan has actively promoted the construction and operation of nuclear power plants as a national policy, with nuclear power as a major energy source. In 1974, the Three Power Supply Laws were enacted to provide greater financial support to municipalities located in and around nuclear power plants (Power Supply Development Promotion Tax Law, Special Accounting Law for Measures to Promote Power Supply Development, and Regional Development Law for the Surrounding Areas of Power Generation Facilities), which accelerated the spread of nuclear power.

Until just before the 2011 Fukushima accident, 54 nuclear reactors provided 25–30% of the country's electricity. Since the accident, all reactors were shut down, and as shown in Fig. 11.5, electricity provision from nuclear fell to 5% in 2018. There remain strong calls for the elimination of nuclear power in Japan, but at this stage, there are no alternative energy sources that do not rely on imported resources, so there are many who want reactors currently under inspection to be restarted.

At the time of the Fukushima disaster, Japan's basic energy plan called for nuclear power to provide about half of total power generation until 2030. At the time, the ruling Liberal Democratic Party (LDP) positioned nuclear power as the "most important power source" for the future. However, after the accident at the Fukushima nuclear power plant, public confidence in nuclear power has wavered greatly, and when the Democratic Party of Japan (DPJ) came to power, it set a scenario of denuclearization in the 2030s as an energy policy goal (Energy and Environment Council 2012).

However, since the LDP took over government again in December 2012, nuclear power is once more positioned as an important energy source. The Fifth Basic Energy Plan (New Energy Basic Plan) was approved by the cabinet in July 2018. Although classifying renewable energy for the first time as a "mainstay energy source" for 2030, nuclear power remains a "baseload power source" with a share of 20–22% of total power generation.

At the end of June 2019, 15 nuclear power plants have been reviewed by the Nuclear Regulatory Commission and approved to resume operation, 12 are still under review, 9 have not yet applied for review, and 17 have abandoned the review process and have been decommissioned, bringing the total number of nuclear power plants in Japan to 36.

As shown in Table 11.2, the accident rate in Japan was the second lowest globally (after the United States), but the level 3.11 Fukushima accident has shattered the myth of nuclear safety in Japan.

11.6 Designing a Nuclear Security System for Japan, China, and Korea

11.6.1 Issues Related to Nuclear Power Plants in Japan, China, and Korea

As shown in Fig. 11.5, the scale of nuclear power problems is no longer that of a single country but extends to regional and global levels. Nuclear power plants in China (48 operational and 10 under construction) are concentrated in the eight coastal provinces of Guangdong, Zhejiang, Jiangsu, Shandong, Liaoning, Guangxi, Fujian, and Hainan. The coastal areas of Japan, South Korea, and China have the highest density of nuclear power plants (number of reactors per unit area) in the world. Because of the close geographical and economic relationship among Japan, China, and South Korea, nuclear issues and accidents in one country can have a significant impact on the others. There is an urgent need to establish a cooperative framework for nuclear safety measures and assessments that accounts for trends in nuclear policy and the increase in risk of future accidents.

For example, the following points pose serious challenges for the Chinese nuclear community.

Shortage of human resources, especially highly skilled engineers According to the nuclear power plant plan, about ten one million kilowatt nuclear reactors will be built simultaneously during the peak period of nuclear power plant construction in China. It is of grave concern to the international community that there is insufficient planning, management, and operational capacity to support this rapid expansion. There is a shortage of human resources, especially high-level engineers, as well as a shortage of manufacturing capacity, construction and installation capacity, and research and development and design capacity for nuclear power generation facilities. There is also a lack of regulatory manpower.

Hidden risks in the scarcity of nuclear fuel resources There is no doubt that China needs nuclear power. However, from a long-term perspective, it may only be able to supply nuclear fuel for approximately 50 years. The rate of uranium utilization in China is extremely low because short- and medium-term nuclear power plants are mainly of the PWR type. Such massive resource consumption and low utilization rates are likely to constrain sustainable development of the nuclear industry due to shortages in nuclear fuel supply.

Lack of technology to treat radioactive waste Despite the fact that China has established a radioactive waste disposal system, the disposal of such large volumes of radioactive waste remains challenging. Treatment of radioactive waste, from research experiments to the production of treatment equipment, requires huge investment and many years of experience. This could provide an excellent business opportunity for companies in other countries that have already developed advanced technology to treat radioactive waste.

11.6.2 Implications of Global Analysis of Nuclear Accidental Events

The following implications are evident from our global analysis of accidental events of INES level 2 or higher at private nuclear power plants since 1992.

By country, the United States has a remarkably low accidental event rate, indicating that significant regulation by the NRC is yielding results, while France has a higher frequency of accidental events but fewer serious events. This reflects differences in safety measures in different countries.

Graphite reactors show a noticeably higher accidental event rate, followed by Russian reactors. Furthermore, since accidental events in Russian reactors are mainly caused by intrinsic phenomena, it is clear that this reactor type is responsible for many accidental events. The IAEA have raised safety concerns with regard to the design of this type of reactor.

When the number of accidental events is categorized by operational period, the number of accidental events significantly increases after 16–20 years of operation. The accidental event rate is less than 1 per-mil (‰) for operational periods of 40 years or less.

The above findings are informative for the safe operation of nuclear reactors and should be taken into account as a minimum in future policy decisions. In particular, Japan, China, and South Korea are heavily dependent on nuclear power generation and are geographically vulnerable to the mutual effects of a nuclear power accident. Therefore, it is extremely important to elucidate the causes of such accidents and accumulate knowledge on security and to construct an East Asia nuclear security system centered on human resource development and exchange, information exchange and sharing, and technology development and transfer.

11.6.3 Proposing an East Asia Nuclear Security System

On May 22, 2011, the leaders of China, Japan, and South Korea met and agreed to strengthen cooperation in nuclear safety and disaster prevention in the wake of the Great East Japan Earthquake and the accident at TEPCO's Fukushima Daiichi nuclear power plant. Mainland China and Taiwan also announced their intention to strengthen cooperation in nuclear security, including the establishment of a mutual notification system for emergencies, coordination of actions, and strengthening technical exchanges. Nuclear security issues are no longer limited to a single country or region but constitute an international issue that transcends national borders and urgently requires international cooperation. We propose the construction of an East Asia nuclear security system through industry-government-academia collaboration for the sound development of nuclear power generation in East Asia. Significant expansion of nuclear power generation in East Asia is expected in future, led by China.

The purpose of the East Asia nuclear security system is to provide a platform for the development and exchange of human resources across countries and regions, the exchange and sharing of information, the transfer and provision of technology, and an intergovernmental panel on nuclear crisis response with the participation of industry, government, and academia. Particularly in relation to nuclear safety, it is important to establish a cooperative framework for mutual monitoring during normal times, building a safety culture and sharing of experiences, developing a cooperative framework during accidents, and sharing response technologies, as well as discussing the exchange of experts in the East Asian region. It is important to identify how to exchange information during emergencies, not only among governments but also among academia, research institutions, and private organizations. Japan should make a proactive contribution to the human resources problems faced by China. In Europe, which has a long history of developing nuclear power, nuclear engineers are aging, and a significant number are expected to retire in the near future. In future, collaborative research among China, Japan, and South Korea will be required on topics such as the social, economic, and environmental impacts of zero nuclear power plants, root cause analysis (RCA) of nuclear accidents, comparison of the existing safety framework for nuclear power plants in East Asia with that of the EU, and the construction of a Japan-China-Korea security framework from the perspectives of human resources, technology, and information. If such a platform is to be established, a key role for Japan, as a technologically advanced country with extensive experience and human resources, is highly anticipated.

References

China National Nuclear Corporation (2014). http://en.cnnc.com.cn/

Energy and Environment Council (2012) Innovative Energy/Environmental Strategy, Sep.14, 2012

IAEA (2011) IAEA Annual Report for (2011). https://www.iaea.org/publications/reports/annual-report-2011. Accessed 03 February 2020

IAEA (2013) International Nuclear and Radiological Event Scale (INES). https://www.iaea.org/topics/emergency-preparedness-and-response-epr/international-nuclear-radiological-event-scale-ines. Accessed 03 February 2020

IAEA (2014) PRIS: Power Reactor Information System. https://pris.iaea.org/PRIS/home.aspx

IAEA-PRIS (2019) Country Statistics. https://pris.iaea.org/PRIS/WorldStatistics/OperationalReactorsByCountry.aspx. Accessed 22 December 2019

IEA (2019) Key World Energy Statistics 2019. https://www.connaissancedesenergies.org/sites/default/files/pdf-actualites/Key_World_Energy_Statistics_2019.pdf. Accessed 22 December 2019

Japan Atomic Energy Agency (2014). https://www.jaea.go.jp/english/

Kainou K (2009) Japan-US comparative analysis of nuclear power plant operating rates and trouble incidence rates. https://www.rieti.go.jp/jp/publications/summary/09120006.html

Korea Electric Power Corporation (2014). http://home.kepco.co.kr/kepco/EN/main.do

LAKA (2020) Documentation and research centre on nuclear energy. https://www.laka.org/docu/ines/location/

Chapter 12
Building a Global Low-Carbon Society Based on Hybrid Use of Natural Clean Energy

Tatsuo Sakai, Shuya Yoshioka, Yoshifumi Ogami, Ryohei Yamada, Yusuke Yamamoto, and Nabila Prastiya Putri

Abstract It is well-known that the extensive emission of greenhouse gases (GHG) such as CO_2 gas and freon gas during a long period after the industrial revolution is a fundamental reason causing the global warming and the climate change on our globe. The average temperature during an entire year at the surface of the globe has been elevated about 1 °C comparing the corresponding temperature before the industrial revolution. Thus, mitigation or reduction of the global warming is one of the most important subjects to keep the continuous development of humankind. In such a circumstance, various kinds of natural energies such as the wind energy, the solar energy, and the water energy have been focused as energy resources to realize the low-carbon society, since these energy resources are clean without emission of any greenhouse gases. In order to use these natural energies as energy resources, the wide variety of the technologies have been developed extensively nowadays. Considering this trend, the authors have attempted to develop new types of power generation systems for the wind energy and the water energy in this work. These new technologies can be effectively combined with solar energy plants and conventional water plants depending on the respective local conditions. In other words, the hybrid use of natural clean energies appropriate to the individual location such as the mountainside, the urban area, and the coast/bay area provides us the distinct advantage to realize the low-carbon society without any emission of the greenhouse gas. In addition, the overall concept to realize the low-carbon society based on these natural energies is finally indicated together with some remaining subjects to be solved in the future.

T. Sakai (✉)
Research Organization of Science and Technology, Ritsumeikan University, Kusatsu, Shiga, Japan
e-mail: sakai@se.ritsumei.ac.jp

S. Yoshioka · Y. Ogami · R. Yamada · Y. Yamamoto · N. P. Putri
College of Science and Engineering, Ritsumeikan University, Kusatsu, Shiga, Japan
e-mail: shuya@fc.ritsumei.ac.jp; ogami@se.ritsumei.ac.jp; rm0071sk@ed.ritsumei.ac.jp; rm0071pp@ed.ritsumei.ac.jp

W. Zhou et al. (eds.), *East Asian Low-Carbon Community*,
https://doi.org/10.1007/978-981-33-4339-9_12

12.1 Introduction

After the industrial revolution in the eighteenth century, fossil fuels such as coal, petroleum, and natural gas have been consumed globally as energy sources in a long sequence of industrial and domestic activities. Extensive emissions of greenhouse gases (GHG), such as CO_2 and freon, since the industrial revolution is a fundamental cause of global warming and climate change (IPCC 2013, Ministry of Environment, Japan 2018, Sakai and Amano 2008). Average annual surface temperatures have increased by approximately 1 °C since the eighteenth century. Thus, mitigation or reduction of global warming is essential for the continued development of humankind.

Energy resources such as fossil fuels are finite resources and do not allow sustainable development. Nuclear power is a clean energy source because it does not use fossil fuels, and CO_2 gas is not emitted during power generation. However, if a critical accident occurs in a nuclear plant, extensive damage may occur over a wide area, in the form of long-term radioactive contamination. Due to this risk, other sources of energy that do not emit greenhouse gases must be explored.

Various natural sources of energy have been focused on in order to realize sustainable development of industrial society. The most typical example is solar energy, the technology for which has progressed rapidly. Wind energy, produced by movement of air, is also common; electric power can be generated using various types of windmill with no emissions of CO_2. Conventional power generation by water mill or water turbine also provides clean energy, because CO_2 is not emitted during power generation. Novel technology for hydraulic power generation systems is a candidate for clean energy.

Solar energy, wind energy, and water energy are typical examples of clean energy that do not lead to emissions of greenhouse gases. Solar power plants have been widely deployed with a range of generation capacities, allowing installation of appropriate capacity according to individual requirements. This chapter presents new types of power generation systems for wind and water energy. These new technologies can be combined with solar energy plants to achieve hybrid use of natural clean energy without GHG emissions.

12.2 Background

12.2.1 Variation in Energy Resources and Consumption in Japan

Domestic annual energy consumption in Japan has varied since 1973, as shown in Fig. 12.1 (Ministry of Economy, Trade and Industries, Japan 2019). Following the Lehman crisis in 2008, total annual energy consumption clearly decreases, which can be mainly attributed to the reduction in industrial energy consumption. This

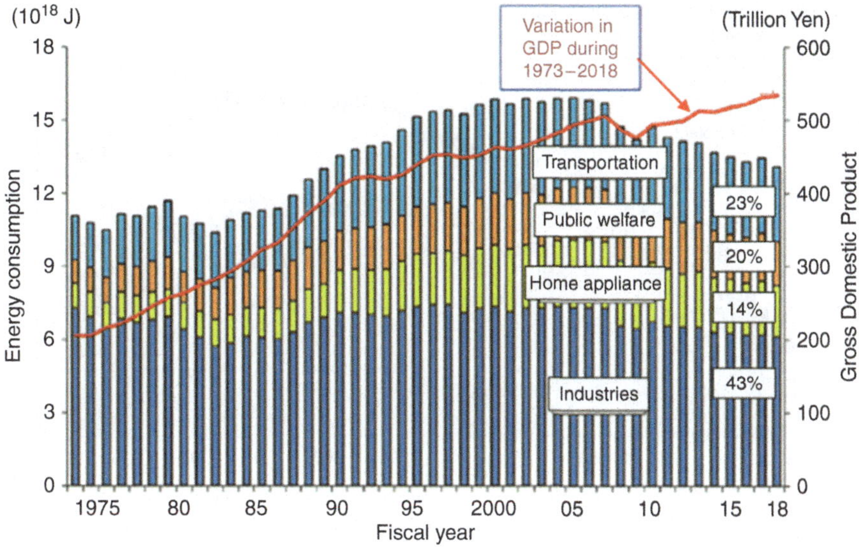

Fig. 12.1 Variation in annual energy consumption in Japan. (Source: Ministry of Economy, Trade and Industries, Japan 2019)

trend was affected by the stagnation of gross domestic product (GDP) in Japan, as shown by the red line in Fig. 12.1. After the Lehman crisis, the percentages of energy consumption for transportation, public welfare, home appliances, and industry remain almost constant at 23%, 20%, 14%, and 43%, respectively, although total energy consumption slightly decreases.

Variation in annual energy supply in Japan since 1965 from petroleum, coal, natural gas, nuclear, water, and other renewable energy sources is depicted in Fig. 12.2 (Ministry of Economy, Trade and Industries, Japan 2019). Total energy supply decreases slightly after the Lehman crisis. This was caused mainly by reductions in petroleum supply and temporary closure of nuclear power plants following the serious accident at the Fukushima nuclear power plant in 2011. In order to compensate for shortages in energy supply, use of petroleum and natural gas was gradually increased, as indicated in Fig. 12.2. The supply of petroleum and natural gas has been further increased since 2012 in order to meet energy requirements in Japan.

Fossil fuels account for more than 85% of the entire energy supply; nonfossil energy accounts for only 11.7%. If the energy supply in Japan is provided only by fossil fuels, emissions of greenhouse gases will significantly increase. Since the risk of nuclear accidents cannot be ignored, alternative energy sources must be prepared to ensure sustainable development.

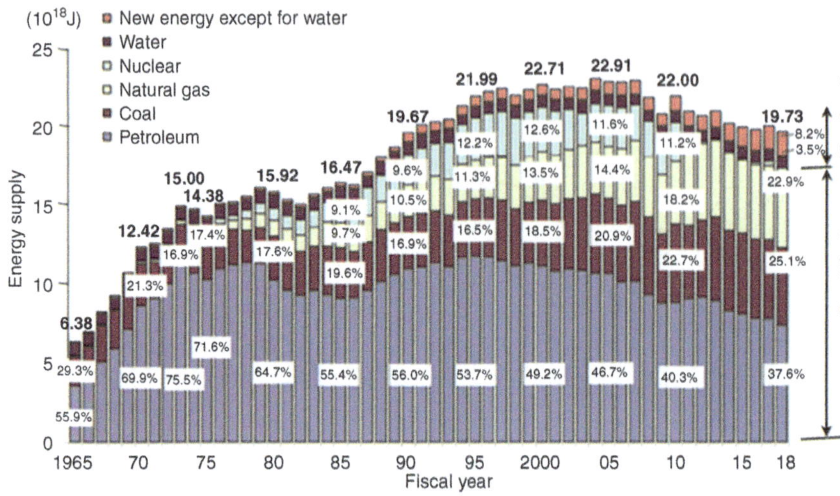

Fig. 12.2 Variation in annual energy supply in Japan. (Source: Ministry of Economy, Trade and Industries, Japan 2019)

12.2.2 New Clean Energy Sources

As described in the introduction, there are many kinds of clean energy that do not emit greenhouse gases. The level of technological development associated with generation of power from these types of energy varies from tentative proposals as new energy resources to mature technology capable of generating large amounts of energy. Table 12.1 provides an overview of conventional and new energy resources.

Among the new energy resources in Table 12.1, technology for solar is already established on a global scale, and large-scale power plants have been constructed in many countries. Given the advanced state of solar technology, it is not a focus of this chapter. Instead, wind (new types), water (new types), and tidal powers are explored in more detail.

Technologies for the other new energy sources listed in Table 12.1 should also be developed. Hybrid use of several different types of clean energy is important, since each type of power plant has both advantages and disadvantages.

12.3 Development of a New Type of Windmill Power Plant

12.3.1 Meteorological Wind Conditions in Japan

Useful applications of windmill power plants to irrigate agricultural land and pump water to the sea have long been present in certain regions such as the Netherlands. Windmills have also been used to mill various cereals and to process wooden

Table 12.1 Overview of conventional and new energy resources

Conventional energy resources			
1	Water power		Traditional
2	Thermal power		Petroleum, coal, gas
3	Wind power		Traditional/Netherland
4	Nuclear power		(Risk for accidents)
New energy resources			
1	**Solar power**		(Almost established)
2	**Wind power**		**New types**
3	**Water power**		**New types**
4	Sea water	①	**Tidal power**
		②	Wave power
		③	Stream power
		④	Temp. difference
5	Geothermal power		(Areas limited)
6	Biomass power		(Developing stage)
7	Wastes power		With thermal power

products, especially in the district of Zaanse Schans. These types of windmill were designed to use drag force rather than lifting force. According to the fundamental concept of fluid dynamics, the drag force acting on a blade is smaller than the lifting force. The most popular type of modern windmill is designed to use lifting force instead of drag force, a design requiring three blades (Matsumiya et al. 2005).

However, superior performance of this windmill is achieved only if the direction of wind remains constant throughout the year, which occurs only in limited areas. In Japan, the wind direction changes with high frequency, depending on variations in meteorological conditions. Therefore, the direction faced by a windmill must be frequently adjusted to match the prevailing wind direction, in order to obtain high efficiency of power generation (Tanaka et al. 2007; Tanaka et al. 2008).

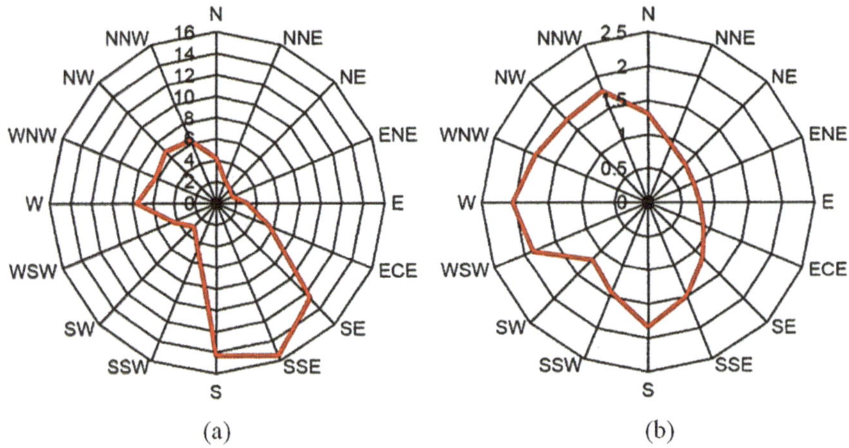

Fig. 12.3 Frequency distributions of wind direction (**a**) and velocity (**b**) on a rooftop on Biwako-Kusatsu campus of Ritsumeikan University, Kusatsu, Shiga, Japan. (Source: Kawai et al. 2010)

Frequency distributions of wind direction and wind velocity on the roof of the Excel III building on Biwako-Kusatsu campus of Ritsumeikan University, Kusatsu, Shiga, Japan, were measured throughout 2010 (Kawai et al. 2010) and are plotted in Fig. 12.3. The frequency distribution of wind direction tends to be governed by surrounding conditions, such as the presence of other buildings and mountains. A preferential direction is observed between S and SE, as shown in Fig. 12.3a, and the frequency is very low in the range of SSW–ECE. This can be attributed to the fact that the observation site is situated between tall buildings located to the NW and ENE.

On the other hand, the wind velocity is highest in the range of WSW–NNW and S, as seen in Fig. 12.3b. Thus, the frequency distribution of wind direction does not correspond to trends in wind velocity. The directions WSW, W, WNW, NW, and NNW give high wind velocity, but the frequency of wind from this direction is fairly low. The most effective combination of wind direction and velocity to generate maximum power should be identified. Even if the frequency of a given direction is low, high velocity can generate high levels of wind power during a short period. The direction faced by the popular three-bladed windmill can be frequently adjusted to obtain the optimum conditions. However, if a non-negligible amount of electricity is consumed in making these adjustments, then this type of windmill power plant becomes less appropriate as a clean energy generator.

Based on the above observations, the one-dimensional distribution of wind velocity during 2010 was analyzed using Weibull statistics (Weibull 1951). In Fig. 12.4, the histogram bars indicate the frequency distribution, and the red line shows the cumulative frequency of wind velocity.

The highest frequency is found at 0.5–1.0 m/s and tends to decrease gradually with increases in wind velocity. Such distribution characteristics can be well

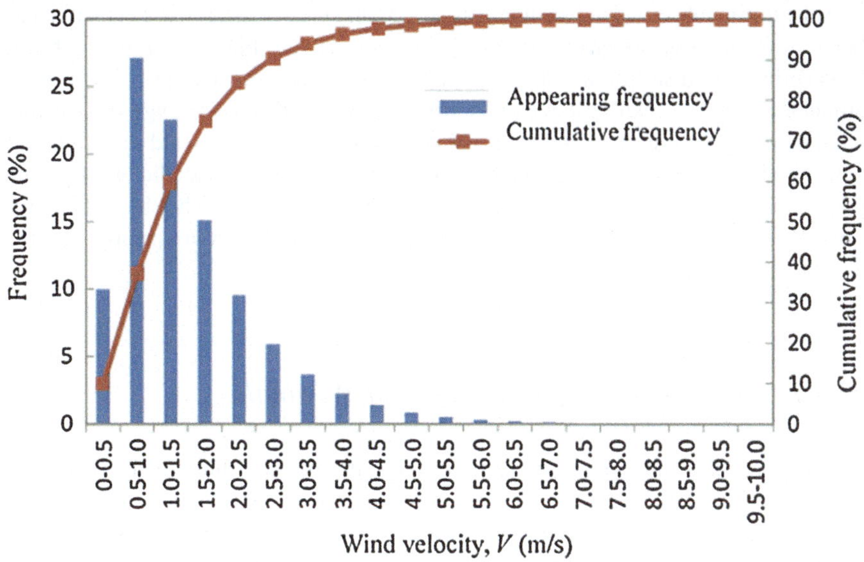

Fig. 12.4 Distribution characteristics of wind velocity during 2010. (Source: Kawai et al. 2010)

represented by the following two-parameter Weibull distribution (Weibull 1951; Sakai and Tanaka 1984):

$$f(V) = \frac{a}{b}\left(\frac{V}{b}\right)^{a-1} \exp\left\{-\left(\frac{V}{b}\right)^{a}\right\} \tag{12.1}$$

$$F(V) = 1 - \exp\left\{-\left(\frac{V}{b}\right)^{a}\right\} \tag{12.2}$$

where Eq. (12.1) is the probability density function and Eq. (12.2) is the cumulative distribution function. Parameter values can be obtained using the least squares method (Sakai and Tanaka 1980, Ninomiya and Yoshioka 2020a, b). The mean wind velocity, V_{m}, is given by:

$$V_{\mathrm{m}} = b\Gamma\left(1 + \frac{1}{a}\right) \tag{12.3}$$

and the power density of wind, PD, is given as follows:

$$\mathrm{PD} = \frac{1}{2}\rho b^{3}\Gamma\left(1 + \frac{3}{a}\right) \tag{12.4}$$

where ρ is the density of air, given as $\rho = 1.225$ kg/m^3, and $\Gamma(\cdot)$ is the gamma function.

Substituting observed values into Eqs. (12.3) and (12.4), mean wind velocity, and wind power density are calculated as $V_m = 1.40$ m/s and PD = 4.09 W/m^3. Here, considering the distribution characteristics of wind velocity in Fig. 12.4, a high efficiency windmill can be developed working within the wide range of 0–6 m/s wind velocity. Variation in meteorological conditions affecting wind direction and velocity should be observed throughout an entire year before installation of power generation facilities based on the common type of windmill. This is particularly important in Japan, where the wind direction tends to vary very frequently, regardless of the season.

12.3.2 Development of a New Type of Windmill

If wind direction and velocity remain constant, the popular three-bladed windmill has the highest efficiency in generating electric power (Matsumiya et al. 2005). However, meteorological wind features are complicated in some countries; wind direction and velocity should be simultaneously evaluated when designing high efficiency windmills, as discussed in the previous section. Here, we present a new type of windmill that works well regardless of variations in wind direction (Kawai et al. 2010) and retains its ability to generate power across a wide range of wind velocities.

Figure 12.5 shows a photograph of the new type of windmill, in which three blades are vertically fixed at the tips of three arms on the top half of a central pole,

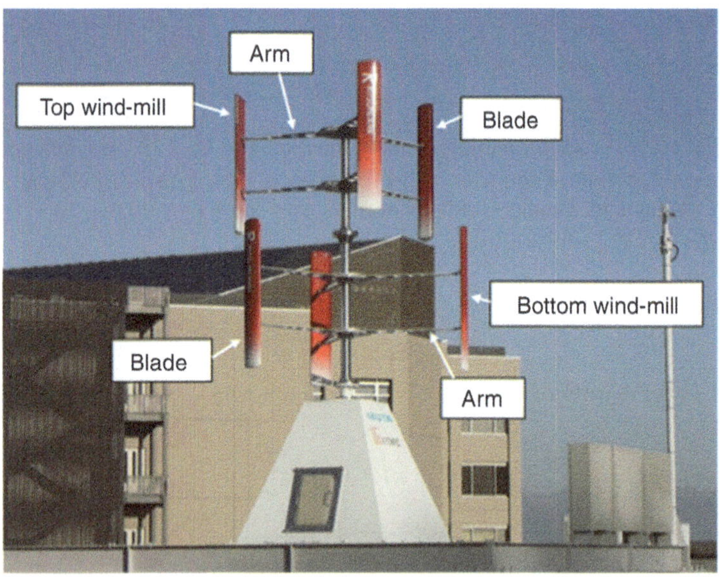

Fig. 12.5 Photograph of proposed new windmill type

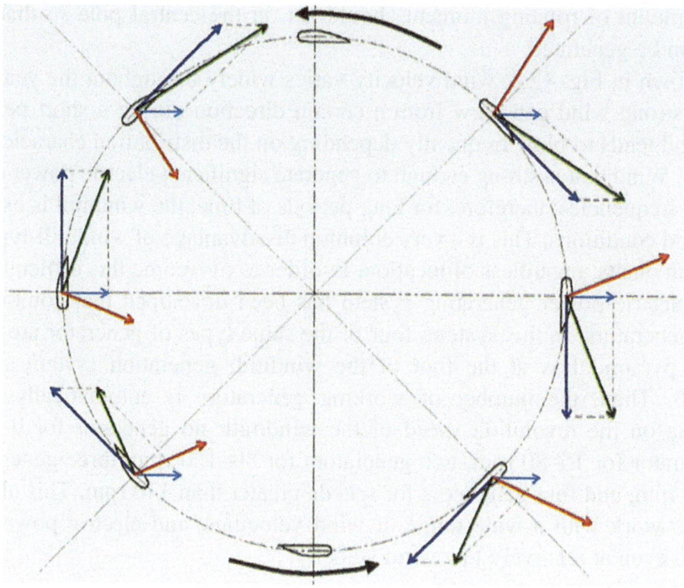

Fig. 12.6 Variation in relative wind velocity and lifting force acting on blades during one revolution of the windmill

and an identical structure is mounted on the bottom half of the central pole. The three blades of the top and bottom windmills are fixed at an equal circumferential angle of 120°, but a circumferential phase gap of 60° is given between the top and bottom windmills. If both windmills are fixed in the same phase, it would be difficult to initiate movement of the windmill; if the wind begins to blow in the direction facing one blade, the remaining two blades are located at both sides in symmetry, and the windmill cannot start to move since the moment of the two symmetrical blades is balanced. Therefore, a phase gap is introduced between the top and bottom windmills. The results in Figs. 12.3 and 12.4 were obtained at the same site as that shown in the photograph in Fig. 12.5.

Figure 12.6 illustrates the variation in relative wind velocity toward the blades and the lifting force acting on the blades during an entire revolution of the windmill, where a constant wind of a given velocity is blowing from the left-hand to the right-hand side. The shape of the blade is based on a popular design for airplanes, NACA4412, and the angle of attack is set as −2°. Wind velocity is represented by a horizontal line (light blue arrow) of identical length at every point of the revolution. During one revolution of the windmill, the blade has a certain velocity in the tangential direction (dark blue arrow), even if the air does not move (i.e., the wind does not blow). In such circumstances, the resultant velocity of the air toward the blade has a different direction (green arrow), depending on its point within one revolution of the windmill. The lifting force on the blade acts perpendicular to the resultant velocity, as indicated by red arrows in Fig. 12.6. The acting direction of the lifting force thus analyzed does not pass through the center of the revolution; a certain amount of rotating moment should act on the central pole so that electric power can be generated.

As shown in Fig. 12.4, wind velocity varies widely throughout the year. Sometimes, a strong wind can blow from a certain direction during a short period, but weak wind tends to blow frequently depending on the distribution characteristics in Fig. 12.4. Wind that is strong enough to generate significant electric power occurs at very low frequencies; therefore, for long periods of time, the windmill is exposed to weak wind conditions. This is a very common disadvantage of windmill-type power generation plants, regardless of location. In order to overcome this difficulty, a new type of electric power generating system has been developed that combines four electric generators. In this system, four of the same types of generator are installed inside a pyramid box at the foot of the windmill generation system shown in Fig. 12.5. Then, the number of working generators is automatically adjusted depending on the revolution speed of the windmill: no generator for 0–15 rpm, one generator for 15–80 rpm, two generators for 80–120 rpm, three generators for 120–140 rpm, and four generators for speeds greater than 140 rpm. This allows the system to work with a wide range of wind velocities, and electric power can be generated even at relatively low wind velocity.

12.3.3 Performance of the Proposed System

As introduced in Sect. 12.3.1, meteorological data were measured during 2010 on the campus of Ritsumeikan University. On November 9, from 6:00 AM to 12:00 AM, wind velocity ranged from 0 to 15.3 m/s with an average velocity of 3.90 m/s. The revolution speed varied from 0 to 154 rpm. This allowed evaluation of the

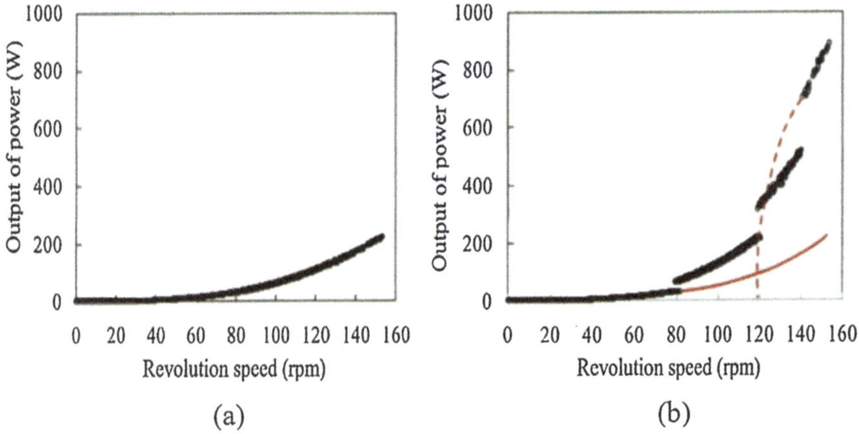

Fig. 12.7 Relationship between power output and revolution speed. (**a**) One generator, (**b**) all generators

performance of the proposed system under any number of working generators, from 1 to 4. In order to maintain system safety, the windmill is designed to stop spinning when revolutions exceed the upper limit (155 rpm).

Figure 12.7a shows the relationship between generator output and windmill revolution speed when only one generator is working. Output tends to increase gradually as revolution speed increases. The relationship between output and speed with all generators working is depicted in Fig. 12.7b. Jumps in output occur at 80, 120, and 140 rpm, due to changes in the number of operational generators. The solid red line shows output at high revolution speeds if only one generator is working, for comparison.

If only one generator is used, output at the maximum speed of 155 rpm is approximately 220 W. However, output at this speed can be increased to approximately 900 W when all generators are operational. Total output of four operating generators is always higher than the output of one generator. This raises the question of whether a large-scale generator with a high capacity can generate significant electric power more efficiently than one set of the proposed multi-type generators. As indicated by the dashed red line in Fig. 12.7b, this type of large-scale generator requires higher wind velocity to initiate windmill revolution due to higher total resistance of the generating system. Wind velocity has to exceed a critical level; in the case of Fig. 12.7b wind velocity must be high enough to cause a revolution speed of 120 rpm. Although the behavior of the large-scale generator is not understood in detail, it is clear that the lower bound of wind velocity to initiate revolution of the windmill is much higher than that required for one small generator in the proposed system. Thus, the electric power generation system presented here can efficiently produce electric power under wind conditions in which a large-scale generator cannot work (0–120 rpm).

12.3.4 Effect of Blade Configuration on Power Generation Performance

It is assumed that the cross-sectional shape of the blade has an important effect on power generation performance. Initially, the cross section of NACA4412, popular as an airplane wing, was accepted for the cross-sectional shape of the blade. In the usual flight of an airplane, wind blows from ahead of the wing. Thus, the wing is put in the optimum conditions to obtain the maximum lifting force for the airplane. However, in the proposed windmill system, the air stream against the blade is always changing direction due to revolution of the windmill axle. Thus, when the circumferential velocity of the blade is less than the wind velocity, the wind direction varies from 0° to 360° during one revolution of the windmill axle. In such a situation, the cross section of NACA4412 would not give the optimum configuration.

The effect of the cross-sectional configuration of the blade on revolution power was investigated. For ease of comparison, continuum fluid dynamics (CFD) analyses

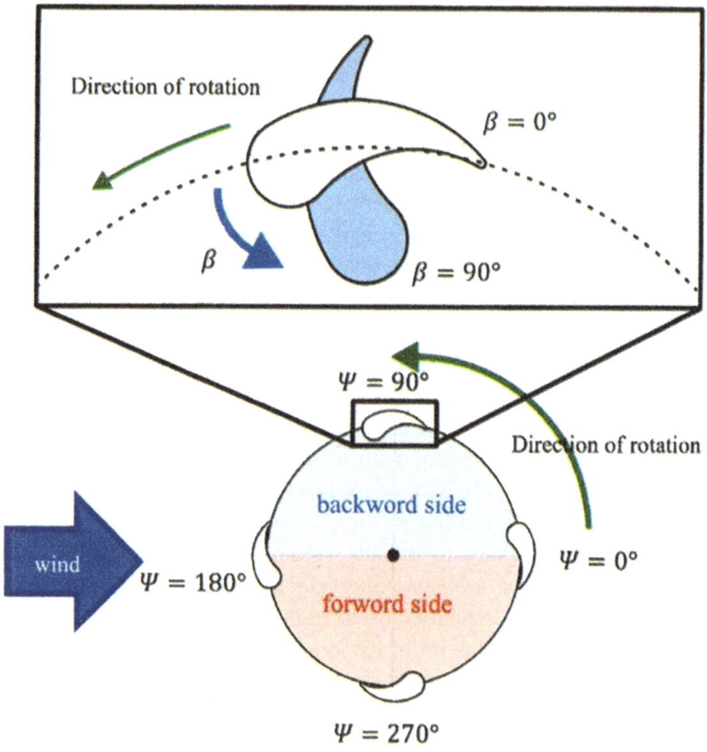

Fig. 12.8 Definition of blade setting angle β and azimuth angle Ψ

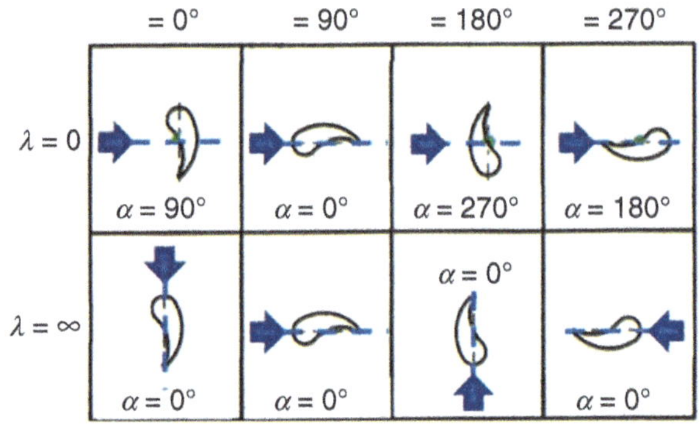

Fig. 12.9 Relation between α and Ψ

were carried out on two typical examples of cross-sectional configuration: NACA0018 without any curvature and Magatama-type (Magatama is an ancient Japanese accessory) (Ninomiya and Yoshioka 2020a, b). Figure 12.8 shows the blade setting angle β and the azimuth angle Ψ. If the blade is set along the circumferential direction, we have $\beta = 0°$, whereas $\beta = 90°$, if the blade is set along

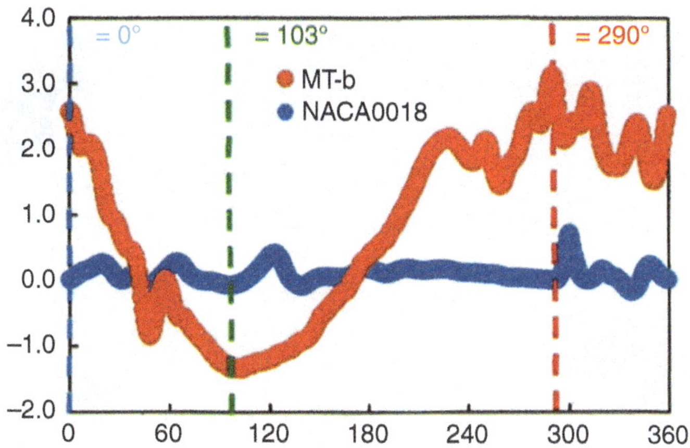

Fig. 12.10 C_F as function of azimuth angle Ψ

the centrifugal direction. The geometrical relation between the attack angle α and the azimuth angle Ψ is depicted in Fig. 12.9, in which parameter λ indicates the ratio of circumferential velocity of the blade to wind velocity. The revolution power is governed by the power coefficient, which corresponds to the nondimensional circumferential component of the force acting from the air. The relationship between the coefficient C_F and azimuth angle Ψ is plotted in Fig. 12.10 (Ninomiya and Yoshioka 2020a, b). Analytical results for both Magatama-type and NACA0018 blades are indicated for the full range of azimuth angles 0–360°. For NACA0018, C_F shows slight variation around 0. However, a positive large C_F can be obtained over a wide range of azimuth angle Ψ, although the coefficient tends to be negative within a limited range of azimuth angle, $\Psi = 60$–$170°$.With the Magatama-type blade, C_F can take a positive value on both forward ($180^\circ \leq \Psi \leq 360^\circ$) and backward ($0^\circ \leq \Psi \leq 60^\circ$) sides.

In Fig. 12.10, azimuth angles of $\Psi = 0^\circ$, $\Psi = 103^\circ$, and $\Psi = 290^\circ$ are selected to give typical positive and negative values of C_F for the Magatama-type blade. Analytical results of pressure and velocity distributions around both types of rotating blade at these three typical azimuth angles are depicted in Fig. 12.11. As shown in Fig. 12.11c, even on the forward side ($\Psi = 290^\circ$), negative pressure on the back surface of the blade can produce a useful driving force to facilitate revolution of the windmill. In addition, as indicated in Fig. 12.11a, even at the transition step of $\Psi = 0^\circ$, high velocity flow on the back surface can produce a useful force to initiate revolution. Accordingly, several synergistic effects producing sufficient power output can be expected when using the Magatama-type blade.

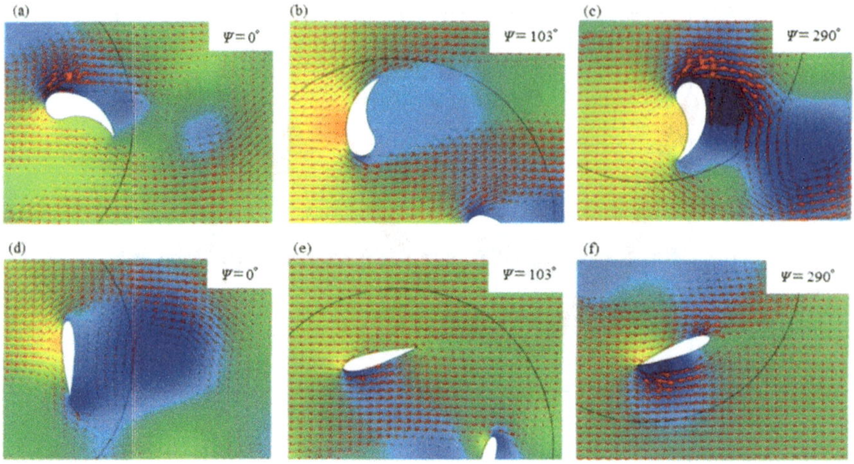

Fig. 12.11 Pressure and velocity distributions around rotating Magatama (**a–c**) and NACA0018 (**d–f**) blades

12.4 Development of a New Type of Micro-Hydropower System

12.4.1 Hydropower Generation Based on Tidal Energy

12.4.1.1 Mechanism Causing Tidal Energy

Sea water levels change depending on the relative locations of the sun, Earth, and moon. At certain points, all three bodies are arrayed linearly as shown in Fig. 12.12a, while at other times they are arranged in an L shape, as illustrated in Fig. 12.12b. Since forces of attraction act between each pair of bodies, sea water is attracted to the nearest celestial body. Thus, sea water tends to either be pulled toward both the moon and the sun, as indicated in Fig. 12.12a, or toward the moon instead of the sun when the bodies are not aligned, because of the shorter distance between the Earth and the moon.

The Earth rotates once daily (during 24 h). Accordingly, any site on the shore passes through both deep and shallow areas of sea water, leading to two high and low tides per day (bottom of Fig. 12.12). In Fig. 12.12a, the difference between high and low water levels becomes more distinct due to the superimposed effect of the sun and the moon, while the difference is smaller in Fig. 12.12b, resulting in spring and neap tides.

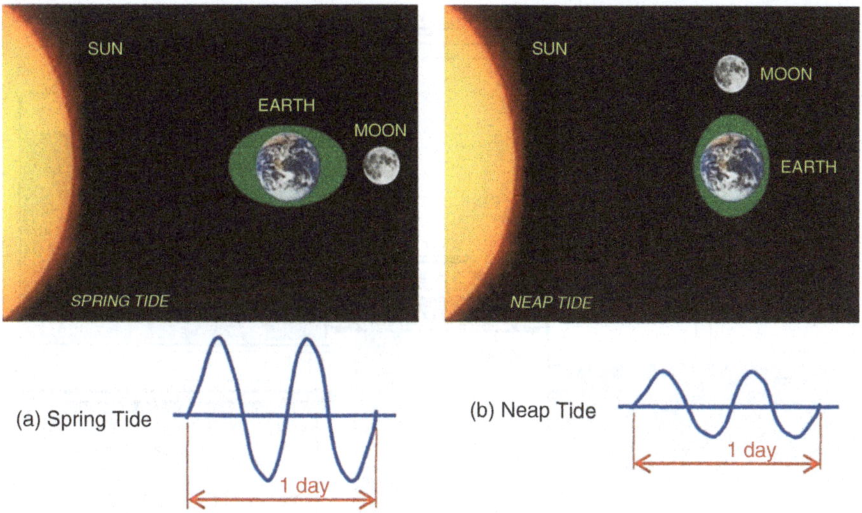

Fig. 12.12 Celestial mechanism causing spring and neap tides

12.4.1.2 Development of Micro-Hydraulic Power System Based on Tidal Energy

If a small bay is enclosed by a wall during high tide, a kind of dam appears after a few hours due to the tidal phenomenon of sea water. Since the potential energy of high-level water is greater than that of low-level water, high-level water tends to flow to areas of lower-level water. If electric power can be generated by such a flow of water, clean energy can be obtained, independent of the weather (Yanabu and Nishikawa 2004, Yamada et al. 2020). In the case of a windmill generator, electric power cannot be generated when the wind does not blow. Solar panels cannot generate electric power at night and on rainy days. In contrast, tidal energy can be generated constantly, regardless of weather and time of day.

A power generation system has been developed in order to extract tidal energy from sea water, as shown in Fig. 12.13 (Lian et al. 2010). Tank 1 and Tank 2 (both approximately 1 m^3) are placed such that Tank 1 is higher than Tank 2 and are connected by a pipe. A spiral screw (Yanabu and Nishikawa 2004) is mounted inside the pipe to drive the generator. Using four pumps, the water in Tank 2 is transported into Tank 1 to achieve a difference in water level between tanks. One of the four pumps has a special function to adjust flow quantity, and the difference in water level in both tanks can be adjusted by changing the water flow of this pump. Tidal phenomena can be simulated in the laboratory by means of this system.

Using this power generation system, electric power output was measured by setting four different water level differences of 60, 75, 90, and 105 cm. Experimental results are depicted in Fig. 12.14. Power output tends to increase with greater differences in water level. Output at 60 cm water level difference is saturated at

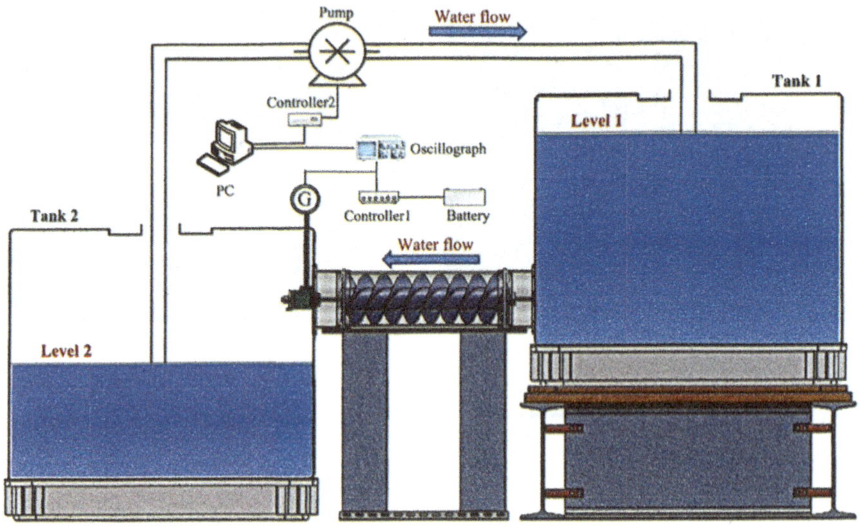

Fig. 12.13 Power generation system based on tidal phenomenon

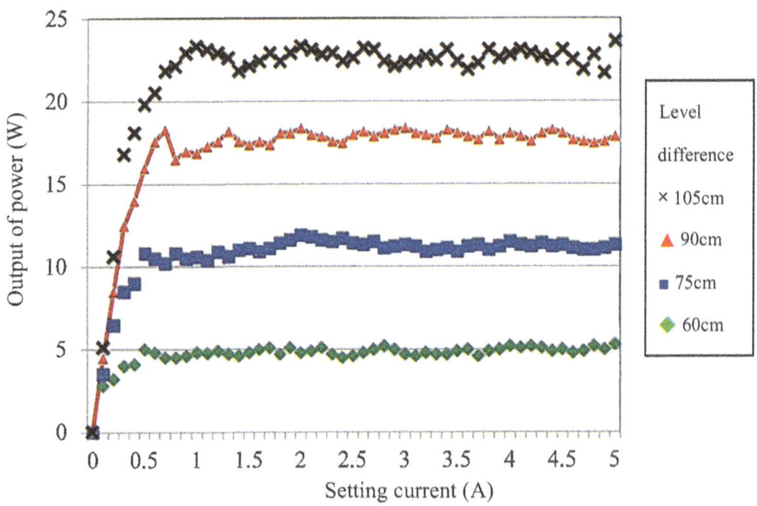

Fig. 12.14 Experimental output of power generated as function of current using potential energy from various water level differences

5 W. Output at 75 cm tends to 11 W, at 90 cm to 18 W, and at approximately 105 cm to 23 W. Thus, 23 W can be generated by the flow of 1 m^3 of water. If sea water along 10 m of the shoreline is used in this type of power generation system, 23 W × 10 = 230 W can be generated. Similarly, if sea water along 100 m of shoreline is used, 23 W × 100 = 2300 W = 2.3 kW can be generated. Using sea water along 100 m of shoreline within a distance of 10 m from the shore to a depth of

1 m provides a volume of 100 m × 10 m × 1 m = 1000 m^3, allowing generation of 23 kW. This is enough to supply 10–20 houses or a small village. Alternatively, this type of micro-hydraulic power generation system could be used to power port facilities.

Although this micro-hydraulic power system was applied here to harness tidal energy, it could equally be applied to extract clean energy from streams and small rivers. In addition, if a certain volume of rainwater is stored in a tank, this water can be transformed into electric power using the same system. This system can be usefully applied to harness different energy sources when the medium is a fluid, such as sea water and fresh water.

12.4.2 Hydropower Generation Using Drainage from Tall Buildings

12.4.2.1 Fundamental Concept

There are many tall buildings in Japan and in cities worldwide. Usually, the height of a single story is approximately 4 m, depending on national building standards. Thus, the height of a 10-story building is 40 m and that of a 20 story building 80 m. Water required for daily utilities in these buildings is supplied to a large tank on the roof by means of an electric pump. Water is distributed to individual stories on demand. After use, the water is discharged from higher to lower stories and eventually into a drainage canal, as shown in Fig. 12.15. Drainage water is treated before being released into the environment. Currently, drainage water is discharged to successively lower stories. The water stores energy, the amount of which depends on its

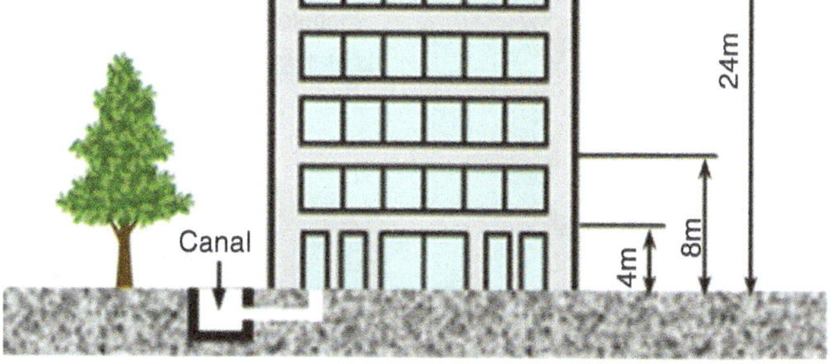

Fig. 12.15 Illustration of standard urban building and heights of several stories

height from the ground. The height of the second story is 4 m, the third is 8 m, and the seventh is 24 m, as indicated in Fig. 12.15. A power generation system based on effective use of the energy of this drainage water is presented here.

12.4.2.2 Preparation of Experimental Drainage Water Power Generation System

In order to extract the potential energy of drainage water, a special type of water mill with bent blades was designed, as depicted in Fig. 12.16. This water mill is installed inside a drain pipe, and connected to a compact electric generator. A prototype of this drainage water power generation system was constructed at ground level of a building on the Biwako-Kusatsu campus of Ritsumeikan University, Kusatsu, Shiga, Japan. In this experimental system, a 5 m^3 tank was set in a dummy space on the third story, at a height from the ground of approximately 8 m. A photograph of the experimental generation system is shown in Fig. 12.17. The volume and velocity of the drainage flow can be adjusted by a special device attached to the outlet gate of the tank on the third story.

12.4.2.3 Analysis of Generation System Performance

In typical cases, the flow of drainage water occupies a limited area inside the drain pipe; the water does not flow through the entire space inside the pipe. This introduces difficulty in harnessing the potential energy of the water because the flow includes several air pockets. In order to overcome this, a special flow adjuster device produces pipe-shaped flow that passes through the edge space inside the pipe. Thus, water flow without air pockets can be provided to the water mill, as indicated in Fig. 12.16. The distribution characteristics of static pressure and the pressure coefficient were analyzed using 3D CFD software. The results are depicted in Fig. 12.18, in which

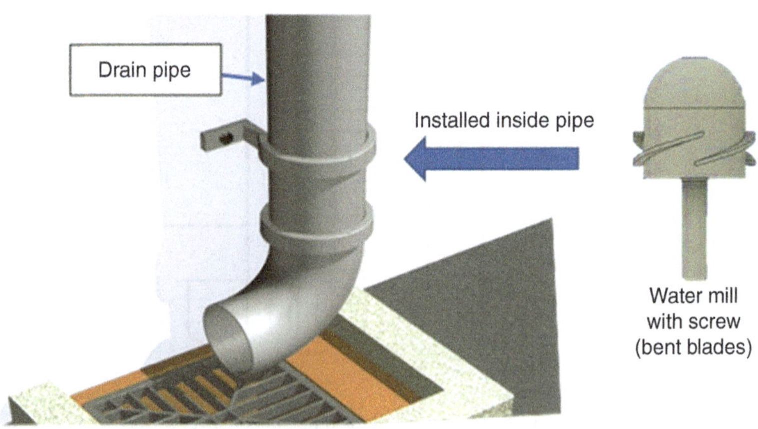

Fig. 12.16 Installation of water mill into drain pipe

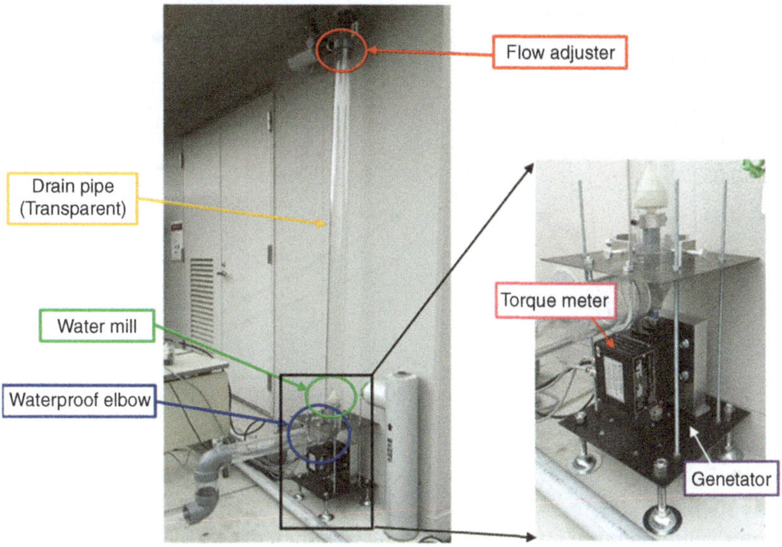

Fig. 12.17 Photograph of experimental water power generation system

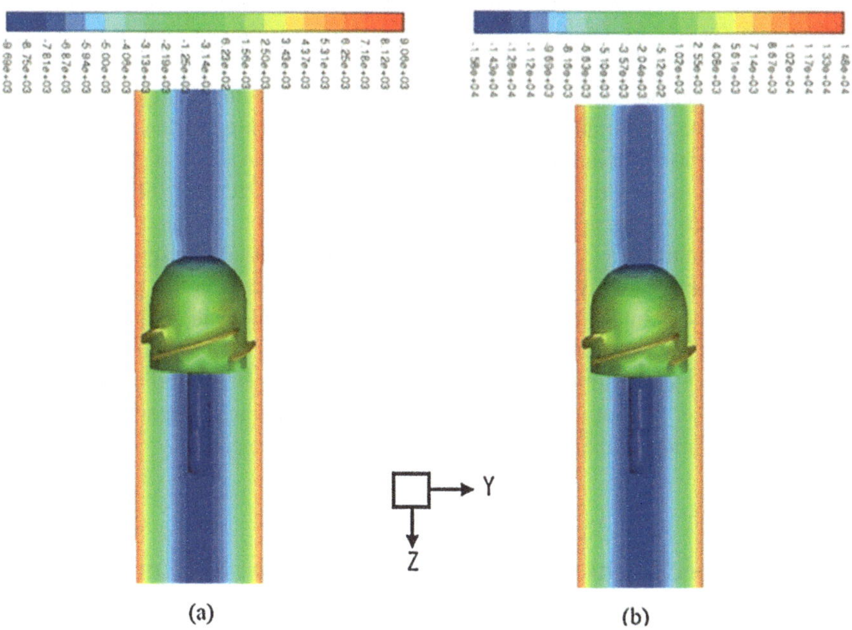

Fig. 12.18 Analytical results of distribution of static pressure and pressure coefficient

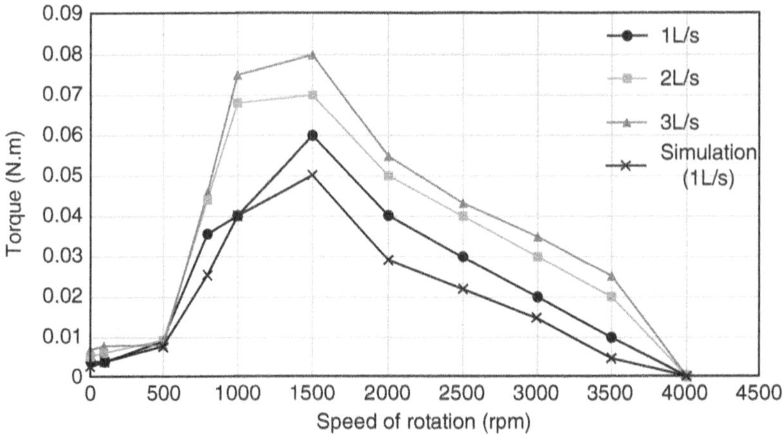

Fig. 12.19 Relationship between torque and speed of rotation of water mill axle

(a) shows static pressure and (b) the pressure coefficient. As seen in both diagrams, the pressure of the water flow tends to give a higher value around the edge, directly facing the blades of the water mill. Water tends to flow along the inner surface of the pipe, whereas the core portion inside the pipe is occupied almost entirely by air. This allows high levels of energy to be harnessed from drainage water with high efficiency.

Finally, the relationships between the torque of the revolution axle of the water mill and the speed of rotation are shown in Fig. 12.19. Three different water flow conditions were used for analysis: 1, 2, and 3 L/s. The torque acting on the water mill axle tends to increase with increased speed of rotation but decreases after peaking at 1500 rpm. Based on these results, maximum torque can be obtained at a speed of 1500 rpm. When the flow volume per unit of time is increased, the speed of rotation tends to increase gradually. However, the torque does not increase proportionally with increases in water flow. Crosses in Fig. 12.19 indicate simulated results obtained using 3D CFD software with a water flow of 1 L/s. The variation in torque is in good agreement with the corresponding analytical result plotted with circles. It is important to set the optimum conditions for the drainage water generation system to provide maximum torque in order to extract sufficient energy.

12.5 Prospects and Challenges of Active Use of Natural Clean Energy

12.5.1 Hybrid Use of Different Types of Natural Energy

The main aim of this chapter is to present a new concept for realization of a low-carbon society based on the transition to natural clean energy sources from

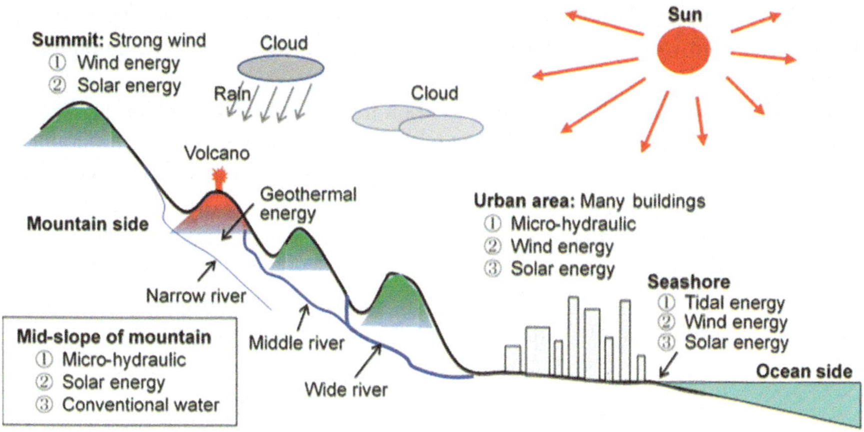

Fig. 12.20 Overview of hybrid use of different types of natural energy

fossil fuels. As shown in Table 12.1, there are many natural clean energies such as solar, wind, water, geothermal, and biomass. Various types of power plant have been developed and their efficiency improved. Using a combination of different types of power generation systems is important to maintain power stability.

Figure 12.20 gives an overview of the use of clean energy generation systems in accordance with local conditions, such as on mountain slopes, in urban areas, and on the seashore. In areas near mountain summits, strong wind tends to blow at high frequencies, and sunshine is less likely to be obstructed. Thus, a combination of windmill and solar panel power plants would be an effective way to obtain sufficient energy. On the mid-slope of mountains, there are many rivers of varying widths, and wide spaces suitable for solar panels are usually available; thus, micro-hydraulic and solar panel power plants would be effective. If large-scale dams can be constructed in convenient places, conventional water power plants can also provide sufficient clean energy. In areas along the coastline, electric power can be generated using tidal energy power plants. Usually, the wind tends to blow strongly on the shore because there are few obstacles to the wind. Solar panel power plants are also appropriate, as the shore receives large amounts of sunshine. Thus, tidal energy, wind energy, and solar energy can be extracted effectively in areas along the coastline. Additional energy generated by waves can be harnessed if wave power plants are constructed.

Many tall buildings have been constructed in urban areas. Since the wind tends to blow more strongly on the roofs of these buildings than at ground level, windmill power plants are superior here for extracting clean energy. If solar panels are installed on the roof, solar energy can be extracted simultaneously. Large volumes of water discharged after use inside the buildings drain to the ground floor through pipes into a drainage canal. The water has potential energy due to its height above the ground. Thus, electric power can be extracted based on the potential energy of the water by means of micro-hydraulic power plants. On rainy days, rainfall drained from the roof can also be harnessed; the sum of the daily drainage water and rainwater can be used as clean energy sources for micro-hydraulic power plants. Retention of rainwater by these buildings will also contribute to reducing flooding in urban areas.

Clean energy should be extracted based on local conditions. Appropriate hybrid use of different types of power plant is preferable to obtain a sufficient amount of clean energy. If each family tries to use a combination of clean energy types, their home can be a "smart house," and if every family in a village tries to have a smart house, their village could be called a "smart village." Furthermore, if every family, every company, every school, and every part of a city tries to use clean energy sources, the city becomes a "smart city." Based on such efforts by every person and every part of society, a low-carbon society can be realized in the future.

12.5.2 Future Challenges

The fundamental performance of a new type of windmill power system and micro-hydraulic power systems have been confirmed experimentally. Electric power generation using solar panels has already been developed, and its applications are expanding. The micro-hydraulic power systems have a wide variety of applications, such as tidal phenomena, narrow rivers, small dams, and drainage water streams in buildings. Annual variation in wind is dependent on local geographical features. The feasibility of harnessing solar energy also depends on geographical features and artificial constructions such as houses and tall buildings. The volume and speed of water flow in rivers vary depending on location, such as in mountains and rural areas. There are many tall buildings in urban areas, and high volumes of used water are discharged from these.

The most preferable method to generate clean energy should be selected considering local conditions, and several different effective methods should be combined as hybrid types of power generation plants in many cases. Each power generation system has both advantages and disadvantages. If different methods are successfully combined, one can obtain sufficient output of clean energy. During the night, solar energy plants cannot generate power. Windmill power plants can only generate energy when the wind is blowing. Effective combination of different power generation systems can provide sufficient electric power when individual power sources are not available.

However, in some cases, there may be no suitable generation method if inconvenient conditions are superimposed. In order to solve this problem, we recommend preparation of an additional artificial generation method combined with various types of natural clean energy, as shown in Fig. 12.21. Among the various types of fossil fuels, CO_2 emissions are lowest for liquefied natural gas (LNG) fuel. Thus, for an artificial power generation system, an energy plant using LNG is preferable from the viewpoint of CO_2 reduction. Power can be generated by consuming LNG when clean energy sources are not available due to inappropriate conditions.

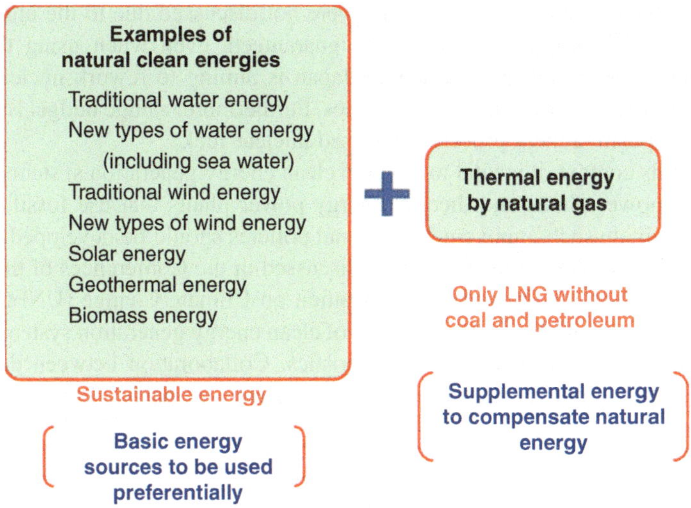

Fig. 12.21 Preferable energy supply to realize a low-carbon society in the future

12.6 Conclusion

This chapter presents new types of clean energy generation systems in the form of a special windmill and micro-hydraulic power generation systems. Since power generation by solar panels is already popular and the technology mature, solar power plants can easily be constructed everywhere. Geothermal and biomass energies are also reliable sources that do not emit CO_2. We present a concept of the hybrid use of several different power generation systems based on local meteorological and geophysical conditions.

The disadvantage of these power generation systems is the fact that power generation ability depends significantly on local conditions. At certain times, all the clean energy plants in a system may be unable to operate. In order to address this problem, a thermal energy generation system using LPG fuel can be combined with the above clean energy generation systems. Thus, we can maintain sufficient energy by using various kinds of clean energy power plants, complemented by additional fossil energy produced by LPG fuel where necessary, although this would result in a small amount of CO_2 emissions.

Electric power generated using natural clean energies can be sold to local electric power companies and supplied using normal electricity transmission networks. However, there are many areas where the transmission networks are not of appropriate scale to transmit power regionally. In such areas, an appropriate size of clean energy power plant can supply power directly to local customers without requiring long-distance transmission networks; clean energy can be created and consumed locally by preparing only a short-distance network system.

In this chapter, nuclear power plants were not discussed due to the high risk of unexpected accidents. Safety cannot be guaranteed, even when using the latest technology. The present government in Japan is aiming to rework nuclear power plants in cooperation with related enterprises. Furthermore, a huge budget is required to establish posttreatment processes for used nuclear fuel.

Given this context, it is vital to develop clean energy generation systems, instead of nuclear power plants and thermal energy power plants that use fossil fuels. In order to facilitate such a movement, national policies should be developed in Japan with reference to international activities discussed in the Conferences of the Parties to the United Nations Framework Convention on Climate Change (UNFCCC). In conclusion, the research and development of clean energy generation systems should be facilitated based on regional characteristics. Collaboration between developed and developing countries will also be vital to realize a global low-carbon society.

References

IPCC (2013) The 5th assessment report, AR5, Working Group I (climate change), The physical science basis summary for policymakers

Kawai Y, Lian B, Shirakawa N, Konishi K, Kawakita H, Yoshioka S, Sakai T (2010) Wind measurement in BKC building roof and fundamental analysis on generation characteristics of a vertical axis wind turbine with straight blades. In: Proceedings of EcoDesign 2010

Lian B, Kawai Y, Sakai T, Ueno A, Shirakawa N, Konishi K, Kawakita H, Nogami H, Okada K (2010) Modeling and some experimental results of micro-hydropower system using tidal energy. In: Proceedings of EcoDesign 2010

Matsumiya H, Aoki S, Iida M (2005) Wind power generation. Kogyo Chosakai Publishing, Tokyo, pp 21–26

Ministry of Economy, Trade and Industries, Japan (2019) White Paper on energy, chap 1

Ministry of Environment, Japan (2018) Annual report 2018, chap 1: Trend of sustainable society that has led to the fifth basic environmental plan

Ninomiya H, Yoshioka S (2020a) Experimental investigation of blade configuration developed for vertical axis wind turbine by genetic algorithm. Adv Exp Mech 5 (in Press)

Ninomiya H, Yoshioka S (2020b) Experimental investigation of aerodynamic characteristics of magatama type vertical axis wind turbine blade using Hele-Shaw cell. J Jpn Soc Exp Mech 19 (1):44–51

Sakai T, Amano K (2008) Realization of resources saving and reduction of global environmental load based on expansion of design life for industrial products. J Environ Conserv Eng 37 (9):39–45

Sakai T, Tanaka T (1980) Estimation of three parameters of Weibull distribution in relation to parameter estimation of fatigue life distribution. J Soc Mater Sci Jpn 29(316):17–23

Sakai T, Tanaka T (1984) Parameter estimation of Weibull-type-fatigue life distributions including non-failure probability. In: Proceedings of FATIGUE'84, vol 2, pp 1125–1137

Tanaka F, Kawaguchi K, Tomioka M (2007) Study on wind measurements and annual energy production of a Darrieus wind turbine. J Jpn Soc Mech Eng Ser B 73(735):111–117

Tanaka F, Kawaguchi K, Sugimoto S, Tomioka M (2008) Influence of wing section and wing setting angle on the starting performance of a Darrieus wind turbine. J Jpn Soc Mech Eng SerB 74(739):624–631

Weibull W (1951) A statistical distribution of wide applicability. J Appl Mech 18:293–297

Yamada R, Yamamoto Y, Yoshioka S (2020) Investigation of performance of magatama-type blade developed for vertical axis wind turbine. J Jpn Soc Exp Mech 19(1):52–59

Yanabu S, Nishikawa H (2004) Energy transformation engineering. Tokyo Denki University/Publication Bureau, pp 225–228

Part IV
Social Innovation for Low-Carbon Community

Chapter 13
Social Innovation Toward a Low-Carbon Society

Xuepeng Qian and Weisheng Zhou

Abstract New demands for social innovation are arising from the quest for a sustainable East Asian Low-Carbon Community (EA-LCC). Innovation refers to the creation of new things that lead society through great changes and is not necessarily limited to the invention of technology. The reformation of a social system can also be counted as an innovation. The social innovation discussed in this chapter seeks to bring society closer to sustainability by changing societal aspects such the economic and business system. In recent years, the spread of the Internet and Internet of Things (IOT) technology has allowed the emergence of a variety of businesses based around the sharing economy, which is a typical and innovative economic phenomenon. At present, the sharing economy is promoting remarkable innovations such as mobility, including travel and movement, as well as lifestyle changes including the sharing of goods. The sharing economy will contribute greatly to the promotion of low carbon emissions through dematerialization, improving convenience, and through a transition from owning things to using services, i.e., shifting from proprietary to sharing, and is expected to develop further in the future. However, issues in the sharing economy associated with existing business systems, market economic systems, and policy remain. In many cases, resources have been wasted due to overinvestment and maintenance. To solve these problems, the legal system must be organized and supply and consumption adjusted.

X. Qian
College of Asia Pacific Studies, Ritsumeikan Asia Pacific University, Beppu, Oita, Japan
e-mail: qianxp@apu.ac.jp

W. Zhou (✉)
College of Policy Science, Ritsumeikan University, Ibaraki, Osaka, Japan
e-mail: zhou@sps.ritsumei.ac.jp

W. Zhou et al. (eds.), *East Asian Low-Carbon Community*,
https://doi.org/10.1007/978-981-33-4339-9_13

13.1 The Sharing Economy as a Low-Carbon Economy

13.1.1 Moving from Proprietary to Sharing

The sharing economy is a newly defined economic model in which access to goods and services is shared on a peer-to-peer basis, such as through community-based and online platforms. The last decade has seen the emergence and rapid growth of the sharing economy, with increasing numbers of people choosing to participate. The sharing economy makes use of Internet services and is currently most active in the mobility sector, such as travel and transportation, through various platforms such as Airbnb, Uber, and DiDi. Digital conversation is made possible via systems that enable the communication and management of all kinds of goods and services through platforms built using ICT and IOT technology. This is the technical background to the rapid evolution and development of the sharing economy.

Meeting growing demands for material and cultural consumption is a fundamental prerequisite for the formation, development, and growth of all industries, with demand for products or services delivered at the lowest possible expense. Satisfaction is a fundamental starting point for consumer behavior. The rapid development of the sharing economy is essentially the result of more people embracing the concept and values of shared consumers. This is the socioeconomic background that has led to the emergence and rapid development of the sharing economy.

The sharing economy is often referred to in the literature as cooperative consumption. Studies of motivations for participating in the sharing economy consider many impact factors, such as sustainability of shared consumption, pleasure, and benefit of shared activities (Hamari et al. 2015). However, Matzler et al. (2015) argue that most people participate in the sharing economy for self-satisfaction, rather than as a subjective promotion of green consumption. That is, most participants in the sharing economy are motivated to meet consumer needs at lower costs, which can create added value from idle assets or time. For example, in a study by Shaheen and Cohen (2007) that used car sharing as an example, the main motivations for participation were cost savings, convenience, and parking guarantees, which all show consumer self-satisfaction. Self-interest is a subjective motivation for participating in shared consumption but has objectively promoted green development of the economy and society as a whole. Enthusiasm for consumption is strengthened, and the concepts of use over ownership and green consumption are gradually formed, further promoting similar concepts.

While growth in the sharing economy has accelerated and many companies are moving toward a shared model, consumer perceptions of past ownership concepts and asset consumption patterns can be difficult to change in the short term (Botsman 2015). In traditional economies, transactions generally require the transfer of product ownership, and the concept of property rights is well established and protected by private property laws that prevent the sharing of tangible goods. Simultaneously, concepts such as materialism, possessive individualism, and self-identity evolve into the concept of property ownership, which hinders development of the sharing economy (Belk 2007).

13.1.2 Classifying the Sharing Economy and Its Policies

The sharing economy can be divided into four categories, according to the content shared: product recirculation, increased utilization of durable assets, exchange of services, and sharing of productive assets (Schor and Attwood-Charles 2017).

The first type of sharing economy is reuse-based recycling. Founded in 1995, eBay and Craigslist provided markets for reusing goods and are now part of the mainstream consumer experience. The consumption methods represented by these sites have been driven by the bulk purchase of cheap, imported goods for nearly two decades, causing a surge in the number of unnecessary items owned by individuals. The evolution of these reuse platforms has reduced the risk of dealing with strangers. By 2010, many similar sites had been launched, including ThredUp and Threadflip for apparel, free exchange sites such as Freecycle and Yerdle, and bartering sites such as Swapstyle.com. Online exchanges are also available for apparel, books, toys, sporting goods, furniture, and household items.

The second type of sharing economy promotes more intensive use of durables and other assets. In wealthy countries, households either buy products or own property that is not used to capacity (e.g., spare rooms or lawnmowers). Zipcar launched as a service that placed vehicles in convenient urban areas and offered hourly rentals. Following the recession in 2009, rental assets became economically attractive, and the popularity of similar initiatives soared, including in the transportation sector car rental sites (e.g., Relay Rides), ridesharing (e.g., Zimride), ride services (e.g., Uber, UberX, Lyft), and bicycle sharing (e.g., Boston's Hubway, Chicago's Divvy Bikes). In the lodging sector, the innovator was CouchSurfing, who began pairing travelers with people offering rooms and couches for free in 1999. CouchSurfing led to Airbnb, with whom more than ten million people have reported staying. There has also been a resurgence of unmonetized initiatives, such as tool libraries, that arose decades ago in low-income communities. These efforts are usually based locally to increase reliability and minimize the cost of transporting bulky items. New digital platforms include the ability to share durable goods as components of neighborhood buildings (e.g., Share Some Sugar, NeighborGoods). These innovations can provide people with low-cost access to goods and space and in some cases offer opportunities to make money to supplement more traditional sources of income.

The third kind of sharing economy is service exchange. Its origins lie in time banking, which began in the 1980s in the United States to provide opportunities for the unemployed. Time banks are community-based, nonprofit, multilateral barter sites that trade on the basis of time spent. They follow the principle that all members'

Table 13.1 Classification of the sharing economy from the stakeholder perspective

Supply	Demand	Basic features	Examples	Effects
Business	Business	Utilization of corporate excess production capacity	Ali TGC	Optimize the allocation of social production resources, promote elimination of surplus capacity, and accelerate the transformation of sustainable industrial development
Business	Person	Rent of corporate assets	Hackerspace, DriveNow	Improve equipment and feature utilization and accelerate service transformation
Person	Business	Utilization of personal skills, extra time, and funds	zbj.com, crowdfunding	Companies save administrative costs and create flexible employment
Person	Person	Personal surplus assets, funds, skills, etc.	Airbnb, DiDi	Improve the use of private assets throughout society and establish new social relationships

Source: Yang (2016)

time is valued equally. In contrast to other platforms, time banks are not growing rapidly. This is due in part to the harsh nature of maintaining the same transaction ratio and due to competition with services such as TaskRabbit and Zaarly, who pair users requiring services with those that can provide them.

The fourth kind of sharing economy consists of initiatives that focus on sharing assets, or space, to facilitate production rather than consumption. Historically, these efforts have taken the form of coops and have been present in the United States since the nineteenth century. Related initiatives include maker spaces, co-working spaces, and co-offices providing shared tools. Other production sites include educational platforms such as Skillshare.com and Peer 2 Peer University, which aim to replace traditional educational institutions by democratizing access to skills and knowledge and facilitating peer instruction.

From the stakeholder perspective of supply and demand, the sharing economy is classified as business-to-business, business-to-person, person-to-business, and person-to-person (Yang 2016). The basic features, typical examples, and effects of these are shown in Table 13.1.

The concept of the sharing economy originated in Europe and the United States but has developed and evolved significantly in China due to favorable conditions, such as the population and market size of China, the information and logistics infrastructure that has been developed in recent years through the spread of e-commerce, and the promotion of policy innovation. Examples of the current sharing economy in China are described in the next section, based on the main points from the *2019 Annual Report on China's Sharing Economy Development* published by the State Information Center of China.

13.1.3 Market Size of the Sharing Economy

The sharing economy in China has continued to grow at a rapid pace in 2018; penetration of the manufacturing sector is accelerating, and capacity sharing is showing rapid development. Since 2017, the rate of development of livelihood services in the sharing economy has slowed. The sharing economy is becoming an important choice for people seeking flexible employment and also offers a wide range of employment opportunities previously unavailable to certain social groups. New business models, such as online car rental, shared accommodation, online takeout, shared medical care, and shared logistics, are updating the structure of the service industry and driving shifts in consumption patterns (Table 13.2).

The sharing of production capacity (capacity sharing) is a highlight of new developments; infrastructure for capacity sharing in large manufacturing companies is becoming increasingly enriched. Service-oriented capacity sharing has become a new model for production services. A "factoryless" manufacturing model has been introduced through capacity sharing.

The evolution and application of artificial intelligence (AI) is helping the sharing economy, while development of the sharing economy has simultaneously provided a wealth of scenarios for innovative applications of AI technology.

In recent years, dramatic changes in the short-term bicycle-sharing market and associated impacts have created controversy for all sectors of the industry and across the sharing economy more generally. From a development perspective, the sharing economy is a combination of technologies, institutions, and organizational innovation methods that can significantly reduce the cost of matching, negotiating, and guaranteeing security between supply and demand during social-industrial transformation processes and can improve the efficiency of resource allocation. Whether in terms of fostering a new momentum of economic growth, facilitating the transformation and upgrading of industries, or meeting the huge potential demand of consumers, the role of the sharing economy is far from being fully realized. The general trend of accelerated penetration and integration of the sharing economy in different sectors will continue.

Table 13.2 Market size of the sharing economy in China

Sector	2017 (billion CNY)	2018 (billion CNY)	Growth rate (%)
Transportation	201	247.8	23.4
Lodging	12	16.5	37.5
Knowledge skills	138.2	235.3	70.3
Living services	1292.4	1589.4	23.0
Medical treatment	5.6	8.8	57.1
Office	11	20.6	87.3
Production capacity	417	823.6	97.5
Total	2077.2	2942	41.6

Source: State Information Center of China (2019)

A milestone in the sharing economy audit process was reached in 2018. Relevant departments adopted a multi-angle approach, combining management, legal systems, and technical supervision methods. The supervision was of unprecedented rigor and scope, and development of standardization was based on stakeholder consensus. The institutional environment for the development of the sharing economy has been further improved, and a regulatory framework has begun to be established with the aim of standardization, institutionalization, and legalization. As a result, platform companies have significantly improved their levels of compliance. Reliability has improved, laying a solid foundation for long-term, rapid, and better development of the sharing economy. Auditing the sharing economy now faces new challenges; establishment of a long-term audit mechanism is far from complete. In addition, improving the level of data sharing between governments and platform companies, emergency response mechanisms, auditing of platform algorithms, and encouragement of platform companies to actively fulfill their social responsibilities are important issues that need to be addressed.

13.2 Current Status of Shared Bicycle and Shared Electric Vehicle Systems in China

13.2.1 Shared Bicycles

Bicycle sharing was first attempted in Amsterdam, the Netherlands, in 1965. Bicycles were painted white and made freely available to everyone in Amsterdam (Runde 2011). The so-called white bikes were also known as free bikes and were the first generation of shared bicycles. Theft and vandalism were so frequent that business in open urban areas could not be sustained, but the use of free bikes became widespread in specific local sites, such as universities. Since the 1990s, there has been an increase in the number of bicycles that can be rented and returned to a fixed location. This began in large cities and tourist cities and then evolved into station-type bicycles with improved freedom of use through being able to return them to stations in different locations. These have been positioned as part of public transportation systems and were expected to contribute to low-carbon transportation in cities.

The latest concept of shared bicycles was formed on the basis that bicycles can be combined with electronic locks and managed on an online platform. Compared to station-type rental bicycles, there is greater freedom of use. In the past 5 years, shared bicycles have experienced a global boom. In 2016, there were more than 1600 public and private bicycle-sharing systems in operation, doubling from 700 in 2013 (Chen 2016).

In recent years, competition in the bicycle-sharing market has undergone significant changes, and rapid development of the industry has caused widespread public concern, instigating reflection on the prospects of development of the sharing economy. Bicycle-sharing systems have developed rapidly due to the fact that components can be built in a short period of time, resulting in a fiercely competitive market. With the support of capital, major companies have completed market development in a relatively short time period, releasing numerous bicycles with high subsidies and rapid expansion. Major companies are growing rapidly in large cities with high levels of cycling and relatively stable user groups. By the end of 2017, Mobike and Ofo accounted for over 90% of the market. However, large upfront investments and long-term price wars have left businesses with huge debts, increased profit pressure, and business models that face great challenges. In addition, since late 2017, issues such as the excessive release of bicycles, which affects public spaces and order, have been widely debated. Many cities have introduced cycling restrictions and strengthened the management and evaluation of cycling operations and maintenance. In the midst of these changes, pressures on business operations have also increased. Against this backdrop, the bicycle-sharing market began to change significantly as capital market attitudes altered and financing in key markets became increasingly difficult to obtain. In April 2018, Mobike was bought outright by Meituan for 3.7 billion USD, including 1.2 billion USD in cash, 1.5 billion USD in shares, and one billion USD in debt. After maintaining brand independence and operations for approximately 10 months in 2019, Ofo plunged into operational difficulties and a debt crisis that led to a sharp reduction in the size of its operations. As a late entry to the market, Harbin Bicycle has adopted a niche market strategy to target regional cities, avoiding large cities, and its users and scale have developed rapidly. Users are beginning to actively consider moving from a single bicycle-sharing service to a combined mobility service that offers a shared bicycle ride service and open entry with operators in other sectors.

In addition to its ambiguous and immature business model, serious flaws in corporate governance structure, irregular user behavior, and inadequate public management in cities, massive entry into the capital markets over a short period of time has been a key cause of upheaval in the bicycle-sharing market. In less than 2 years from 2016 to the beginning of 2018, the bicycle-sharing industry has received huge amounts of funding. The Ofo bicycle platform has undergone an estimated ten rounds of financing, raising approximately 1.5 billion USD. Participating investors include well-known companies and investment institutions such as Alibaba, Ant Financial, DiDi, and CITIC. Financing frequency, large sums of money, and high popularity have been very rare in the history of Internet development.

The enormous entry of operators to the massively capital-backed shared bicycle market, excess vehicle production, and excessive competition across the market have resulted in rollercoaster development of the industry, which has severely harmed stakeholders, especially suppliers. Since 2016, short-term market distortion and high levels of prosperity resulted in an overexpansion of production capacity in the design, research and development, manufacturing, and marketing businesses associated with bicycle sharing. As the market shrank rapidly in 2018, the overcapacity of shared bicycle-related manufacturers and companies became apparent. The number of bicycle manufacturers in Tianjin's Wuqing town, known as the "first town of bicycles," has dropped from more than 500 at its peak in 2017 to less than 300 in

2018. Shanghai Phoenix, a renowned bicycle supplier, experienced overproduction and underproduction of bicycles and parts in the first half of 2018, with revenue and profits decreasing by more than 50% year on year. Affected by the bankruptcy of some bicycle-sharing companies, the operating profit of Cronus Bike fell by 82.3% in the first half of 2018 compared to the previous year. The company was unable to collect a large amount of its receivables, resulting in a significant decrease in liquidity and profits. It was down 220.4% from the same period in the previous year.

The impact of dramatic changes in the bicycle-sharing market over such a short period of time has triggered questions about the industry and the sharing economy itself. The sharing economy has been the subject of a deluge of pseudopropositions, short-lived, cold winter theories, and death theories. Other negative stories and comments have directly impacted public and market confidence in the development of the sharing economy. These controversies are inevitable in the growth of new business formats. Objectively speaking, however, "survival of the fittest" is a fundamental principle of market competition. The dilemmas faced by particular applications are equated with the overall state of the sharing economy rather than the failure of particular models. This is a failure of the sharing model as a whole, and developmental issues may be seen as overstated.

13.2.2 Shared Vehicles

Another important market in the area of shared mobility is shared vehicles. A typical company is DiDi Chuxing, which focuses on an objective of "making the journey better" and has enhanced its innovation and business layout. Several widely reported incidents in the past 2 years caused a considerable loss of confidence among its users. Since then, the company has comprehensively integrated its safety infrastructure, improved its service quality, and gradually restored user trust. In 2018, DiDi showed steady development. More than ten million car drivers operate on the DiDi platform each year, transporting more than ten billion passengers over 48.8 billion kilometers (Tables 13.3 and 13.4).

Table 13.3 Size of shared vehicle market in China

	Taxi reserved online				
Year	Amount of use (billion times)	Transportation volume (billion people)	Street taxi transportation volume (billion people)	Combined transportation volume (billion people)	Market share of taxis reserved online (%)
2015	2.09	4.18	39.67	43.85	9.5
2016	3.76	7.52	37.74	45.26	16.6
2017	7.85	15.7	36.54	52.24	30.1
2018	10	20	35.07	55.07	36.3

Source: State Information Center of China (2019)

Table 13.4 Legislation relating to vehicle sharing market in China

Date	Department responsible	Name of ordinance
June 2018	Ministry of Transport	Taxi service quality reputation assessment method
September 2018	Ministry of Transport, Ministry of Public Security	An urgent notice on further strengthening the safety management of online taxi booking and private passenger car sharing
September 2018	Ministry of Transport	Notice of search work for information platform security terms between ride-hailing platform companies and private passenger cars

Source: State Information Center of China (2019)

DiDi is a technology company with global influence and a one-stop travel platform. Over the past year, DiDi has continuously increased investment in science and technology, expanded its strategic layout in terms of cutting-edge technology and innovative business, and become a world-class technology company leading transformation of the automotive and transportation industry. In 2018, DiDi founded DiDi AI Labs, which primarily investigates technical issues in AI and focuses on cutting-edge research in areas such as machine learning, natural language processing, computer vision, operations research, and statistics. The application actively deploys next-generation technology to build a new ecology of intelligent mobility. Based on innovation, DiDi has released the Transportation Brain, a smart transportation strategy product that combines cloud computing, AI technology, big data from transportation, and transportation engineering.

Continued investment in innovation has improved the company's products and service offerings, greatly improved the efficiency of shared travel, and resulted in significant energy savings and emissions reductions. DiDi reduced carbon dioxide emissions by 1.57 million tons in 2017, which equates to the average annual emissions from 800,000 vehicles covering 10,000 km.

DiDi has designed a business ecology that enables industry partners to transform and upgrade. Travel is a basic human need. As a global industry leader, DiDi is actively deploying its business ecosystem to accelerate change across the industry. Efforts are being made to open up technical capabilities and accelerate the creation of a one-stop travel platform. In 2018, DiDi integrated information on buses, subways, taxis, and bicycles, providing users with a combination of multiple travel options in a single interface. Meanwhile, a one-stop auto service platform has been established to provide complete auto services to all car owners, including DiDi drivers. The platform seeks to broaden its reach to business related to car ownership such as refueling, car services, charging, and finance. In addition, DiDi is addressing future travel needs and aggressively pushing reforms in automobile manufacture. DiDi and 31 automotive industry chain companies have collaborated to establish the Trent Alliance to promote the development of new energy, intelligence, and shared industries, jointly building a service platform for future travel users and car owners.

In 2018, DiDi accelerated implementation of its international business layout. It has acquired the company Brazil 99; launched online car rental services in Mexico, Australia, and elsewhere; and expanded into the taxi service market in countries and regions such as Taiwan, Hong Kong, and Japan. In this process, DiDi has exported Chinese capital, technology, and models to the global market, as well as Chinese experience and solutions to transportation problems to countries and regions worldwide.

13.3 Shared Bicycles as Low-Carbon Mobility

13.3.1 Bicycle Sharing in Hangzhou City

From the launch of bicycle-sharing businesses on May 1, 2008, until October 23, 2019, 4253 cycle ports have been installed in Hangzhou City. Over 100,000 bicycles are available, and up to 470,000 journeys are made daily. Most bicycles are rented more than four times each day, and total usage since the launch of the scheme has exceeded one billion rentals, surpassing the volume of public bicycle rental in developed countries. Hangzhou City has the largest community bicycle system in the world (Fig. 13.1).

According to data from the Hangzhou City Public Bicycle Transportation Service Development Corporation, there is one cycle port every 300 m on average in the central city of Hangzhou. In 2013, there were 135 ports in tourist areas, 422 in West Lake District, 385 in Shangcheng District, 418 in Xiacheng District, 431 in Jianggan District, and 386 in Gongshu District. Across these areas, 41.2% of ports are near bus

Fig. 13.1 Cycle port in Hangzhou. (Source: Hangzhou City Public Bicycle Transportation Service Development Corporation)

Table 13.5 Changes in shared bicycle usage in Hangzhou City over time

Time	Number of bicycles	Number of cycle ports	Service areas	Number of daily rentals (times/day)	Daily rental frequency (times/ bicycle)
May 2008	2500	61	West Lake Tourist Area, west and north of the city center	5000	0.93
May 2009	17,000	800	Tourist areas, Gongshu District, Binjiang District, Jianggan District, West Lake District, Shangcheng District, Xiacheng District	99,798	5.6
May 2010	50,000	2000	Tourist areas, Gongshu District, Binjiang District, Jianggan District, West Lake District, Shangcheng District, Xiacheng District, Xiaoshan District	301,647	6.04
May 2011	60,600	2411	Eight districts of Hangzhou	363,600	6
May 2013	69,750	2962		378,500	5.43
October 2019	100,000	4253		470,000	More than 4

stations, 31.9% are in residential areas, 5.8% are in commercial areas, and 2.4% are near schools and hospitals.

The number of bicycles, cycle ports, service areas, and usage throughout Hangzhou City from May 2008 to May 2013, as well as targets for 2015, is shown in Table 13.5.

The public bicycle system in Hangzhou City will continue to develop, the scale of inputs will expand, and rental volume will increase. Large facility inputs and convenience of the system service network are seen as the main reasons for increases in rental volume.

13.3.1.1 Business Entity

Hangzhou City initially invested 150 million CNY (1 CNY is approximately 19.2 JPY); outsourcing operation and management to Hangzhou Public Transport Co., Ltd.; Hangzhou Public Bicycle Transport Service Development Co., Ltd.; and Hangzhou Public Bicycle Technology Development Co., Ltd. (in collaboration with Hangzhou Dianzi University). This investment supported the establishment of two venture companies, each of which is responsible for the operation and development of the public bicycle business in Hangzhou City. The public bicycle system is deployed as a government-led, business-managed, and citizen-based system.

Table 13.6 Hangzhou City shared bicycle rental fee system

Duration of use	Usage fee
0–1 h	Free (1.5 h free if connecting from bus travel)
1–2 h	1 CNY/h (approximately 19.2 JPY; one use of Kyoto City bus is roughly equivalent to 12 uses of a Hangzhou shared bicycle)
2–3 h	2 CNY/h (approximately 38.4 JPY)
Over 3 h	3 CNY/h (approximately 57.6 JPY)

Source: Hangzhou City Public Bicycle Transportation Service Development Corporation

13.3.1.2 Rental Procedures

To rent a bicycle, a Hangzhou public transportation IC (integrated circuit) card or citizen IC card with a balance of more than 200 CNY is necessary. People who do not own such a card, e.g., tourists, can obtain a public bicycle IC card called a Z-Card at the cycle port. This costs 300 CNY, of which 200 CNY is a deposit and 100 CNY is a temporary deposit.

13.3.1.3 Fee System

The fee system is shown in Table 13.6. For users, the public benefit of a shared bicycle is that it is low-cost or free (if using for less than 1 h).Over 90% of bicycle rentals are for less than 1 h. This means that free use has maximized active use of this low-carbon green transportation tool by Hangzhou citizens and tourists.

13.3.1.4 Anti-Theft Measures

All bicycles are equipped with GPS. If a bicycle is not returned or its loss reported within 24 h, users are fined 10 CNY per day in addition to the rental fee. Malicious users are reported to the police and put onto a credit blacklist.

13.3.1.5 System Characteristics

In addition to stationary cycle ports, 30 mobile cycle ports mounted on trucks can be placed to respond flexibly to demand (Fig. 13.2).

13.3.1.6 Central Dispatch System

In August 2009, a central dispatch system was introduced. When a cycle port is at or above 80% capacity, 20 bicycles are reassigned to ports at less than 20% capacity.

Fig. 13.2 Hangzhou mobile cycle port. (Source: Hangzhou City Public Bicycle Transportation Service Development Corporation)

13.3.2 Analysis of Low-Carbon Effects

Bicycles have a low-carbon impact as environmentally friendly vehicles because they do not emit carbon dioxide or harmful substances when used. Such effects are important to consider in terms of building low-carbon cities and transportation (Qian et al. 2017).

Driving 10 km in a private car emits 2.73 kg of carbon dioxide. Assuming that the average rental time of public bicycles in Hangzhou is 0.56 h and average cycling speed is 12 km/h, each trip covers approximately 6.7 km. If on average there were 500,000 daily rentals in 2019, annual CO_2 reductions would be approximately 334,000 tons. As shown in Fig. 13.3, the public bicycle system has significantly reduced carbon dioxide emissions since the beginning of 2008. The combination of this initiative of the Hangzhou municipal government and the introduction of taxes led to relatively rapid adoption of public bicycles. Cooperation with public transportation such as subways and buses, active introduction of new technologies and systems, close collaboration between the public and private sector, and optimized cycle port design are important components of a successful community bicycle scheme.

However, introduction of this scheme in Hangzhou has shown that despite its large scale and wide range of users, there is no resultant decrease in the number of cars and city buses (Fig. 13.4). The rate of uptake of bicycle use has not slowed since bicycle sharing was launched, demonstrating the potential for further carbon

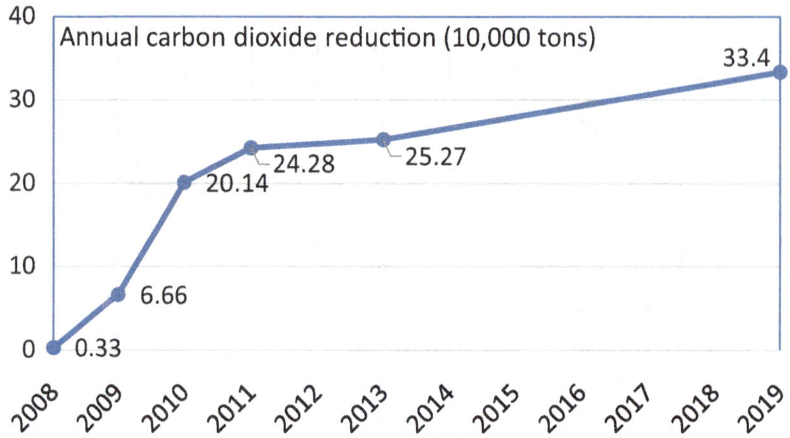

Fig. 13.3 Annual carbon dioxide reduction due to replacement of private cars with shared bicycles in Hangzhou. (Source: Hangzhou City Public Bicycle Transportation Service Development Corporation)

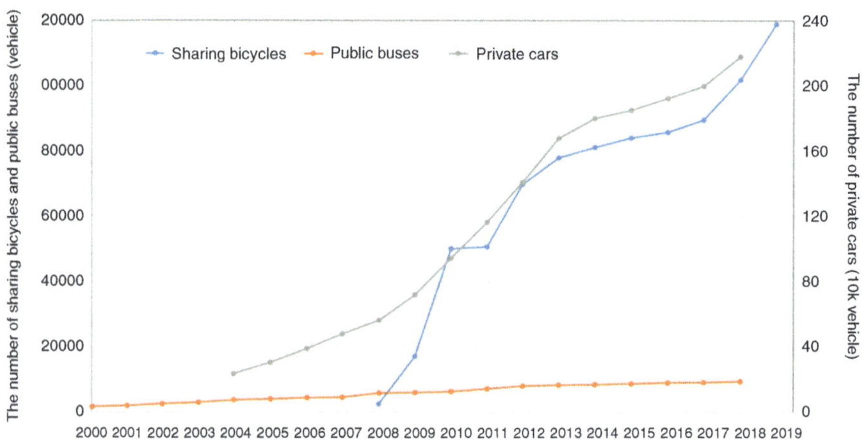

Fig. 13.4 Changes in vehicle use in three transportation systems in Hangzhou. (Source: Hangzhou City Public Bicycle Transportation Service Development Corporation, Hangzhou Public Transport Group Co., Ltd., Hangzhou City Statistical Yearbook)

reductions through increased bicycle use. Although the current use of shared bicycles in Hangzhou City alone has shown finite effects of environmental protection and low-carbon effects, the increase in low-carbon traffic and environmental awareness among citizens is clear.

13.3.3 Lessons Learned from Shared Bicycle Scheme

Although the shared bicycle scheme in Hangzhou has been a success, this has not been replicated nationwide.

In August 2019, there were 19.5 million shared bicycles in 360 cities in China, more than 300 million registered accounts, and an average of 47 million daily rides. However, this has fallen dramatically compared to previous years. In 2017, up to 77 companies entered the market, which is now composed of fewer than 10. In Shanghai, there were up to 1.7 million shared bicycles in 2017, but by 2019 this had decreased by approximately 30% to 500,000.

What has happened to all the shared bicycles that have been discarded in China due to overproduction? They are stored by different companies and visible in aerial photographs, where their bright colors resemble fields of flowers. By the end of 2020, nearly ten million bicycles are expected to be discarded in China, with over-competition resulting in a saturated mass of bicycles flooding the streets, destined to be removed and taken to storage and waste disposal sites (Fig. 13.5).

These graveyards of shared bicycles are spreading across China. All aspects of shared bicycle systems, from investment to disposal, are progressing in China at a rate that is unimaginable in Japan. Problems such as exaggerated investment due to excessive competition, proliferation of abandoned bicycles on the streets, and the uncertainty of deposit management are common. New investment is decreasing and risks to cash flow have been reported. However, these issues are expected to be resolved in the future.

13.4 Sustainability in the Sharing Economy

13.4.1 Outlook for the Sharing Economy

The following trends in technological progress and social system improvement can be observed toward sustainable development (Graedel and Allenby, 2003).

1. Dematerialization: Reducing the use of materials as much as possible to deliver the same or better functions or services, e.g., the compactness of personal computers, weight reduction of product packaging, and shifting from product to service, and from proprietary to sharing
2. Material substitution: Replacing currently used materials in order to reduce damage to the environment, e.g., development of biological pesticides, renewable energy use, and improving recycling
3. Decarbonization: Mainly improving energy efficiency, e.g., through the introduction of new transportation systems, popularization of shared bicycles
4. Coupling with information and communication technology (ICT): Using ICT to enable efficient planning, monitoring, and management of various business

Fig. 13.5 Discarded shared bicycles. (Source: people.cn)

activities, leading to more efficient social systems, e.g., mobile payment, smart society

In this section, the sharing economy will be reviewed from those aspects. Over the next 3 years, the overall average annual growth rate of the sharing economy will exceed 30%. Government audits and corporate compliance have been strengthened, the behavior of market participants has become more rational, and development imbalances in various areas have decreased. Capacity sharing combines the two main advantages of large manufacturing countries and Internet giants. Improving resource utilization, restructuring supply and demand structures and industrial organization, and empowering small and medium-sized enterprises (SMEs) can create superimposition, aggregation, and multiplication effects. As levels of digitization, networking, and intelligence in manufacturing continue to increase, the role of capacity sharing in stimulating "double innovation," fostering new models, and driving supply-side structural reforms will become more prominent.

The role of reservoirs and stabilizers in employment in the sharing economy becomes more pronounced, with an increasing number of people participating as flexible workers depending on interest, skills, time, and resources. Through the sharing of various economic activities, the sharing economy has become an important source of growth in the employment sector and can promote consumption. It can continue to stimulate novel consumer demand as well as meet the limited consumer demand of traditional service models. With changes in attitudes and priorities for personal fulfillment, shared services become an important force to accelerate penetration into key areas of life and promote consumption. The sharing economy is a catalyst for innovation, including employment, working, and consumption.

The sharing economy will be an important scenario for technological innovation in the field of AI. In travel, lodging, and healthcare, AI is used for identity verification, content governance, decision support, risk prevention and control, service assessment, and network and information security auditing. It plays an important role in other aspects and has great application potential. The pace of application of blockchain technology will also be accelerated to provide technical support for the establishment of social trust systems and credit guarantee systems in the sharing economy.

Following various corrective and regulatory actions in 2018 and the formal implementation of the Electronic Commerce Act, the sharing economy will continue its strong regulatory developments. The development of standardization systems in the sharing economy is accelerating, and industry-wide service standards and specifications are expected to be introduced in areas such as shared offices, cloud creation platforms, shared medicine, and online takeout.

13.4.2 The Sharing Economy and the Internet: A Future Economy

The Internet combined with various platforms based on time and space provide technical support for accelerating the prosperity of the sharing model. Before the advent of the Internet, human activities such as production and leisure were limited by the physical world and geographically dispersed in urban and rural areas. The arrival of the Internet has removed the limitations of time, space, and social class. Consumers from different countries and regions can form different virtual communities online, initiating economic activities that are different from those in traditional markets. The Internet is the technological foundation of the sharing economy Belk (2014a, b) attributes the phenomenon of collaborative consumption and the sharing economy to the Internet age, where networks have opened a new era of sharing. Sach (2015) sees the sharing economy as an IT-driven phenomenon. While these views attribute varying degrees of importance to Internet technology in the sharing economy, there is no doubt that the sharing economy supports it. Matzler et al. (2015) claim that the Internet provides a convenient channel for sharing items, and development of social media on the Internet is stimulating the sharing economy. According to Leismann et al. (2013), the spread of the Internet and the networking opportunities it provides play an important role in changing consumption patterns. Schor (2014) argues that from a technical point of view, complex software technology in the Internet age has reduced the high cost of traditional trading in the distribution market, improved the transparency of transactions in a shared model, and reduced the risk of trading with strangers. Undoubtedly, in a shared platform, the basic premise is to use Internet technology to provide participants with the right support to bring the masses together, promote growth, increase trust, and ensure security. In the future, the evolution of technologies such as IOT, blockchain, and AI will contribute significantly to the creation of more diverse sharing business models and further development of the sharing economy.

13.4.3 A Finite Earth: Maximizing Resource Use and Minimizing Environmental Impact

In literature on the sharing economy, many discuss the environmental impact of sharing consumption. Thus far, most studies have argued that the sharing economy contributes to energy savings, emission reductions, and green and low-carbon development, though some have disagreed.

The mainstream view is that the sharing economy promotes green development. For example, Rifkin (2015) argues that the sharing economy, which promotes a switch in consumption from ownership to use rights, can significantly reduce sales of new products and reduce resource consumption and greenhouse gas emissions. An empirical analysis using car-sharing survey data by Belk (2014a, b) and Martin and Shaheen (2011) shows that the average level of car use in North America and Western Europe is only 8%. Vehicle sharing can help reduce greenhouse gas emissions by reducing vehicle purchases, delaying vehicle disposal, and reducing vehicle fuel consumption. By shifting from owning things to using them to meet needs, the sharing economy can reduce the burden of consumption and stimulate long-term green changes in consumer behavior. In addition, centralized management can be expected to improve environmental efficiency, including of energy, due to increased ease of circulation, maintenance, and introduction of the best performing products. In Shanghai, bicycle sharing reduced carbon dioxide and nitrogen oxide emissions by 25,000 and 64 tons, respectively, in 2016 (Zhang and Mi 2018).

Many researchers have questioned the ability of the sharing economy to promote emissions reductions. Schor (2014), for example, argues that for most people, car sharing increases emissions by improving the ease of car use. In fact, this effect is analyzed using basic economic theory, namely, the substitution effect of consumption and the income effect. There are antagonistic effects of car sharing: on one hand, the substitution effect leads to reduced emissions through the use of new, more efficient cars and reduced fuel consumption. On the other hand, the income effect promotes increases in demand, mileage, and emissions due to reduced costs. It is widely believed that the substitution effect should be greater than the income effect. For example, Leismann et al. (2013) view the economy as a resource-saving model. Schor (2014) states that lenders may purchase products with a more severe environmental impact due to the money they earn, and the use of old objects stimulates the purchase of new objects. Reduction of expense also stimulates increased consumption, with a range of other negative ramifications of the sharing economy.

These discussions are based on the form of the sharing economy up to the present and demonstrate advantages and disadvantages of its impact on the economy, environment, and sustainability. At this stage, many sharing businesses are being targeted by capital as new markets, and economic value is emphasized over environmental value. Therefore, national and local administrations should adjust their policies as soon as possible to promote corporate environmental responsibility and social responsibility (CSR) and introduce environmental assessments, such as life cycle assessments of goods and services, during the development of proposals for sharing businesses (Mi and Coffman 2019).

Human society has transitioned from inland civilizations in the non-globalized era (agricultural civilization, river civilization): the hunting society (Society 1.0) and agricultural society (Society 2.0), to marine society plus inland civilization, with the industrial society (Society 3.0) and information society (Society 4.0). Japan was the first to advocate Society 5.0 as the fourth industrial revolution in its Fifth Science and Technology Basic Plan. This is described as a society where economic development and the resolution of social issues are both achieved through a system that fuses cyber space (virtual space) and physical space (real space) (Cabinet Office 2019). Society 5.0 is a smart society (maximizing utility and minimizing cost and risk) that aims to optimize social, economic, and technological systems, with innovation in all fields as the driving force. In the future, the sharing economy is expected to evolve to a more mature and sound stage, allowing it to make a significant contribution to maximizing use of the Earth's finite resources and minimizing environmental impact.

References

Belk R (2007) Why not share rather than own. Ann Am Acad Polit Soc Sci 61:126–140

Belk R (2014a) You are what you can access: sharing and collaborative consumption online. J Bus Res 67:1595–1600

Belk R (2014b) Sharing versus pseudo-sharing in Web 2.0. Anthropologist 18(1):7–23

Botsman R (2015) Defining the sharing economy: what is collaborative consumption -and what isn't. Fast Company 27:2015

Chen M (2016) Hangzhou abuzz over bike sharing. http://www.chinadaily.com.cn/business/2016hangzhoug20/2016-09/01/content_26665873.htm. Accessed 22 Jan 2020
Graedel TE, Allenby BR (2003) Industrial ecology, Prentice-Hall international series in industrial and systems engineering. Prentice Hall, Upper Saddle River
Hamari J, Sjöklint M, Ukkonen A (2015) The sharing economy: why people participate in collaborative consumption. J Assoc Sci Technol 67(9):2047–2059
Leismann K, Schmitt M, Rohn H et al (2013) Collaborative consumption: towards a resource-saving consumption culture. Resources 2:184–203
Martin E, Shaheen S (2011) Greenhouse gas emission impacts of carsharing in North America. IEEE Trans Intell Transp Syst 12(4):1074–1086
Matzler K, Veider V, Kathan W (2015) Adapting to the sharing economy. MIT Sloan Manage Rev 56(2):71–77
Mi Z, Coffman D (2019) The sharing economy promotes sustainable societies. Nat Commun 10:1214
Qian X, Zhou L, Zhou W, Nakagami K (2017) Japan-China comparative study on low-carbon transportation: case studies on improving public bus service in Kyoto and Hangzhou. Policy Sci 24(3):57–74
Rifkin J (2015) Towards internet of things and shared economy. Corp Res 2:14–21
Runde S (2011) Readers Digest Deutschland (in German) 06/11, pp 74–75
Sach A (2015) IT-user-aligned Business Model Innovation (ITUA) in the sharing economy: a dynamic capabilities perspective. In: ECIS 2015 Completed Research Papers
Schor J (2014) Debating the sharing economy. Great Transition Initiative. http://www.greattransition.org/publication/debating-the-sharing-economy
Schor J, Attwood-Charles W (2017) The sharing economy: labor, inequality, and social connection on for-profit platforms. Sociol Compass 11:e12493. https://doi.org/10.1111/soc4.12493
Shaheen S, Cohen A (2007) Growth in Worldwide Carsharing: An International Comparison. Transp Res Rec 1992(1):81–89
Yang S (2016) Sharing economy types, factors and influences: the perspective of literature research. Rev Ind Econ 2:35–45
Zhang Y, Mi Z (2018) Environmental benefits of bike sharing: a big data-based analysis. Appl Energy 220:296–301

Chapter 14
Achievement of Nationally Determined Contributions (NDCs) Through Emissions Trading in China, Japan, and South Korea

Xuanming Su and Weisheng Zhou

Abstract Carbon emissions trading is a market-based approach used to lower abatement costs for both sellers and purchasers of carbon. By simulating a scenario in which emissions trading takes place among China, Japan, and Korea in a low-carbon community, this chapter explores the possible impact of carbon emissions trading. The results show that developed countries, such as Japan and Korea, and developing countries, like China, can lower their total abatement costs through carbon trading, considering the costs of buying and selling carbon credits.

14.1 Introduction

The Paris Agreement adopted in 2015 aims to restrict global warming to 1.5–2 °C above pre-industrial levels by the end of this century. To achieve this target, the Paris Agreement requires each Party to outline their post-2020 climate actions, which are known as nationally determined contributions (NDCs). An important part of the global economy, China, Japan, and South Korea account for approximately 23.5% of global gross domestic product (GDP) (The World Bank 2020, calculated using current USD). They exhibit different levels of socioeconomic development: China is currently a developing country, Japan is a mature developed country, and South Korea can be considered semi-developed. These three countries can play an important role in addressing climate change. Under the Paris Agreement, China has pledged to reach peak carbon dioxide emissions by 2030 and to lower CO_2 emissions per unit of GDP by 60–65% from 2005 levels. Japan plans to reduce

X. Su
Research Institute for Global Change/Research Center for Environmental Modeling and Application/Earth System Model Development and Application Group, Japan Agency for Marine-Earth Science and Technology (JAMSTEC), Yokohama, Japan
e-mail: suxuanming@jamstec.go.jp

W. Zhou (✉)
College of Policy Science, Ritsumeikan University, Ibaraki, Osaka, Japan
e-mail: zhou@sps.ritsumei.ac.jp

W. Zhou et al. (eds.), *East Asian Low-Carbon Community*,
https://doi.org/10.1007/978-981-33-4339-9_14

greenhouse gas (GHG) emissions by 26.0% by fiscal year (FY) 2030 compared to FY 2013 (a 25.4% reduction compared to FY 2005; approximately 1042 $MtCO_2$-eq as 2030 emissions). South Korea has pledged to reduce emissions by 37% compared to the business-as-usual (BAU) level (850.6 $MtCO_2$-eq) by 2030 (UNFCCC 2020).

Some studies have assessed the use of emissions trading among countries or local regions to achieve NDCs. Fujimori et al. (2016) showed that achieving NDCs with emissions trading could decrease reductions in global welfare stemming from changes in household consumption by 75%. Caciagli (2018) reviewed existing emissions trading schemes (ETS) and suggested ways to combine them with international carbon credits to contribute to achieving NDCs. Schneider et al. (2020) analyzed how to allow the transfer and use of mitigation not covered by NDCs to facilitate the identification of mitigation potential and reduce the costs of achieving NDCs. Gao et al. (2019) assessed the opportunities and challenges for China to participate in international carbon markets. Inspired by these studies, we use a case study to assess the possible economic benefits or losses, for example, GDP losses or carbon prices changes, if emissions trading were to occur under NDCs among three important East Asian countries, China, Japan, and Korea. We utilize the Glocal Century Energy Environment Planning (G-CEEP) model (Su et al. 2010, 2012b, 2014) for our analysis, assuming a balanced scenario, the shared socioeconomic pathway 2 (SSP2; O'Neill et al. 2014, Fricko et al. 2016) as the BAU scenario.

In Chap. 4, carbon abatement and its associated co-benefits are assessed in the context of achieving emission reduction targets domestically. For developed countries, such as Japan and Korea, existing carbon intensities are relatively low, and it is costly to further reduce carbon emissions. Developing countries do not have compulsory emission reduction obligations according to the "common but differentiated responsibilities" of the UNFCCC; carbon emissions originating in developing countries will grow to meet social and development needs. A market-based approach is used to abate carbon emissions by providing economic incentives for achieving reductions, allowing countries to sell excess capacity of their allowable emissions to countries that have exceeded their targets. Thus, countries with strict climate policies are able to meet their reduction targets at significantly lower than projected costs.

China, Japan, and Korea lead the economic development of Asia, covering developing and developed countries, and can form a representative economic community in East Asia. This chapter introduces a comparative scenario with emissions trading occurring only among China, Japan, and Korea as a "Glocal Low-Carbon Community" and focuses on the following questions:

- What occurs when emissions trading is allowed?
- How does emissions trading affect the cost of carbon emission reduction, and what is the price of traded carbon credits?
- How sensitive are different abatement measures to emissions trading?

14.2 Methodology

We assume the same BAU scenario as in Chap. 4; the balanced SSP2 scenario (O'Neill et al. 2014; Fricko et al. 2016) is used to represent future socioeconomic projection. We considered the emission reduction scenario (TAR) and an emissions trading scenario (TRD) as described above. The G-CEEP model (Su et al. 2010, 2012a, b) was used to optimize results. In addition, we distinguished the effects and costs of various abatement options.

There are many kinds of abatement options to reduce CO_2 emissions. The contribution of different abatement options to CO_2 emissions are divided into three terms shown in Eq. (14.1) (Akimoto et al. 2004):

- Fuel switching among fossil fuels.
- Fuel switching to nonfossil fuels—the contribution of the shift to nonfossil fuels is decomposed into that of each of the nonfossil fuels.
- Energy saving in both the energy supply side and the end-use sectors.

$$\begin{aligned} E_{\mathrm{ref}} - E_{\mathrm{tar}} &= P_{\mathrm{tar}}\left(\frac{E_{\mathrm{ref}}}{P_{\mathrm{ref}}} - \frac{E_{\mathrm{tar}}}{P_{\mathrm{tar}}}\right) + \frac{E_{\mathrm{ref}}}{P_{\mathrm{ref}}}(P_{\mathrm{ref}} - P_{\mathrm{tar}}) \\ &= P_{\mathrm{tar}}\frac{P^{\mathrm{f}}_{\mathrm{tar}}}{P_{\mathrm{tar}}}\left(\frac{E_{\mathrm{ref}}}{P^{\mathrm{f}}_{\mathrm{ref}}} - \frac{E_{\mathrm{tar}}}{P^{\mathrm{f}}_{\mathrm{tar}}}\right) + P_{\mathrm{tar}}\frac{E_{\mathrm{ref}}}{P^{\mathrm{f}}_{\mathrm{ref}}}\left(\frac{P^{\mathrm{f}}_{\mathrm{ref}}}{P_{\mathrm{ref}}} - \frac{P^{\mathrm{f}}_{\mathrm{tar}}}{P_{\mathrm{tar}}}\right) + \frac{E_{\mathrm{ref}}}{P_{\mathrm{ref}}}(P_{\mathrm{ref}} - P_{\mathrm{tar}}) \end{aligned} \tag{14.1}$$

where E is carbon emissions and P is primary energy consumption. The subscript ref denotes the BAU scenario, and tar is used for both the TAR and TRD scenarios.

14.3 Results and Discussion

14.3.1 Carbon Trading

Carbon trading lowers domestic carbon emission reductions in countries that are purchasing carbon credits, namely, Japan and Korea in this study. It increases the demand for carbon trading and then lowers the total reduction costs. The sellers of carbon credits, namely, China in this study, gain revenue from carbon trading and increase domestic carbon emission reduction levels. Carbon emissions abatement and trading for China, Japan, and Korea in 2030 are given in Fig. 14.1. China sells 176.4 Mt carbon, while Japan and Korea reduce 87.8 and 88.7 Mt carbon, respectively, by carbon trading. By means of carbon trading, abatement costs are reduced significantly (Fig. 14.2). Without carbon trading, Japan would need to spend 70.7 billion USD (2005) to reduce its GHG emissions to 1042 $MtCO_2$-eq. With carbon trading, domestic reduction costs drop to 61.5 billion USD (2005); even when the 38.0 billion USD (2005) cost of carbon trading is included, total abatement cost is

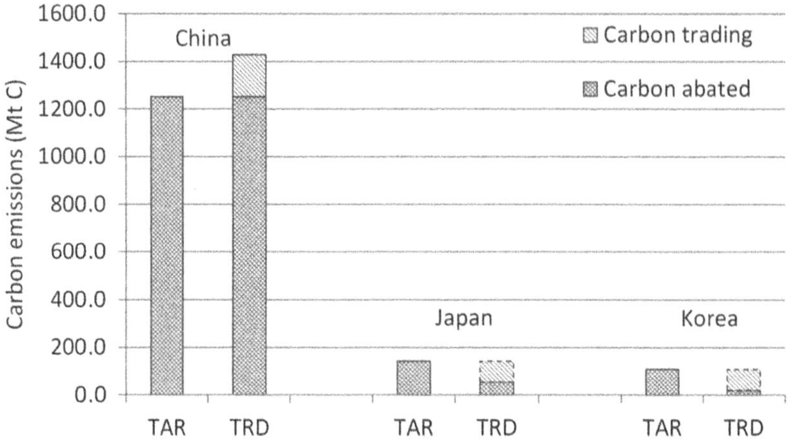

Fig. 14.1 Carbon emissions abatement and trading for China, Japan, and Korea in 2030, under emissions reduction (TAR) and emissions trading (TRD) scenarios

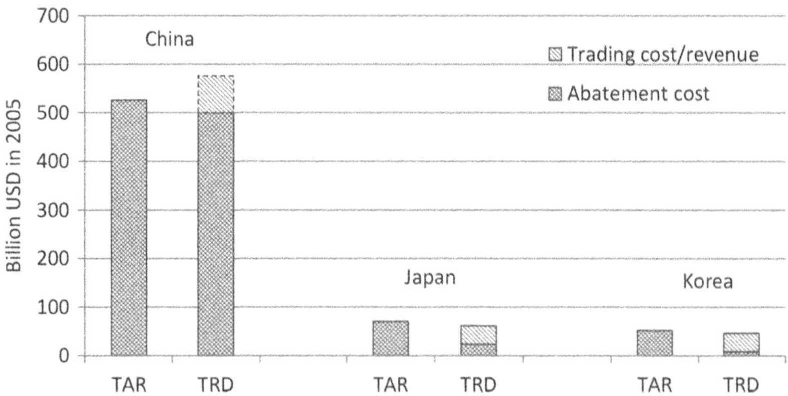

Fig. 14.2 Carbon abatement and trading cost for China, Japan, and Korea in 2030, under emissions reduction (TAR) and emissions trading (TRD) scenarios

reduced by 13.0%. For Korea, the total abatement cost to achieve the 2030 NDC target reduces from 52.2 to 46.0 billion USD (2005). China gains 76.4 billion USD (2005) revenue from selling carbon credits. The total abatement cost for China is reduced by 5%, although fewer reductions are made than in the no trading scenario. Generally speaking, reducing carbon emissions domestically is costly and inefficient for developed countries because of existing low carbon intensity. Carbon emissions trading is a market-based approach used to control climate change at relatively lower abatement costs. Developing countries can sell their carbon credits, profit from carbon trading, and still lower their total abatement costs; it is a win-win solution for carbon emissions abatement.

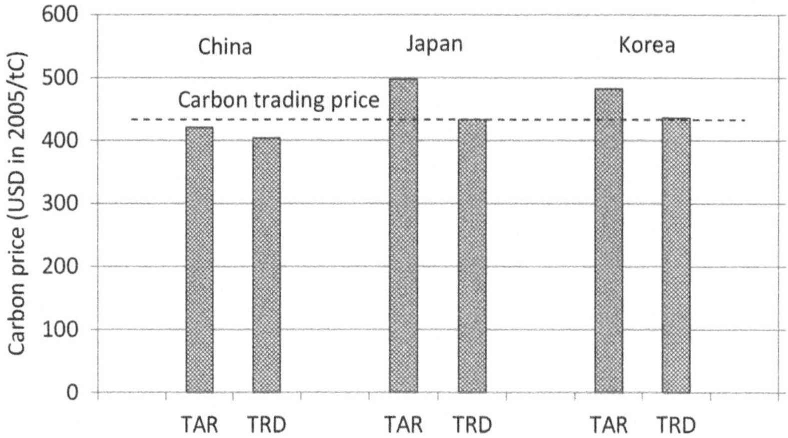

Fig. 14.3 Carbon abatement and trading prices for China, Japan, and Korea in 2030, under emissions reduction (TAR) and emissions trading (TRD) scenarios

We considered an international carbon price (carbon trading price) and a domestic carbon price (marginal abatement cost (MAC) of carbon). The international carbon price is determined using MAC curves, which derive from carbon credit demand and supply curves. The domestic carbon price is determined by domestic reductions and abatement costs. Prices in 2030 are given in Fig. 14.3. The international carbon price under trading among the specified countries is 433.2 USD (2005)/tC. Domestic carbon prices in Japan decrease from 497.5 to 432.8 USD (2005)/tC and in Korea decrease from 482.4 to 435.9 USD (2005)/tC. China decreases its domestic carbon price from 420.5 to 403.6 USD (2005)/tC, considering revenue from carbon credit sales. The domestic abatement cost will increase in China due to the consequent reduction in additional carbon for trading. However, the cost is partly compensated by the revenue from selling carbon credits, and the improved "cleaner" environment produced by additional carbon abatement is also valuable. Similar analysis can be seen in den Elzen et al. (2011), in which the international carbon price under full emissions trading is 132 USD in 2005/tC. All countries in the trading community benefit from selling or buying carbon credits, because sellers can gain revenue from higher international carbon prices (relative to domestic prices), and buyers can lower their abatement cost through lower international carbon prices (relative to domestic prices). Our results reveal a relatively large carbon price for China when attaining its NDC target, which implies that it will be costly for China to reduce carbon intensity by 65% in 2030 compared to 2005 levels.

MAC curves for the carbon trading scenario in China, Japan, and Korea in 2030 are given in Figs. 14.4, 14.5, and 14.6. The trading cost or revenue is not included; the MAC curves show the domestic marginal abatement costs under carbon trading.

China raises its average MAC to produce additional carbon emissions for sale. However, the revenue from selling carbon credits will decrease the average MAC from 420.5 to 403.6 USD (2005)/tC. The most significant change is energy saving,

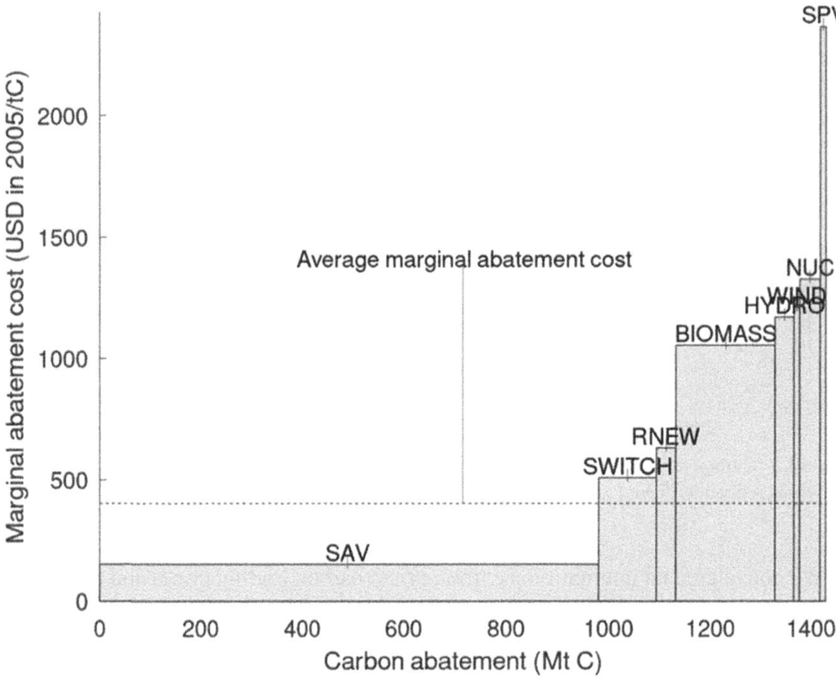

Fig. 14.4 Marginal abatement cost in China in 2030

for which the MAC increases from 95.2 to 152.5 USD in 2005/tC. The abated carbon emissions produced by energy saving account for 63.7% of the total (or 68.6% in the no trading scenario), which indicates that low-cost energy saving cannot produce sufficient carbon emission reductions to both achieve emission targets and sell extra carbon emissions internationally. Thus, the abatement cost of energy saving increases. Switching among fossil fuels is an effective measure for China to reduce carbon emissions. The MAC of the nonelectric renewable energy increases from 603.4 to 632.8 USD (2005)/tC. Other renewable energy accounts for up to 2.6% compared to 2.9% in the emission target scenario without trading, though the absolute value is slightly greater in the emission target scenario. Biomass, switching among fossil fuels, and energy saving account for most of the total abatement costs, at 36.2%, 25.9%, and 9.9%, respectively, in the carbon trading scenario. The total abatement cost of hydro-power accounts for a small proportion (7.4%) because of its relatively low capacity in the carbon trading scenario. Carbon trading raises domestic carbon prices in China and the reduction efficiencies of different abatement measures are improved to meet the extra carbon reduction requirements.

Energy saving becomes the most important abatement measure, if Japan is to achieve partial carbon emission reductions by purchasing carbon credits from developing countries such as China. The MAC of energy saving decreases from 157.6 to 153.4 USD (2005)/tC in the carbon trading scenario. The abatement cost of

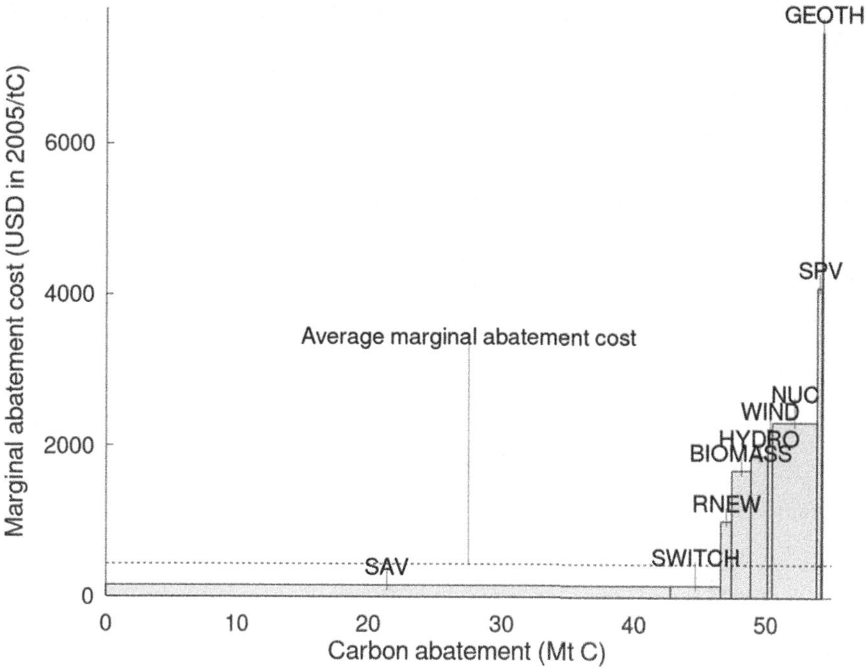

Fig. 14.5 Marginal abatement cost in Japan in 2030

energy saving accounts for 27.9% of total abatement costs but 78.8% of abated carbon because of its relatively low MAC. Nuclear reduces 6.3% of total carbon abatement in Japan in the carbon trading scenario, due to nuclear limits in Japan. Its total abatement cost accounts for 33.7% of the total. Introduction of biomass, wind, and solar power reduces emissions by 2.6%, 0.7%, and 0.6%, respectively, and accounts for 10.3%, 3.4%, and 6.1%, respectively, of total abatement costs. Abatement measures with low MAC are likely to substitute abatement measures with higher MAC in Japan under a carbon trading scenario. Thus, domestic abatement costs are lowered.

In Korea, energy saving is the most important abatement measure under the carbon trading scenario, accounting for 63.6% of total carbon abatement, followed by nuclear, switching among fossil fuels, and biomass. The total abatement cost of nuclear accounts for the majority of total abatement costs, at 68.0%. The total abatement cost of energy saving, biomass, and switching among fossil fuels accounts for 15.5, 2.7, and 1.9, respectively. The average MAC of Korea is 435.9 USD (2005)/tC, which is a significant decrease from the MAC with no carbon trading (482.4 USD (2005)/tC). Abatement measures with low MAC are apt to substitute abatement measures with high MAC, in order to lower domestic abatement costs in Korea.

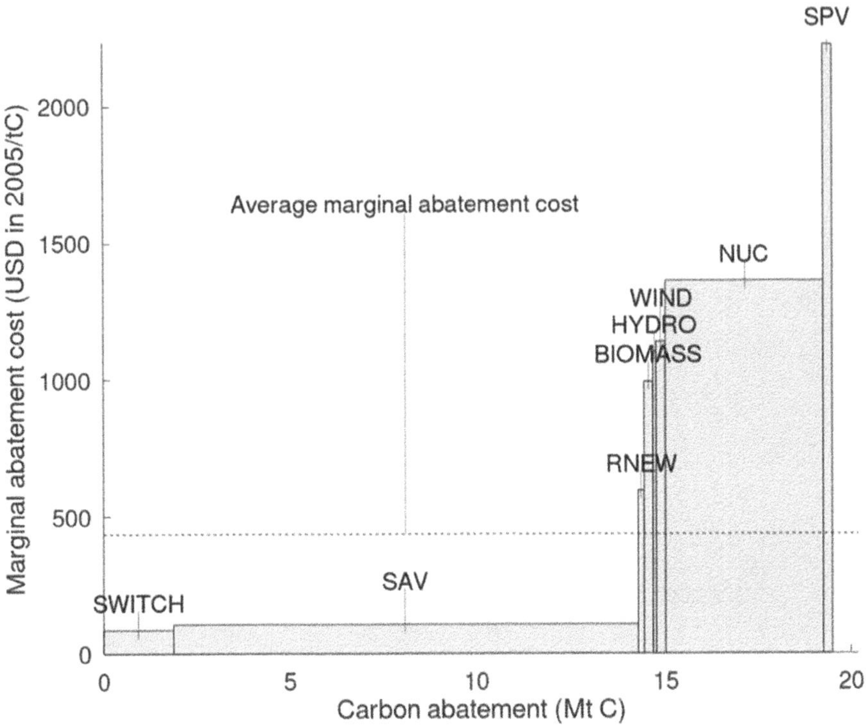

Fig. 14.6 Marginal abatement cost in Korea in 2030

14.4 Conclusion

Carbon emissions trading is a market-based approach used to lower abatement costs in countries selling and purchasing carbon. This chapter explores the possible impact of emissions trading by simulating a scenario in which emissions trading takes place among China, Japan, and Korea in a low-carbon community. Developed countries such as Japan and Korea can lower their total reduction costs through carbon trading, even when the cost of purchasing carbon credits is factored in. Japan and Korea reduce their total abatement costs by 13.0% and 10.1%, respectively, compared to the no trading scenario. Developing countries (China in this study), benefit from the sale of carbon credits. China faces increased domestic abatement costs to produce extra carbon emissions for sale but gains revenue from carbon credit sales and reduces total abatement costs by 5.0%. In addition, cleaner environments provided by relatively stringent constraints on carbon emissions provide additional benefits for China. The international carbon price in this carbon trading scenario is 433.2 USD (2005)/tC, within the range of domestic carbon prices of countries buying and selling carbon. Analyses of MAC curves reveal that carbon trading raises the domestic carbon price in China, and China needs to improve the reduction efficiency of different abatement measures to achieve the additional carbon reduction. For countries purchasing carbon credits, such as Japan and Korea, abatement measures with low MAC are apt to substitute abatement measures with higher MAC under a carbon trading scenario, leading to lower domestic abatement costs.

References

Akimoto K, Tomoda T, Fujii Y, Yamaji K (2004) Assessment of global warming mitigation options with integrated assessment model DNE21. Energy Econ 26:635–653

Caciagli V (2018) Emission trading schemes and carbon markets in the NDCs: their contribution to the Paris agreement. In: Alves F, Leal Filho W, Azeiteiro U (eds) Theory and practice of climate adaptation. Springer International Publishing, Cham, pp 539–571. https://doi.org/10.1007/978-3-319-72874-2_31

den Elzen MGJ, Hof AF, Beltran AM, Grassi G, Roelfsema M, van Ruijven B et al (2011) The Copenhagen Accord: abatement costs and carbon prices resulting from the submissions. Environ Sci Policy 14(1):28–29

Fricko O, Havlik P, Rogelj J, Klimont Z, Gusti M, Johnson N et al (2016) The marker quantification of the shared socioeconomic pathway 2: a middle-of-the-road scenario for the 21st century. Glob Environ Chang 42:251. https://doi.org/10.1016/j.gloenvcha.2016.06.004

Fujimori S, Kubota I, Dai H, Takahashi K, Hasegawa T, Liu J, Hijioka Y, Masui T, Takimi M (2016) Will international emissions trading help achieve the objectives of the Paris agreement? Environ Res Lett 11(10):104001. https://doi.org/10.1088/1748-9326/11/10/104001

Gao S, Li M, Duan M, Wang C (2019) International carbon markets under the Paris agreement: basic form and development prospects. Adv Clim Chang Res 10(1):21–29. https://doi.org/10.1016/j.accre.2019.03.001

O'Neill BC, Kriegler E, Ebi KL, Kemp-Benedict E, Riahi K, Rothman DS et al (2014) The roads ahead: narratives for shared socioeconomic pathways describing world futures in the 21st century. Glob Environ Chang 42:169. https://doi.org/10.1016/j.gloenvcha.2015.01.004

Schneider L, La Hoz Theuer S, Howard A, Kizzier K, Cames M (2020) Outside in? Using international carbon markets for mitigation not covered by nationally determined contributions (NDCs) under the Paris agreement. Clim Pol 20(1):18–29. https://doi.org/10.1080/14693062.2019.1674628

Su X, Ren H, Zhou W, Mu H, Nakagami K (2010) Study on future scenarios of low-carbon Society in East Asia Area, part 1: development of Glocal century energy and environment planning model and case study. Policy Sci 17(2):85–96

Su X, Zhou W, Nakagami K, Ren H, Mu H (2012a) Capital stock-labor-energy substitution and production efficiency study for China. Energy Econ 34(4):1208–1213. https://doi.org/10.1016/j.eneco.2011.11.002

Su X, Zhou W, Ren H, Nakagami K (2012b) Co-benefit analysis of carbon emission reduction measures for China, Japan and Korea. Policy Sci 19(2):99–112

Su X, Zhou W, Sun F, Nakagami K (2014) Possible pathways for dealing with Japan's post-Fukushima challenge and achieving CO_2 emission reduction targets in 2030. Energy 66:90–97. https://doi.org/10.1016/j.energy.2014.02.002

The World Bank (2020) World development indicators. https://data.worldbank.org/

UNFCCC (2020) Nationally determined contributions (NDCs) Interim Registry. https://www4.unfccc.int/sites/ndcstaging/Pages/Home.aspx

Chapter 15
Design and Analysis of a Carbon Emissions Trading System for Low-Carbon Development in China

Yishu Ling, Weisheng Zhou, and Xuepeng Qian

Abstract Tackling climate change is one of the most significant challenges for human beings today. China, the largest emitter of CO_2, has taken carbon pricing as a market-oriented policy for reducing its emissions. At the end of 2017, the National Development and Reform Commission (NRCD) announced the implementation of a national carbon trading scheme (ETS) in the power sector, which has initiated eight pilot projects since 2013. What have the pilot projects devoted to low-carbon development? What is the present status of its emissions reduction schemes, and what issues remain to be solved? This study first provides a comprehensive analysis of each pilot project from the perspectives of policies, transactions, and market performance. Then, we explore the impact of China's pilot ETSs on low-carbon development with targets of carbon emissions, carbon per person carbon intensity, energy consumption, and energy intensity from 2008 to 2017 using a panel data applying a difference-in-differences method. The main conclusions are as follows: (1) overall, China's pilot ETSs have a powerful impact on low-carbon development. (2) Pilot projects have offered lessons that could be learned and applied to the national trading market.

Y. Ling
Graduate School of Policy Science, Ritsumeikan University, Ibaraki, Osaka, Japan
e-mail: ps0290ih@ed.ritsumei.ac.jp

W. Zhou (✉)
College of Policy Science, Ritsumeikan University, Ibaraki, Osaka, Japan
e-mail: zhou@sps.ritsumei.ac.jp

X. Qian
College of Asia Pacific Studies, Ritsumeikan Asia Pacific University, Beppu, Oita, Japan
e-mail: qianxp@apu.ac.jp

W. Zhou et al. (eds.), *East Asian Low-Carbon Community*,
https://doi.org/10.1007/978-981-33-4339-9_15

15.1 Introduction

Carbon pricing is considered the most cost-effective tool for reducing greenhouse gas (GHG) emissions (Mehling and Tvinnereim 2018). As the primary type of carbon pricing, emissions trading schemes (ETS) can have significant impacts on reducing carbon dioxide and energy consumption. As China is the world's largest CO_2 emitter, accounting for 30% of global emissions, it is taking action to reduce emissions. Like the European Union, the United States, Japan, South Korea, and other countries that have already adopted ambitious policies to reduce carbon emissions through implementation of carbon pricing, China has been developing pilot carbon ETS projects in Shenzhen City (hereafter Shenzhen), Shanghai City (hereafter Shanghai), Beijing City (hereafter Beijing), Guangdong Province (hereafter Guangdong), Tianjin City (hereafter Tianjin), Hubei Province (hereafter Hubei), Chongqing City (hereafter Chongqing), and Fujian Province (hereafter Fujian) since 2013. Analysis of the effect of an ETS on emission reductions can support promotion of the policy. In 2015, China submitted a national voluntary reduction draft, claiming it would reach its maximum CO_2 emissions target by 2030, and CO_2 emissions per unit of gross domestic product (GDP) would be reduced by 60–65% compared with 2005 levels. Cap-and-trade ETS pilot projects, including more than 2000 companies, factories, offices, and institutions, were established. CO_2 emission quotas were approximately 3.3 billion tons coal equivalent by June 2019, with the aim of achieving national targets (Ministry of Ecology and Environmental of the People's Republic of China 2019). Lessons learned from pilot ETS markets will determine the future of a nationwide ETS.

At the end of 2017, the NDRC announced the implementation of a national carbon trading scheme (ETS) in the power sector, providing a pilot model in an additional sector that functions in parallel with the regional pilot model. Before a national market is launched, it is pertinent to analyze the outcome of the eight pilot projects to ascertain whether the ETS policy has a practical impact on CO_2 emissions and energy consumption reduction.

In this chapter, we first focus on the problems mentioned above using comparative analysis to clarify the current situation of each ETS pilot project based on published statistical data. Then, we use a difference-in-differences (DID) method with panel data from 2008 to 2017 to analyze the impacts of ETS policy.

It has been approximately 6 years since the first pilot market launched in Shenzhen City. Although the carbon trading market pilot schemes have been running for a comparatively long time, many issues are unresolved, including lack of well-designed guidelines for implementation, of a strategy for building the capacity essential for constructing an active carbon market, and of an overall strategy for integrating pilot projects into a national scheme (Kong and Freeman 2013).

In China, ETS policy is still being explored, and implementation is a lengthy process requiring adjustment and refinement. How effectively are the pilot projects working as an emission reduction tool? Researchers have recently focused on the impact of ETS using both qualitative and quantitative analyses. Overviews summarizing lessons learned, including comprehensive details of the seven pilots, are given chronologically in (Wang 2016). China was not the first country to build a carbon market, and market planners can learn from the experiences of many international schemes (Dong et al. 2016). The specific aspects of ETS pilot projects have been discussed, including the choices of allocation methods of carbon emissions trading (Feng et al. 2018), management of the carbon emissions trading market (Chang et al. 2018), and reference to the experiences of developed countries (Jotzo and Löschel 2014).

Most quantitative analyses have used computable general equilibrium (CGE) models (Tang et al. 2016; Mu et al. 2018) to clarify the impacts of ETS in terms of economic and environmental performance. CGE models are useful for policy forecasting, allowing analysis of future impacts of ETS policy. As pilot ETS policies have already been running for several years, it is essential to understand current implementation impacts. Comparative analysis of changes in carbon emissions and carbon intensity in pilot areas or industries over time has allowed discussion of the effects of pilot ETS in China (Zhang et al. 2017a, b). Concerns have been raised about using DID methods based on panel data from pilot regions to estimate the effect of ETS policy (Jin et al. 2018, Zhang et al. 2017b, 2019). However, due to 2-year delays in data collection, ETS policies must be revised before the launch of the national market in the power sector, based on only limited results from the pilots. Furthermore, prevailing trend hypothesis and robustness tests have not previously been conducted.

In this chapter, we combine qualitative and quantitative analyses to clarify the impact of ETS policy from 2013 to 2017, based on provincial level panel data from 2008 to 2017. We use carbon emissions, carbon intensity, energy consumption, and energy intensity as target indicators to evaluate whether the ETS policy is attaining its low-carbon goals. As Shenzhen is a part of Guangdong, we take the 6 pilot regions as the treatment group and the remaining 24 provinces as the control group (excluding Tibet, Hong Kong, Macao, and Taiwan).

The rest of this chapter is organized as follows: Section 15.2 describes the eight pilot regions from policy and market performance perspectives. Section 15.3 describes data collection and the methodology applied. Section 15.4 shows the empirical results, discussion, and a robustness test of the DID model. Section 15.5 concludes with policy implications for promoting the development of a national carbon trading market based on the experiences of the pilot schemes.

15.2 ETS Pilot Projects in China

There are many reasons that the eight regions were chosen to serve as pilots. The regions cover the north (Beijing, Tianjin), the mid-west (Hubei, Chongqing), and the southeast coastal area (Shanghai, Shenzhen, Guangdong, Fujian) of China, so can reflect geographical disparities. Furthermore, they include the national political center (Beijing), the national economic center (Shanghai), and a city with prior

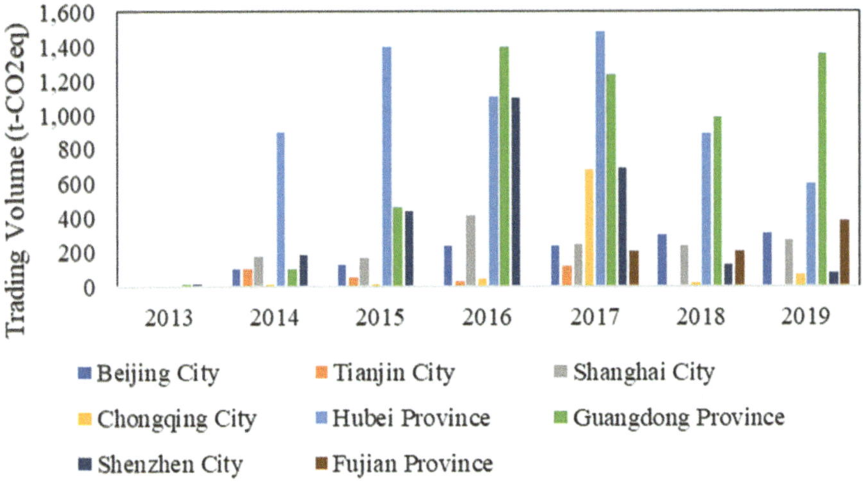

Fig. 15.1 Trading volume of eight emissions trading scheme (ETS) pilot projects in 2013–2018. (Sources: Beijing Environment Exchange (2020), Chongqing Emission Exchange (2020), Fujian Emission Exchange (2020), Guangdong Emission Exchange (2020), Hubei Emission Exchange (2020), Shanghai Environment and Energy Exchange (2020), Shenzhen Emission Exchange (2020), Tianjin Emission Exchange (2020))

experience of participating in pilot projects (Chongqing). Therefore, implementation was relatively easy, and sufficient funds could be raised to support the pilot projects. Experience gained from the pilots is expected to inform a national market. Fujian Province, the most recent pilot to adopt national standards and guidelines for carbon verification, is an actual experiment for the future nationwide market. Finally, energy consumption in China is concentrated in secondary and tertiary industries; pilot regions whose industrial structure is of crucial national status thus offer an opportunity to reduce emissions both in the pilot regions and nationally.

The trading volume (Fig. 15.1), trading value (Fig. 15.2), and allowance prices (Fig. 15.3) of each ETS pilot are vital signals indicating how the markets are functioning. The performance of the ETS pilot project trading markets can be considered in two phases. The first is from 2013 to 2014, which is the initial period, when the pilots show similar features, such as low trading volumes and high allowance prices. The second phase is from 2015 to 2018, when the pilots begin to perform differently from one other.

The absolute trading volume varies among pilots; from 2014 to 2019, trading volume in Hubei ranked first three times and that in Guangdong ranked first twice. In contrast, the trading volume in Chongqing ranked last twice and that in Tianjin ranked in last place four times. In the ETS in Beijing and Fujian, the trading volume continues to increase. The ETS in Shanghai and Shenzhen both reach a peak in 2016 and then gradually decline. There is a particular anomaly in the ETS in Chongqing in 2017, when trading volume increases massively before sharply decreasing in 2018.

Patterns in absolute trading value differ from those of trading volume. Due to the relative advantage of a larger trading volume, trading value in Hubei ranked in first

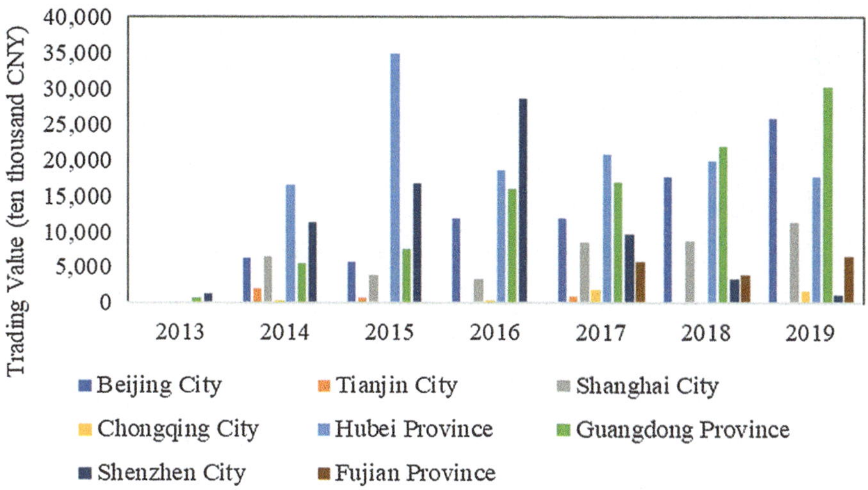

Fig. 15.2 Trading volume of eight emissions trading scheme (ETS) pilot projects in 2013–2019. (Sources: Beijing Environment Exchange (2020), Chongqing Emission Exchange (2020), Fujian Emission Exchange (2020), Guangdong Emission Exchange (2020), Hubei Emission Exchange (2020), Shanghai Environment and Energy Exchange (2020), Shenzhen Emission Exchange (2020), Tianjin Emission Exchange (2020))

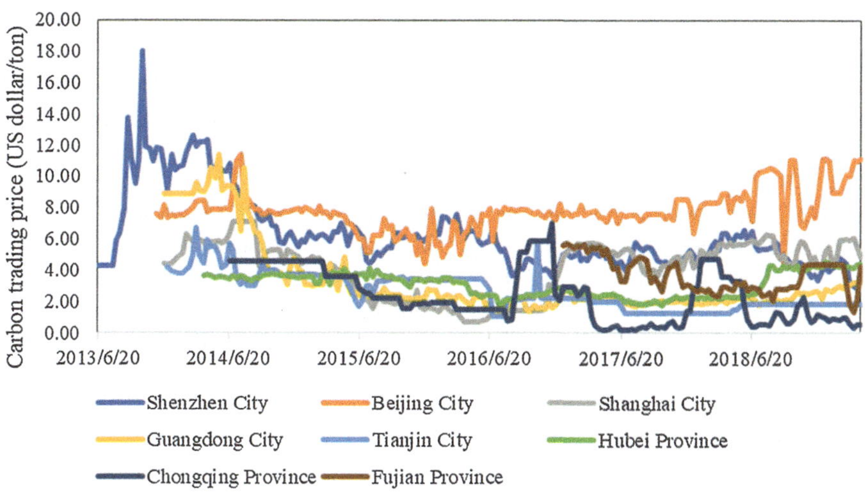

Fig. 15.3 Carbon price of emissions trading scheme (ETS) pilots in China from June 2013 to December 2019 (USD/ton). (Sources: Beijing Environment Exchange (2020), Chongqing Emission Exchange (2020), Fujian Emission Exchange (2020), Guangdong Emission Exchange (2020), Hubei Emission Exchange (2020), Shanghai Environment and Energy Exchange (2020), Shenzhen Emission Exchange (2020), Tianjin Emission Exchange (2020))

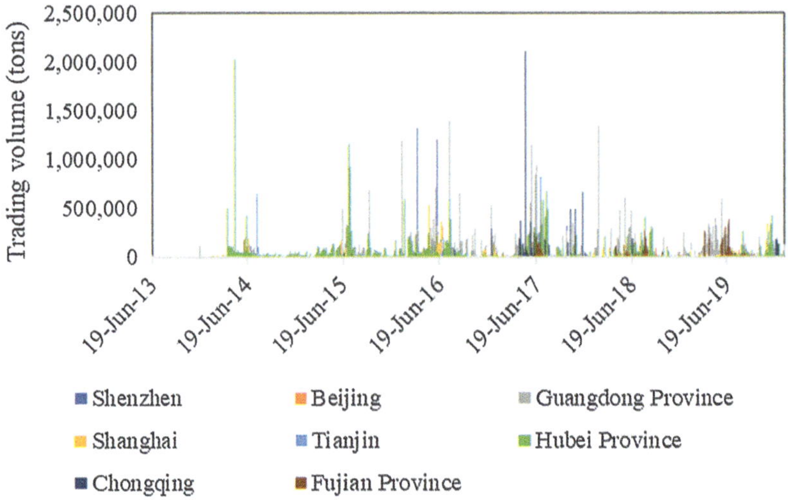

Fig. 15.4 Trading volume of emissions trading scheme (ETS) pilots in China from June 2013 to December 2019. (Sources: Beijing Environment Exchange (2020), Chongqing Emission Exchange (2020), Fujian Emission Exchange (2020), Guangdong Emission Exchange (2020), Hubei Emission Exchange (2020), Shanghai Environment and Energy Exchange (2020), Shenzhen Emission Exchange (2020), Tianjin Emission Exchange (2020))

place three times and that in Guangdong twice from 2014 to 2019. In 2016, trading value in Shanghai was highest, despite Guangdong having a larger trading volume. In contrast, because of small trading volumes, trading values in Chongqing and Tianjin are low throughout the period. There is a dramatic increase in trading value in the Beijing ETS from 2014 to 2019. The trading value of the Shenzhen ETS peaked in 2016 before gradually decreasing. The Shanghai ETS reached its lowest point in 2016, after which the trading value gradually increases.

The allowance price is vital for clarifying the relationship between trading volume and value. After higher prices during the initial period, prices in each pilot remained at relatively low levels (Fig. 15.3). Considering trends in allowance prices, Beijing ETS is the only pilot that retained a stable price, showing a slight increase from 2019. Shenzhen ETS allowance price decreased from the initial stage, from its highest price of 122.97 CNY/ton in 2013 to its lowest price of 7.89 CNY/ton in 2019. Shanghai ETS experienced a slump in prices in 2016 when trading volume increased, meaning that the trading value did not increase. Guangdong ETS experienced a similar trend, although prices after 2017 increased more slowly than those in Shanghai. In Tianjin and Chongqing, prices slumped from the initial stage; prices in Chongqing ETS in 2017 fell almost to zero. The huge increase in trading volume in 2017 did not translate into a similar increase in trading value; the trading value remained constant and almost fell to zero in 2018.

The compliance periods for each pilot fall around the beginning of June. All pilots have compliance rates close to 100%, which is very high. However, observation of daily trading volumes (Fig. 15.4) reveals that there is a concentration in trading during June and July. To ensure participants can comply with annual targets, Beijing, Tianjin, Shanghai, and Hubei have all extended their compliance periods at least once, leading to a high concentration of trading before and after June. Since liquidity directly affects prices, the price decreases due to this concentration of trading, and there is a higher risk to market security. It follows that the higher the historical volatility is, the riskier the security is. The current pilot markets tend to trade within a short term, so the liquidity of each pilot market remains low.

15.3 Methodology

15.3.1 *Research Design*

The market performance analysis in Sect. 15.2 shows that the ETS pilots have different scheme designs due to differences in developmental stages and emission situations. However, they share some common characteristics, such as "seizing large enterprises and releasing medium and small ones." Overall, pilot ETS in China can be considered a quasi-experiment. Therefore, we take Beijing, Tianjin, Shanghai, Chongqing, Guangdong (including Shenzhen), Hubei, and Fujian as the treatment group and the remaining 24 provinces of China (excluding Tibet, Hong Kong, Macao, and Taiwan) as the control group. We defined the treatment period as 2013–2017. The impact of the pilot ETS can be analyzed by comparing changes in carbon emissions, carbon intensity, energy consumption, and energy intensity between the treatment and control groups before and after the treatment period.

15.3.2 *Methodology*

15.3.2.1 Difference-in-Difference Model

Changes in the indicators of carbon emission reduction of the treatment group and control group are estimated before and after ETS implementation, allowing calculation of the difference between the changes, namely, the difference-in-difference. The original DID model is given as follows:

$$\mathrm{CAE}_{\mathrm{it}} = \beta_0 + \beta_1 \mathrm{ETSpilot}_{\mathrm{it}} + \beta_2 \mathrm{T}_{\mathrm{it}} + \beta_3 (\mathrm{ETSpilot}_{\mathrm{it}} * T_{\mathrm{it}}) + \beta_4 \mathrm{CON}_{\mathrm{it}} + \gamma_t + \mu_i + \varepsilon_{\mathrm{it}} \tag{15.1}$$

where CAE is the dependent variable representing low-carbon development; i and t represent the ith region and tth year, respectively; and $\mathrm{ETSpilot}_{\mathrm{it}}$ is the policy dummy variable. If i belongs to the treatment group, then $\mathrm{ETSpilot}_{\mathrm{it}} = 1$; otherwise $\mathrm{ETSpilot}_{\mathrm{it}} = 0$. T is the time dummy variable. If t belongs to the treatment period (2013–2017), then $T = 1$, otherwise, $T = 0$. The coefficient β_3 of $\mathrm{ETSpilot}_{\mathrm{it}} \times T_{\mathrm{it}}$

Table 15.1 Observable variable in DID models

Variables	Symbols	Categories of variables	Calculation methods
CO_2 emissions	CAE	Dependent	Calculation according to Table 1, taking natural logarithm for it
Carbon intensity	CAI	Dependent	Emitted CO_2 per CNY10,000 GDP
Energy consumption	ENC	Dependent	Derived from statistical yearbooks, taking natural logarithm for it
Energy intensity	ENI	Dependent	Consumed energy in standard coal per CNY 10,000 GDP
ETS piloting	ETSpilot	Policy	Treatment group, ETSpilot = 1; control group, ETSpilot = 0
Time	T	Policy	2010–2017, if time > =2014, $T = 1$; if time < 2014, $T = 0$
Population	POP	Control	Derived from statistical yearbooks
GDP	GDP	Control	Derived from statistical yearbooks, taking natural logarithm for it
Industrial structure	SEI	Control	The proportion of secondary industry's added value to GDP
Technical level	TEC	Control	The ration of R&D put to GDP
Wages	WAG	Control	Average wages of employees of provinces
Urbanization	URB	Control	The ratio of urban population to the total population

represents the estimator of difference-in-differences and measures the impacts of pilot ETS on low-carbon development. γ_t represents time fixed effects, μ_i represents region fixed effects, and ε_{it} represents random disturbances. Reductions in CO_2 emissions, CO_2 emissions per capita, carbon intensity, energy consumption, and energy intensity are estimated as dependent variables to comprehensively assess carbon emissions reduction.

Control variables are included to ensure the model is robust. Variables of industrial structure, technology, R&D, population, GDP per capita, and energy intensity are often selected in studies (Zhang et al. 2017a, b and Wang et al. 2019). We consider population and the scale of the economy to be the main factors affecting carbon emissions. Industrial structure, especially secondary industry, has a significant impact on production. Regional technological progress is also related to the effectiveness of emission reductions. Due to incomplete data on disposable income, we use residents' wages to measure the effect of income levels on carbon emissions. As urbanization has a significant positive effect on carbon emissions in China, the proportion of urban population to the total population at each year end is used to measure the role of urbanization. In summary, we use GDP, population, proportion of secondary industry to total industry, technical level, wages, urbanization, and environmental regulation as control variables (Table 15.1).

Thus, Eq. (15.1) can be improved as follows:

$$\ln \mathrm{CAE_{it}} = \beta_0 + \beta_1 \,\mathrm{ETSpilot_{it}} + \beta_2 T_{\mathrm{it}} + \beta_3 (\mathrm{ETSpilot_{it}} * T_{\mathrm{it}}) + \beta_4 \ln \mathrm{POP_{it}} + \beta_5 \ln \mathrm{GDP_{it}} + \beta_6 \mathrm{SEI_{it}} + \beta_7 \mathrm{TEC_{it}} + \beta_8 \mathrm{WAG_{it}} + \beta_9 \mathrm{URB_{it}} + \beta_{10} \mathrm{ENV_{it}} + \gamma_t + \mu_i + \varepsilon_{\mathrm{s}} \quad (15.2)$$

$$\mathrm{CAEPP_{it}} = \beta_0 + \beta_1 \,\mathrm{ETSpilot_{it}} + \beta_2 T_{\mathrm{it}} + \beta_3 (\mathrm{ETSpilot_{it}} * T_{\mathrm{it}}) + \beta_4 \ln \mathrm{POP_{it}} + \beta_5 \ln \mathrm{GDP_{it}} + \beta_6 \mathrm{SEI_{it}} + \beta_7 \mathrm{TEC_{it}} + \beta_8 \mathrm{WAG_{it}} + \beta_9 \mathrm{URB_{it}} + \beta_{10} \mathrm{ENV_{it}} + \gamma_t + \mu_i + \varepsilon_{\mathrm{it}} \quad (15.3)$$

$$\mathrm{CAI_{it}} = \beta_0 + \beta_1 \,\mathrm{ETSpilot_{it}} + \beta_2 T_{\mathrm{it}} + \beta_3 (\mathrm{ETSpilot_{it}} * T_{\mathrm{it}}) + \beta_4 \ln \mathrm{POP_{it}} + \beta_5 \ln \mathrm{GDP_{it}} + \beta_6 \mathrm{SEI_{it}} + \beta_7 \mathrm{TEC_{it}} + \beta_8 \mathrm{WAG_{it}} + \beta_9 \mathrm{URB_{it}} + \beta_{10} \mathrm{ENV_{it}} + \gamma_t + \mu_i + \varepsilon_{\mathrm{it}} \quad (15.4)$$

$$\ln \mathrm{ENC_{it}} = \beta_0 + \beta_1 \,\mathrm{ETSpilot_{it}} + \beta_2 T_{\mathrm{it}} + \beta_3 (\mathrm{ETSpilot_{it}} * T_{\mathrm{it}}) + \beta_4 \ln \mathrm{POP_{it}} + \beta_5 \ln \mathrm{GDP_{it}} + \beta_6 \mathrm{SEI_{it}} + \beta_7 \mathrm{TEC_{it}} + \beta_8 \mathrm{WAG_{it}} + \beta_9 \mathrm{URB_{it}} + \beta_{10} \mathrm{ENV_{it}} + \gamma_t + \mu_i + \varepsilon_{\mathrm{it}} \quad (15.5)$$

$$\mathrm{ENI_{it}} = \beta_0 + \beta_1 \,\mathrm{ETSpilot_{it}} + \beta_2 T_{\mathrm{it}} + \beta_3 (\mathrm{ETSpilot_{it}} * T_{\mathrm{it}}) + \beta_4 \ln \mathrm{POP_{it}} + \beta_5 \ln \mathrm{GDP_{it}} + \beta_6 \mathrm{SEI_{it}} + \beta_7 \mathrm{TEC_{it}} + \beta_8 \mathrm{WAG_{it}} + \beta_9 \mathrm{URB_{it}} + \beta_{10} \mathrm{ENV_{it}} + \gamma_t + \mu_i + \varepsilon_{\mathrm{it}} \quad (15.6)$$

15.3.2.2 Data Collection

Fossil fuel-related CO_2 emissions by energy type were calculated following the IPCC (2006):

$$\mathrm{CE}_i = \mathrm{EN}_i * \mathrm{EF}_i \quad (15.7)$$

where CE_i is the CO_2 emissions from energy type i; EN_i is the annual consumption of the ith energy source according to the China Statistical Yearbook from 2011 to 2018; and EF_i is the emission factor for the ith fossil fuel. According to IPCC (2006), the CO_2 emission factors are estimated as follows:

$$\mathrm{CE}_i = \mathrm{EN}_i * \mathrm{EF}_i \quad (15.8)$$

where C_i is the carbon content, O is the oxidation rate, and J_i is the average low calorific value of energy i. The emission factors of each energy type are shown in

Table 15.2 Overall descriptive statistics

Variables	Mean	Std. Dev.	Min	Max
CAE (10,000 tons)	32,440.27	21,875.91	2387.95	94,897.22
CAI (ton/CNY 10,000)	2.32	0.51	1.14	4.32
ENC (10,000-ton tce)	13,668	8475.34	1135.33	40,837.34
ENI (ton tce/CNY 10,000)	0.90	0.49	0.25	2.90
ETSpilot	0.23	0.42	0	1
POP (10,000 people)	4583.89	2797.03	554.30	11,169
GDP (billion CNY)	19,750.37	16,378.97	961.53	89,705.23
SEI (%)	46.43	8.85	19	68.08
TEC (%)	1.51	1.07	0.22	6.03
WAG (CNY)	48,321.01	18,982.04	20,597.00	13,1700.00
URB	0.54	0.14	0.12	0.89

Table 15.1. By summing the emissions from different energy types, the total CO_2 emissions for a single province from 2008 to 2017 are given by Eq. (15.9).

$$\mathrm{CE} = \sum \mathrm{CE}_i \tag{15.9}$$

Data on population, GDP, added value of secondary industry, R&D input, and SO_2 emissions are collected from the China Statistical Yearbook (2009–2018). The overall descriptive statistics are shown in Table 15.2. Data are presented in their raw form, although some were transformed before being put into the DID models.

15.4 Results and Discussion

15.4.1 Regression Results

The impact of ETS pilots in China on CO_2 emissions and reductions in CO_2 emissions per capita, carbon intensity, energy consumption, and energy intensity were estimated based on Eqs. (15.2)–(15.6). Table 15.3 shows the ordinary least square regression of the dynamic impacts of ETS pilots, controlling for population, GDP, the ratio of secondary industry, and technical level. The analysis confirms that the role of the ETS pilots causes reductions in carbon emission and carbon intensity. The reduction in carbon intensity consists with the target of its Intended Nationally Determined Contribution (INDC) submitted to the UN. Therefore, there is a clear relationship between carbon emission trading schemes and low-carbon development.

Based on this model, we used different emission measurement standards to estimate the average and dynamic effects of carbon emission rights trading on carbon emissions. Besides, carbon intensity, energy consumption, and energy intensity are explanatory variables, highlighting the impacts of ETS policy on low-carbon

Table 15.3 Empirical test on the impact of carbon emission trading scheme on carbon mitigation

	(1) lncae	(2) cai	(3) lnenc	(3) eni
DID	−0.182** (−2.667)	−0.337*** (−3.652)	−0.027 (−0.462)	0.114 (1.784)
Treated	−0.347*** (−5.947)	−0.199* (−2.486)	−0.271*** (−5.161)	−0.300*** (−5.429)
Time	0.108 (1.429)	0.251** (2.831)	−0.013 (−0.232)	−0.081 (−1.410)
GDP	0.741*** (4.856)	0.184 (0.940)	0.631*** (5.472)	−0.368** (−3.156)
POP	0.016 (0.091)	−0.104 (−0.480)	0.079 (0.578)	0.025 (0.182)
SEI	0.016*** (3.827)	0.005 (1.003)	0.015*** (4.999)	0.016*** (5.004)
TEC	−0.149*** (−4.767)	−0.118** (−2.819)	−0.074** (−2.734)	−0.018 (−0.652)
WAG	−0.466** (−2.892)	−0.495* (−2.288)	−0.163 (−1.278)	0.037 (0.276)
URB	0.289 (0.652)	0.048 (0.094)	0.271 (0.837)	0.148 (0.468)
Region fixed effects	Yes	Yes	Yes	Yes
Time fixed effects	Yes	Yes	Yes	Yes
Adj. R^2	0.766	0.196	0.817	0.612

Note: t statistics in parentheses, $^{*}p < 0.05$, $^{**}p < 0.01$, $^{***}p < 0.001$

development. After controlling for fixed effects and time fixed effects of individual provinces and cities, the crossover term DID has significant negative impacts on carbon emission and carbon intensity and possesses the significance at the 1 or 5% level (Table 15.3). This indicates that the carbon emission trading market promotes carbon emission reductions through the establishment of property rights and a market mechanism. In line with the target of lowering carbon dioxide emissions in China per unit of GDP by 60–65% from 2005 levels, ETS will be a crucial policy instrument in low-carbon development as it lowers CO_2 intensity.

In addition, the coefficient of wages has significantly negative impacts on carbon emissions, carbon intensity. In contrast, the coefficient of population has positive impacts on carbon emissions and energy consumption. The increasing population leads an insufficient in energy consumption and carbon emission theoretically. Besides, GDP and secondary industry also have positive impacts on carbon emission and energy consumption and possess the significance at the 1% level. It is clarified that economic development has massive demand for energy consumption and carbon emission. China is progressing toward industrialization, which is an important source of carbon emissions as industry continues to develop with high fossil fuel consumption.

It is particularly noteworthy that in the regression results, the coefficient of the technological level is significantly negative at the 1% level, reducing carbon dioxide emissions and energy consumption. This shows that while developing a low-carbon economy and improving technology levels, carbon dioxide emissions can be effectively controlled and reduced. As a result of investment in technology development projects, energy efficiency is improved, thereby reducing energy consumption. In summary, most control variable coefficients are significant and in line with expectations.

15.4.2 Discussion

15.4.2.1 Trends in CO_2 Emissions, Carbon Intensity, Energy Consumption, and Energy Intensity

The compelling premise of the difference-in-differences method adopted in this chapter is that if the ETS pilot policy is implemented, trends in the growth of carbon dioxide emissions and energy consumption in pilot provinces and cities and non-pilot regions will be broadly similar. Carbon emissions, carbon intensity, energy consumption, and energy intensity in pilot and non-pilot regions from 2008 to 2017 are plotted in Figs. 15.5, 15.6, 15.7, and 15.8, respectively. Both pilot and non-pilot regions show roughly parallel upward trends before 2013. After 2013, emissions from non-pilot regions maintain an upward trend, while those in pilot regions begin to decrease, which shows a clear positive effect of the carbon emissions trading policy.

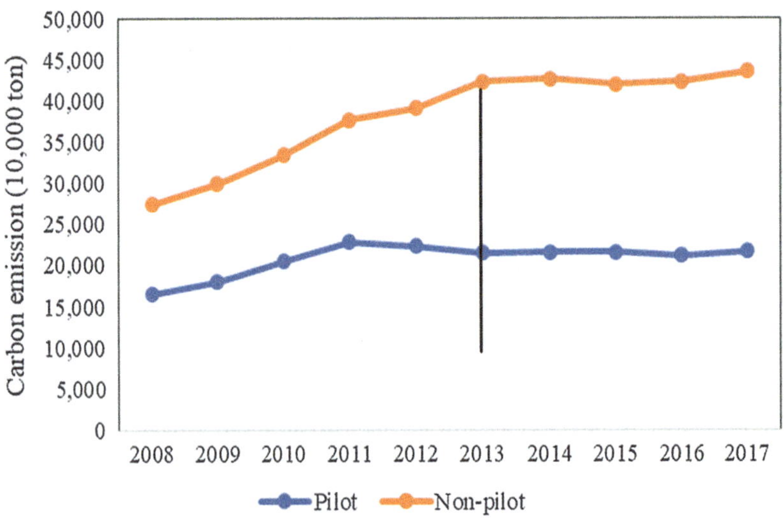

Fig. 15.5 CO_2 emissions in pilot and non-pilot regions

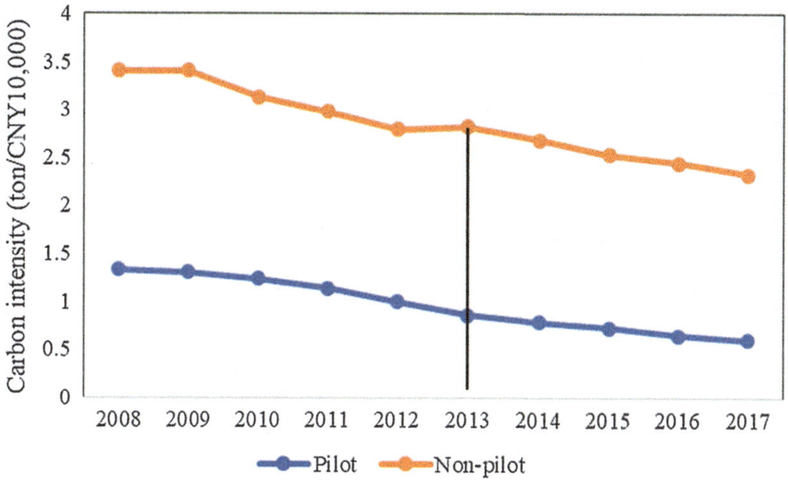

Fig. 15.6 CO_2 intensity in pilot and non-pilot regions

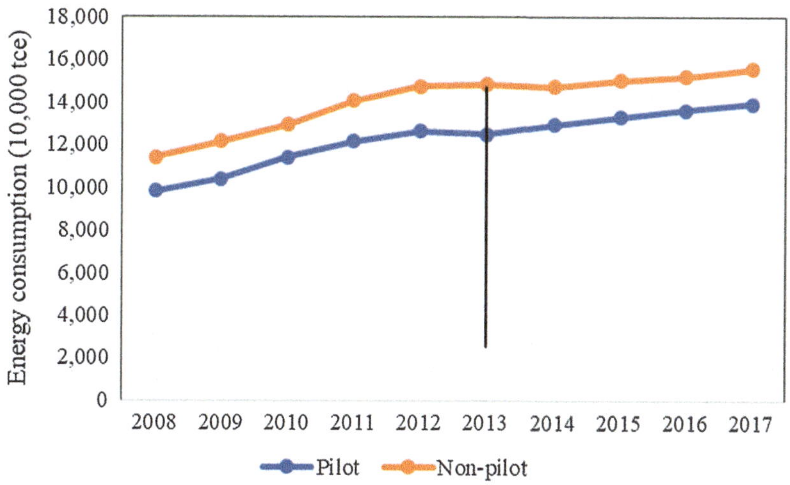

Fig. 15.7 Energy consumption in pilot and non-pilot regions

ETS policy has a positive impact on reducing CO_2 emissions and carbon intensity. The assumption of parallel trends is satisfied, indicating that our conclusions are robust.

15.4.2.2 Pilot ETS Issues and Implications for Nationwide ETS

Analysis of pilot ETS has confirmed that ETS policy beneficially reduces CO_2 emissions and energy consumption. However, several issues remain. The

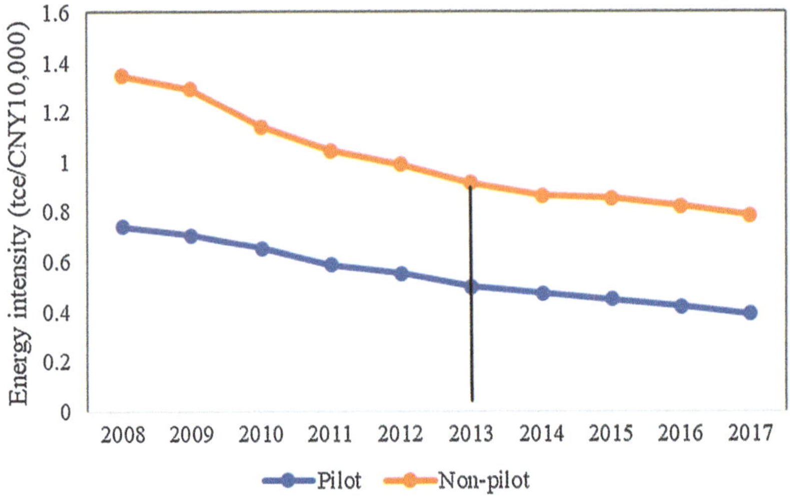

Fig. 15.8 Energy intensity in pilot and non-pilot regions

performance of pilot ETS varies (Figs. 15.1 and 15.2). While Shenzhen, Guangdong, and Hubei are actively trading, stagnation of markets in Tianjin and Chongqing removes incentives for participation. After initial high prices, the prices of each pilot remain at relatively low levels (Fig. 15.3); allowance prices in the second period are low expect for in Beijing. Since liquidity directly affects price, periods during which transactions are concentrated (Fig. 15.4) often lead to higher risks to market security.

Which lessons learned from pilot ETS can be applied to a nationwide ETS? Due to different geographical areas, stages of development, and industrial structure of pilots, different characteristics are exhibited, including the industrial sectors covered by each scheme and the allocation of allowances. For example, Guangdong adopts an allowance auction system, and Chongqing implements an allocation model for self-declaration of participants. These differences provide valuable experience for the establishment of a unified national market. There are large differences between regions in China, so the carbon market must be developed based on regional characteristics. The low liquidity of trading has been a prominent problem during the pilot period. In particular, pilots have been limited to the province level, and similarity of participants has led to low tradability, further reflecting the necessity of establishing a national market. Therefore, based on the outcome of the pilots, the following should be considered when building the national carbon market: equality between industries in different regions; efficiency of carbon prices; information symmetry between central and local governments and between government and enterprises; the balance of government intervention; and making use of market mechanisms.

15.5 Conclusion

Through analysis of policies, market performance, and compliance of each ETS pilot, schemes are constantly being refined in an adaptive fashion.

In this chapter, we analyzed the market performances of ETS pilots in China and calculated panel data for 30 provincial regions from 2008 to 2017 to examine the impacts of ETS policy on carbon emission reduction. Using a DID model, we show that ETS policy has had a positive effect on pilot regions after a trial period of 6 years. Controlling for GDP, population, the ratio of secondary industry, and technology level reveals that development of the economy has a positive impact on decreasing both carbon intensity and energy intensity. China aims to reduce its carbon intensity by 40% by 2020 and 80% by 2050. The 2020 goal was reached in 2017, and the goal for 2050 may also be achieved early. Technology level also shows a positive impact on decreasing both carbon emissions and energy consumption, and ETS policy can be implemented in combination with technology improvements to reduce carbon emissions more effectively.

The pilot markets and national market will continue to operate in parallel for a certain period. The pilot projects have certainly provided useful experiences to inform design of the national market. Important considerations for building the national carbon market include equality between industries in different regions; efficiency of carbon prices; information symmetry between central and local governments and between government and enterprises; the balance of government intervention; and making use of market mechanisms.

References

Beijing Environment Exchange (2020) Trading information. http://www.cbeex.com.cn/. Accessed 9 April 2020

Chang K, Chen R, Julien C (2018) Market fragmentation, liquidity measures, and improvement perspectives from China's emissions trading scheme pilots. Energy Econ 75:249–260

Chongqing Emission Exchange (2020) Trading information. https://tpf.cqggzy.com/. Accessed 9 April 2020

Dong J, Ma Y, Sun H (2016) From pilot to the National Emissions Trading Scheme in China: international practice and domestic experiences. Sustainability 8:522

Feng S, Stephen H, Liu Y, Zhang K, Yang J (2018) Towards a national ETS in China: cap-setting and model mechanisms. Energy Econ 73:43–52

Fujian Emission Exchange (2020) Trading information. https://carbon.hxee.com.cn/. Accessed 9 April 2020

Guangdong Emission Exchange (2020) Trading information. http://www.cnemission.com/. Accessed 9 April 2020

Hubei Emission Exchange (2020) Trading information. http://www.hbets.cn/. Accessed 9 April 2020

IPCC (2006) IPCC Guidelines for National Greenhouse Gas Inventories Volume 2 Energy. https://www.ipcc-nggip.iges.or.jp/public/2006gl/vol2.html. Accessed 9th April 2020

Jin Z, Mizuno Y, Liu X (2018) Development situation of the emissions trading system in China and prospects. (in Japanese). https://pub.iges.or.jp/pub/china_ets_2018. Accessed 29 Oct 2018
Jotzo F, Löschel A (2014) Emissions trading in China: Emerging experiences and international lessons. Energy Policy 75:3–8
Kong B, Freeman C (2013) Making sense of carbon market development in China. Carbon Climate Law Rev 194–212
Mehling M, Tvinnereim E (2018) Carbon pricing and the 1.5° C target: near-term decarbonization and the importance of an instrument mix. Carbon Clim Law Rev 12(1):50–61
Mu T, Evans S, Wang C, Cai W (2018) How will sectoral coverage affect the efficiency of an emission trading scheme? A CGE-based case study of China. Appl Energy 227:403–414
Ministry of Ecology and Environmental of the People's Republic of China (2019). United Nations Climate Action Summit: China's position and actions.http://www.mee.gov.cn/ywgz/ydqhbh/syqhbh/201909/t20190917_734045.shtml (in Chinese). Accessed on 18th Jan. 2020
Shanghai Environment and Energy Exchange (2020) Trading information. http://www.cneeex.com/ . Accessed 9 April 2020
Shenzhen Emission Exchange (2020) Trading information. http://www.cerx.cn/. Accessed 9 April 2020
Tang L, Shi J, Bao Q (2016) Designing an emissions trading scheme for China with a dynamic computable general equilibrium model. Energy Policy 97:507–520
Tianjin Emission Exchange (2020) Trading information.https://www.chinatcx.com.cn/. Accessed 9 April 2020
Wang H (2016) Evaluation of emissions trading scheme pilot project in 5 cities and 2 provinces of China (in Japanese). AGI working paper series, 2016-02, pp 1–47
Wang H, Chen Z, Wu X, Nie X (2019) Can a carbon trading system promote the transformation of a low-carbon economy under the framework of the porter hypothesis? Empirical analysis based on the PSM-DID method. Energy Policy 129:930–938
Zhang J, Wang Z, Du X (2017a) Lessons learned from China's regional carbon market pilots. Econ Energy Environ Policy 6(2):1–20
Zhang M, Liu Y, Su Y (2017b) Comparison of carbon emission trading schemes in the European Union and China. Climate 5:70
Zhang W, Zhang N, Yu Y (2019) Carbon mitigation effects and potential cost savings from carbon emissions trading in China's regional industry. Technol Forecast Soc Change 141:1–11

Chapter 16
An Empirical Analysis of International Carbon Transfer

Hirotaka Haga and Weisheng Zhou

Abstract This chapter quantified international carbon flow as a result of economic globalization based on input-output analysis using the World Input-Output Database (WIOD). Trends in CO_2 emissions induced from 7 regions and 35 industries were analyzed as a case study of the production of transportation equipment sector. Total CO_2 emissions induced in China were larger than in the other six regions. Production-based emissions tend to be larger than consumption-based emissions in China and East Asia, while the inverse is true elsewhere. The ratio of production- to consumption-based CO_2 emissions induced by a region to itself is close to 1. CO_2 emissions from manufacturing transportation equipment such as automobiles are larger than emissions from purchasing in this sector in East Asia. Developing countries tend to have higher consumption- than production-based emissions. Furthermore, consumption-based calculations tend to underestimate CO_2 emission in developing countries and emerging countries compared with production-based estimates. It is important to compile further analysis on global carbon flows in consumption-based cases. In considering burdens of responsibility for global CO_2 emissions, ratios must be taken into account, allowing identification of the hidden, larger amount of emissions in RoW (developing countries), and BRIIAT (emerging countries). High volumes of CO_2 emissions flow to RoW from the Euro-zone, NAFTA, RoW, East Asia, and BRIIAT; CO_2 emissions must be reduced in both developed and developing countries. The East Asian area emits more CO_2 from manufacturing transportation equipment than it does from purchasing in this sector, while in developing countries, CO_2 emissions in consumption were higher than those of production. This analysis highlighted the importance of assessing trends in emissions from emerging and developing countries when considering global burdens of responsibility for CO_2 emissions.

H. Haga
Department of Regional Design and Development, Department of Business Economics, University of Nagasaki, Nagasaki, Japan
e-mail: hhaga@sun.ac.jp

W. Zhou (✉)
College of Policy Science, Ritsumeikan University, Ibaraki, Osaka, Japan
e-mail: zhou@sps.ritsumei.ac.jp

W. Zhou et al. (eds.), *East Asian Low-Carbon Community*,
https://doi.org/10.1007/978-981-33-4339-9_16

16.1 Introduction

An international framework to address climate change has been constructed on the basis of the United Nations Framework Convention on Climate Change (UNFCCC) in 1992 and the Kyoto Protocol, which was adopted at the third Conference of the parties (COP3). After that, UNFCCC adopted COP decision with regard to the Paris Agreement and its implementation on 12 December 2015. The Paris Agreement aims to hold the increase in the global temperature to well below 2 °C above pre-industrial levels and to pursue efforts to limit the temperature increase to 1.5 °C by achieving net zero greenhouse gas emission in the second half of this country (Takamura 2016, p. 11).

Climate change is a global issue; thus international cooperation to mitigate it is essential. However, there is intense debate between developed and developing countries concerning the burden of responsibility (Fujikawa et al. 2015, p. 130). Due to industrialization of emerging nations and globalization of the economy, the division of labor to provide goods and services spans international borders. Thus, environmental burdens are internationally distributed. There is an "embodied environmental load"; consumption of goods and services induces production of goods in multiple countries, and corresponding environmental loads are embodied in each country (Fujikawa et al. 2015, p. 129). Domestic policies addressing climate change mitigation must consider the carbon leakage that occurs when emission sources are transferred overseas, as this greatly increases CO_2 emissions intensity.

CO_2 emissions can be generated by production. The *IPCC Greenhouse Effect Gas Inventory* provides a method for estimating production-based emissions, which are the total CO_2 emissions generated inside a country as a result of production in that country. Another method allows estimation of consumption-based CO_2 emissions, or "trade embodied CO_2 emissions." These are calculated by summing the emissions from production of goods in the country of origin with emissions in the country consuming the goods. The conventional method for calculating national emissions is production-based. Considering trade embodied CO_2 emissions includes emissions generated internationally.[1]

There has been much research seeking to determine whether the producer or consumer should be responsible for CO_2 emissions, through exploration of trade embodied CO_2 emissions (Hoshino and Sugiyama 2012), attributed emissions (Na 2006), and other concepts (e.g., Wyckoff and Roop 1994; Bastianoni et al. 2004; Peters and Hertwich 2008; Davis and Caldeira 2010; Peters et al. 2012; Mózner 2013; Pang et al. 2016; Fujikawa and Ban 2016). Yamazaki et al. (2012) estimated production-based and consumption-based CO_2 emissions of Japan, China, Korea, and the European Union through international carbon flow with an input-output model, concluding that consumption-based estimation was theoretically and quantitatively more appropriate. Yoshinaga (2013) examined CO_2 emission trends

---

[1]For detailed information on international carbon flows, especially empirical research on trade embodied CO_2 emissions; see Hoshino and Sugiyama (2012) and Fujikawa et al. (2015).

and factors influencing household and industrial emissions across all 27 countries of the European Union, calculating CO_2 emissions induced by final demands in each country.

National CO_2 emissions are usually estimated based on levels of energy consumed during production of goods and services in producing districts. If consumption of goods is also included in CO_2 emission calculations, results vary considerably (Fujikawa et al. 2015). The CO_2 emissions of countries that are net importers are hugely underestimated. In contrast, net exporters of goods (producing countries) overestimate their CO_2 emissions. Responsibility for environmental conservation can be attributed more fairly to producing and consuming countries by using embodied environmental load (Fujikawa et al. 2015, p. 130). Responsibility for regional CO_2 emissions can be shifted from producing to consuming countries. The continuing growth of regional international trade is likely to support this distinction, e.g., using the international input-output structure in the Asia-Pacific region based on analysis of Global Trade Analysis Project (GTAP) models (Lee et al. 2016, pp. 349). If the trade deals allow for the transfer of low-carbon technology (e.g., from Japan and Korea to China), then it would be possible to simultaneously achieve higher production levels and lower emission levels (Lee et al. 2016, p. 348).[2]

Research on responsibility for environmental issues has also been conducted in the field of air pollution.[3] Consumption-based accounting is necessary to support international policies that distribute the burden of mitigation among all beneficiaries according to their ability and willingness to pay (Guan et al. 2014, p. E2631).

Rapidly developing regions such as China have been generating huge emissions, leading to environmental destruction. As Asia increasingly becomes the "world's factory" and the "world's market," emissions will continue to increase, highlighting the political, economic, and societal importance of low-carbon development (Ueta 2010, pp. 2, 4). Economic globalization leads to international carbon flow; this international carbon transfer must be analyzed in order to understand how CO_2 emissions can be reduced.

In this chapter, we quantitatively assess international carbon flow due to economic globalization based on input-output analysis using the World Input-Output Database (WIOD). In Sect. 16.2, we discuss global trends in CO_2 emissions. The WIOD and computational methods are described in Sect. 16.3. Section 16.4 presents

[2]Lee et al. (2016) quantitatively assessed the macroeconomic and environmental effects of various free trade agreements (eliminating tariffs) among East Asian countries using the E3ME-Asia and GTAP-E models. Their results show that free trade can modestly improve the macroeconomic situation (GDP and employment), but higher production levels will lead to higher CO_2 emissions (Lee et al. 2016, p. 348).

[3]For example, as the United States outsourced manufacturing to China, sulfate pollution in 2006 increased in the western United States but decreased in the eastern United States, reflecting the competing effects from enhanced transport of Chinese pollution and reduced US emissions (Lin et al. 2014).

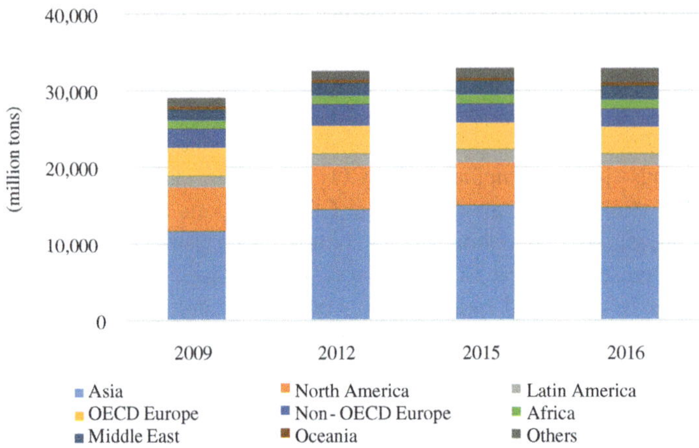

Fig. 16.1 Global trends in CO_2 emissions. (Source: EDMC 2012, 2015, 2018, 2019)

the quantitative analysis of CO_2 emissions induced across different areas. The results are discussed in Sects. 16.5 and 16.6 concludes.

16.2 Global Trends in CO_2 Emissions

Figure 16.1 shows regional trends in CO_2 emissions. Total global CO_2 emissions are approximately 30 billion tons. Asia was the highest CO_2 emitting region in 2008 and 2016, with 11.7 and 14.8 billion tons CO_2, respectively. Emissions in North America, European countries in the Organization for Economic Co-operation and Development (OECD), and non-OECD Europe have slightly decreased, though emissions from the United States are notable.[4]

CO_2 emissions from Asia accounted for more than 40% of global emissions from 2009 to 2016 (Fig. 16.2). CO_2 emissions in Asia have increased to a high level, reaching 45% of global emissions in 2016. The proportion of emissions from North America, OECD Europe, and non-OECD Europe slightly decreased from 2009 to 2016, to 16.5%, 10.6%, and 7.4%, respectively.

Figure 16.3 indicates trends in CO_2 emissions in countries and regions in Asia from 2009 to 2016. China has the highest emissions, emitting approximately 7.5 times as much CO_2 as Japan. Korea, Taiwan, the ASEAN,[5] and other countries in Asia[6] have increased their emissions; in particular, the sum of emissions from

[4] CO_2 emissions in the United States were approximately 4.8 billion tons in 2015 and 2016, accounting for 89% of North American emissions.

[5] Indonesia, Malaysia, Myanmar, the Philippines, Thailand, and Vietnam.

[6] "Other countries" refers to all countries in Asia except China, Japan, Korea, Taiwan, and the countries of the ASEAN.

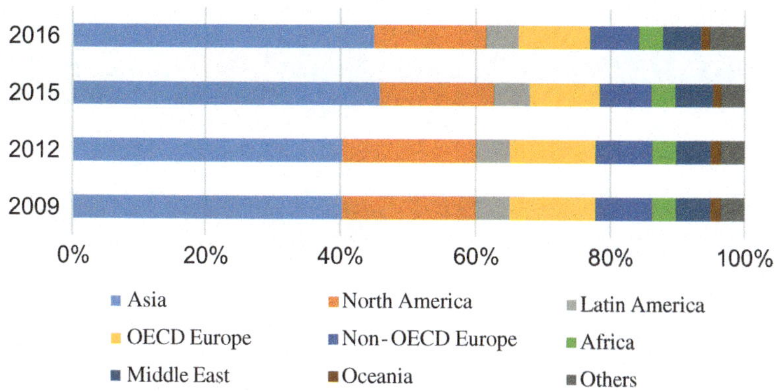

Fig. 16.2 Regional contributions to global CO_2 emissions. (Source: EDMC 2012, 2015, 2018, 2019)

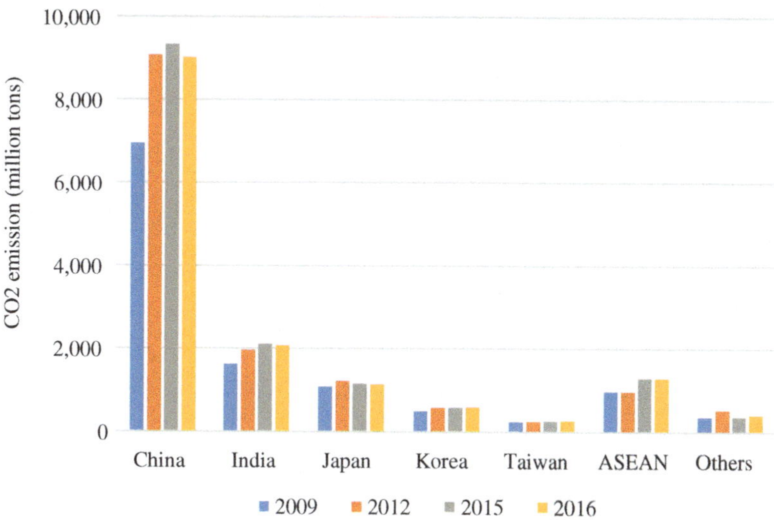

Fig. 16.3 Trends in CO_2 emissions in countries and regions of Asia (2009–2016). (Source: EDMC 2012, 2015, 2018, 2019)

ASEAN and other countries is greater than that of Japan. Figure 16.4 shows the ratio of CO_2 emissions by country and region in Asia. Within the Asian region, China accounts for approximately 60% of total emissions from 2009 to 2016, India for 14%, Japan for approximately 8.2%, Korea 4.1%, and Taiwan 1.8%. CO_2 emissions from Asia account for more than 40% of global emissions from 2009 to 2016. Over this 7 year period, CO_2 emissions from China increased from 23.9% of the global total to 27.9%.

To summarize, trends in overall CO_2 emissions have remained high, especially in Asia, mainly accounted for by China.

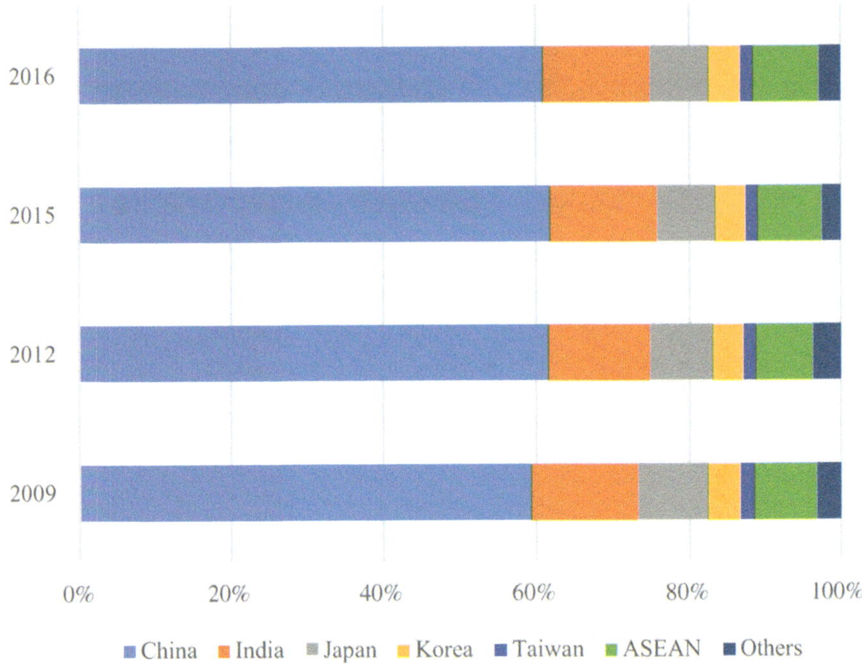

Fig. 16.4 Contribution of individual countries and regions to CO_2 emissions in Asia. (Source: EDMC 2012, 2015, 2018, 2019)

16.3 Methodology

16.3.1 World Input-Output Database

The World Input-Output Database (WIOD) was developed to analyze the effects of globalization on trade patterns, environmental pressures, and socioeconomic development across a wide set of countries. The database covers 27 countries of the European Union and 13 other major countries over the period from 1995 to 2009 (Timmer 2012, p. 3). The structure of a national input-output table (IOT) is shown in Fig. 16.5. An important component of the IOT is that total output by domestic industry is equal to the use of output from domestic industry, such that all flows in the economic system are accounted for (Timmer 2012, p. 4).

The world input-output table (WIOT) is an extension of the WIOD concept.[7] Each product is explicitly tracked in the WIOT. For example, for a country, flows of

[7]The full WIOT contains data for 40 countries from the WIOD, including the largest, in order to cover more than 85% of global GDP. Nevertheless, to complement the WIOT and make it suitable for our modeling purposes, we added a region called the rest of the world (RoW) that provides a proxy for all other countries.

	Industry	Final Use		Total
Industry	Intermediate Use	Domestic Final Use	Exports	Total Output
	Imports			
	Value Added			
	Total Output			

Fig. 16.5 Structure of national input-output table. (Source: Timmer 2012, p. 63)

products both for intermediate and final use are split into domestically produced and imported. The WIOT shows in which foreign industry a product was produced (Timmer 2012, pp. 4–5). This database has been combined with detailed data on production activity in each country and international trade to illustrate supply of 59 products and use of 35 industries. "Use" can refer to issues related to fragmentation, socioeconomic aspects (e.g., labor and the creation of added value), and environmental aspects (e.g., energy use, atmospheric emissions, and water consumption) (Dietzenbacaher et al. 2013, pp. 91–92).

In order to reflect globalization, an effective database for policymaking and analysis must be global, use time-series data to evaluate previous development, include several socioeconomic and environmental indicators, and base all data on a consistent framework (e.g., using the same industrial classification). In this respect, WIOD is useful for policy analysis.

The core of the environmental database consists of atmospheric emission accounts. Atmospheric emissions resulting in impact categories covered in the WIOD (global warming, acidification, and tropospheric ozone formation) originate from gases emitted in energy use processes. Inventory data from the UNFCCC and the Convention on Long-Range Transboundary Air Pollution (CLRTAP) are compiled using extended energy balances from the International Energy Agency (IEA) (Timmer 2012, p. 15). We use CO_2 emissions data from the WIOD 2009 release.

Increasing numbers of studies are using WIOD. For example, Boitier (2012) used input-output analysis to calculate consumption-based emissions of greenhouse gases (GHG) and applied this to environmental issues (Leontief 1970). The WIOD database was used to calculate a consumption-based inventory of national GHG emissions for 41 countries from 1995 to 2009, which were then compared with production-based GHG inventories. The structures of CO_2 emissions in the European Union (Yoshinaga 2013) and Germany (Yoshinaga 2004) have been analyzed using data from WIOD. The former study examines trends in domestic and industrial CO_2 emissions in all European Union countries, including Central European and Baltic countries that joined the European Union in 2005.

16.3.2 Analysis

First, CO_2 emissions were divided into production-based and consumption-based. The former were considered to be CO_2 emissions from production for which

Table 16.1 Regional classification in a World Input-Output Table (WIOT)

Euro-zone	Non-Euro EU	NAFTA	China	East Asia	BRIIAT	RoW
Austria	Bulgaria	Canada	China	Japan	Australia	
Belgium	The Czech Republic	Mexico		Korea	Brazil	
Cyprus	Denmark	United States (USA)		Taiwan	India	
Estonia	Hungary				Indonesia	
Finland	Latvia				Russia	
France	Lithuania				Turkey	
Germany	Poland					
Greek	Rumania					
Ireland	Sweden					
Italy	United Kingdom (UK)					
Luxembourg						
Malta						
Netherland						
Portugal						
Slovakia						
Slovenia						
Spain						

Source: Timmer (2012, p. 42)

manufacturing, or supply-side, countries were wholly responsible. The latter were considered as wholly the responsibility of the countries demanding the product. The production of transportation equipment in various sectors was selected as a case study. CO_2 emissions from 35 sectors in both production-based and consumption-based countries were considered across 7 areas (Euro-zone, other European countries (hereafter non-Euro zone), North American Free Trade Agreement area (NAFTA), China, East Asia, BRIIAT (Brazil, Russia, India, Indonesia, Australia, Turkey), and the rest of the world (RoW)). Table 16.1 shows the regional classification.

Production-based and consumption-based emissions are defined as follows. The former case indicates CO_2 emissions from production as the final demand, while the latter case shows CO_2 emissions from the production of transportation equipment for the country from which demand stems. Production-based emissions are calculated as the total sum of CO_2 emissions induced in each country. Consumption-based emissions estimate the final demand in each area and then calculate the CO_2 emissions induced by this demand. The emissions are calculated as follows:

$$\mathbf{Y} = \mathbf{C} \cdot \mathbf{X} \quad \text{(production-based)} \tag{16.1}$$

$$Y = \mathbf{C} \cdot (\mathbf{I} - \mathbf{A})^{-1} \cdot \mathbf{F} \quad \text{(consumption-based)} \tag{16.2}$$

where, **Y** is the estimated matrix of CO_2 emissions; **X** is the matrix of CO_2 emitted; **C** is a diagonal matrix containing the CO_2 emission coefficient in each area; **I** is a unit matrix; **A** is an input coefficient matrix consisting of 7 areas and 35 industries; and **F** is the final demand vector in each region.

16.4 Quantitative Analysis of Induced Regional CO_2 Emissions

Figures 16.6 and 16.7 show total production-based and consumption-based CO_2 emissions, respectively.

Table 16.2 shows production-based CO_2 emissions induced in each region by the production of transportation equipment sector. The highest CO_2 emissions in the Euro-zone were induced by demand from the Euro-zone itself (64,057 kt CO_2) followed by demand from RoW. CO_2 emissions from NAFTA induced by NAFTA are 108,061 kt CO_2, those induced by RoW are 6586 kt CO_2, and those induced by the Euro-zone are 5950 kt CO_2. Most CO_2 emissions in China were induced by its own demand (218,294 kt CO_2), followed by that of East Asia (150,867 kt CO_2), NAFTA (22,596 kt CO_2), RoW (22,314 kt CO_2), and the Euro-zone (17,115 kt CO_2).

Table 16.3 shows CO_2 emissions induced in each region in the production of transportation equipment sector based on consumption. For the Euro-zone, the highest CO_2 emissions (45,263 kt CO_2) are due to demand from the Euro-zone itself, followed by demand from RoW. CO_2 emissions from NAFTA are mostly due

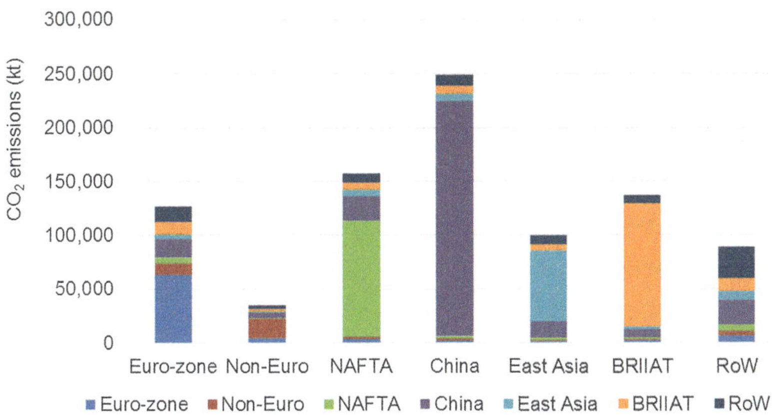

Fig. 16.6 Regionally induced production-based CO_2 emissions from sectors involved in production of transportation equipment

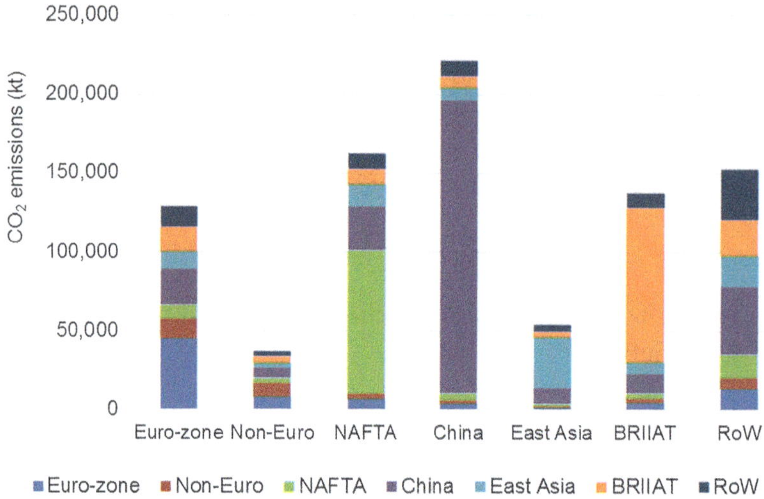

Fig. 16.7 Regionally induced consumption-based CO_2 emissions in sectors involved in production of transportation equipment

Table 16.2 CO_2 emission induced in an each region in a transport and equipment sector (production-based) (unit: kt-CO_2)

	Euro-zone	Non-Euro	NAFTA	China	East Asia	BRIIAT	RoW
Euro-zone	64,057	4930	3648	2358	1374	2080	6586
Non-Euro	9567	17,149	1874	1107	733	1044	3708
NAFTA	5950	1502	108,061	2976	2487	1980	6586
China	17,115	4487	22,596	218,294	15,087	7269	22,314
East Asia	4245	1158	5953	6614	65,626	2281	8298
BRIIAT	11,597	2826	6692	7639	6350	114,873	11,719
RoW	14,304	3029	8482	9720	7906	7422	29,985

Table 16.3 CO_2 emission induced in an each region in a transport and equipment sector (consumption-based) (unit: kt-CO_2)

	Euro-zone	Non-Euro	NAFTA	China	East Asia	BRIIAT	RoW
Euro-zone	45,263	8298	6973	4272	1557	4747	13,926
Non-EU	12,308	8844	2991	1608	633	2221	6607
NAFTA	8931	3147	91,621	4685	2123	3733	15,303
China	23,098	6618	27,154	185,541	9752	12,269	42,729
East Asia	10,545	2642	14,376	8163	32,037	7090	19,322
BRIIAT	16,068	4305	9211	7592	3595	98,105	22,819
RoW	12,908	3355	9820	9206	4288	9273	31,997

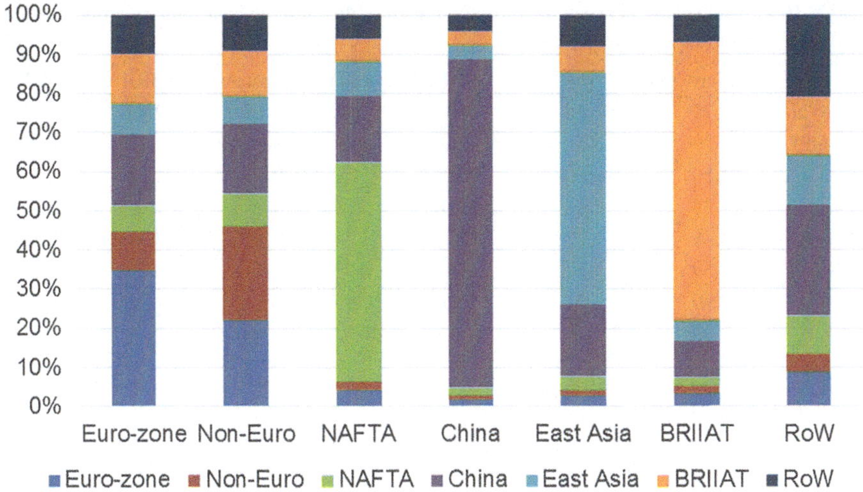

Fig. 16.8 Ratio of CO_2 emissions induced in each region in production of transportation equipment sector (consumption-based)

to demand from NAFTA (91,621 kt CO_2), followed by RoW (15,303 kt CO_2), and the Euro-zone (8931 kt CO_2).The highest CO_2 emissions induced in China are from demand in China itself (185,541 kt CO_2), followed by NAFTA (27,154 kt CO_2).

Figure 16.8 shows the proportion of consumption-based CO_2 emissions in the production of transportation equipment sector induced in each region. Demand from the Euro-zone accounts for 35.05% of Euro-zone emissions, and China, RoW, and the non-Euro zone account for 17.89%, 10%, and 9.53%, respectively. Demand from NAFTA accounts for 56.51% of its own emissions, while demand from China, East Asia, and RoW accounts for 16.75%, 8.87%, and 6.06%, respectively, of NAFTA emissions. China has the highest self-induced CO_2 emissions, accounting for 83.93% of its total emissions. Additional demand stems from RoW (4.16%), East Asia (3.69%), BRIIAT (3.43%), NAFTA (2.12%), and the Euro-zone (1.93%). CO_2 emissions in East Asia, including Japan, are mainly a result of East Asian demand (59.35%), while carbon flow to China accounts for 18.07%, and that to RoW, BRIIAT, NAFTA, and the Euro-zone accounts for 7.94%, 6.66%, 3.93%, and 2.88%, respectively. Most CO_2 emissions from RoW are induced by China (27.98%), followed by demand from RoW (20.95%), then BRIIAT, East Asia, NAFTA, and the Euro-zone and non-Euro zone (14.94%, 12.65%, 10.02%, 9.12%, and 4.33%, respectively).

Table 16.4 and Fig. 16.9 provide the ratio of consumption-based to production-based CO_2 emissions in each region. Many regions induce CO_2 emissions in East Asia, resulting in ratios of consumption-based to production-based CO_2 emissions greater than 1. For BRIIAT, this ratio is 3.11, those for the Euro-zone, NAFTA, RoW, and the non-Euro zone are all greater than 2, and that for China is 1.23. The ratio of CO_2 emissions induced in a region by its own demand tends to be below 1, except for RoW. The ratio of consumption to production-based emissions in

Table 16.4 Comparison between regions with regard to the ratio the consumption-based divided by the production-based CO_2 emission

	Euro-zone	Non-Euro	NAFTA	China	East Asia	BRIIAT	RoW
Euro-zone	0.71	1.68	1.91	1.81	1.13	2.28	2.11
Non-EU	1.29	0.52	1.60	1.45	0.86	2.13	1.78
NAFTA	1.50	2.09	0.85	1.57	0.85	1.89	2.32
China	1.35	1.48	1.20	0.85	0.65	1.69	1.91
East Asia	2.48	2.28	2.41	1.23	0.49	3.11	2.33
BRIIAT	1.39	1.52	1.38	0.99	0.57	0.85	1.95
RoW	0.90	1.11	1.16	0.95	0.54	1.25	1.07

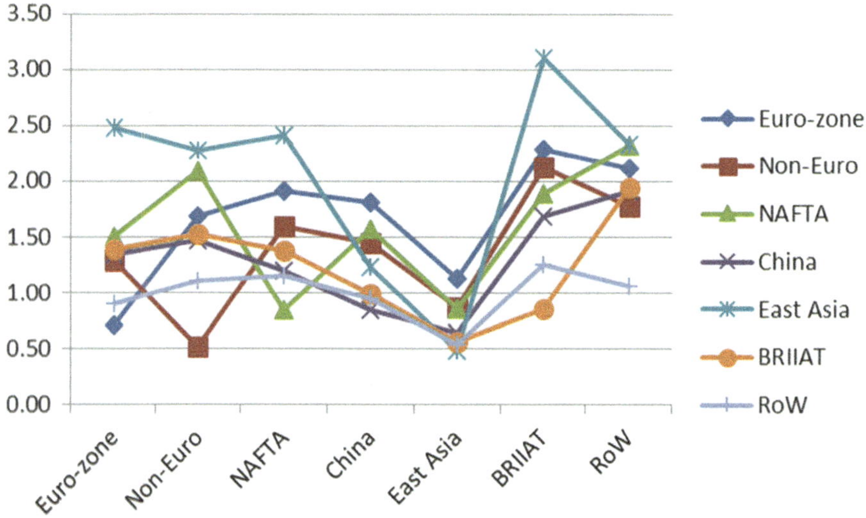

Fig. 16.9 Trends in the ratio of consumption-based CO_2 emissions divided by production-based CO_2 emissions

China is greater than 1 for RoW, BRIIAT, and both Euro and non-Euro zones. For emissions induced by China, ratios are greater than 1 for the Euro and non-Euro-zone, and East Asia, while those for BRIIAT, RoW, and China are below 1. The ratios of emissions induced from BRIIAT to RoW, the Euro and non-Euro zone, and NAFTA are all greater than 1, while those to East Asia and BRIIAT are less than 1. The ratio of emissions induced in the Euro-zone from the Euro-zone itself, the non-Euro zone, and East Asia are greater than 1, while those from BRIIAT, RoW, and China are less than 1. Emissions induced in BRIIAT have a ratio greater than 1 for all regions excluding BRIIAT itself. The ratio of emissions induced by the Euro-zone in BRIIAT and RoW is greater than 2, while emissions induced in the Euro-zone from itself and RoW have a ratio of less than 1. The ratio of emissions induced in NAFTA by RoW and the non-Euro zone are greater than 2, and those of the Euro-zone, China, and BRIIAT are greater than 1. Induction of emissions by

Table 16.5 Difference of CO_2 emission (unit: Mt-CO_2)

	Euro-zone	Non-Euro	NAFTA	China	East Asia	BRIIAT	RoW
Euro-zone	18.79	−3.37	−3.32	−1.91	−0.18	−2.67	−7.34
Non-Euro	−2.74	8.30	−1.12	−0.50	0.10	−1.18	−2.90
NAFTA	−2.98	−1.64	16.44	−1.71	0.36	−1.75	−8.72
China	−5.98	−2.13	−4.56	32.75	5.33	−5.00	−20.41
East Asia	−6.30	−1.48	−8.42	−1.55	33.59	−4.81	−11.02
BRIIAT	−4.47	−1.48	−2.52	0.05	2.76	16.77	−11.10
RoW	1.40	−0.33	−1.34	0.51	3.62	−1.85	−2.01

Note: Difference indicates (CO_2 emission (production-based)-CO_2 (consumption-based))

NAFTA is greater than 1 everywhere except in NAFTA itself. With regard to emissions induced by RoW (especially developing countries), ratios for BRIIAT, the non-Euro zone, and RoW are greater than 1 and are greater than 2 for East Asia, NAFTA, and the Euro-zone.

16.5 Discussion

International carbon transfer can be interpreted by examining the results of our analysis. For example, for transportation equipment, estimated production-based CO_2 emissions in China are 17,115 kt (17.1 Mt), whereas consumption-based emissions are 23,099 kt (23.1 Mt). The ratio of consumption-based[8] to production-based emissions is 0.85, so, approximately 15% of Chinese CO_2 emissions occur as a result of imports.

To investigate the transfer of CO_2 emissions, inflow and outflow can be calculated by subtracting consumption-based emissions from production-based emissions (Table 16.5). In the Euro and non-Euro zones, NAFTA, BRIIAT, and RoW, there is overseas inflow, while there is outflow to East Asia. Figure 16.10 shows carbon flow among the seven regions, taking the example of China and East Asia from Table 16.5, rather than showing overall flow among the seven areas. In many developing countries in RoW, CO_2 is emitted to China, NAFTA, and Asia through trade, while inflows come from the non-Euro zone, NAFTA, BRIIAT, and RoW itself.

About 90% of world direct emissions and responsibilities (consumer, producer, and total responsibility) result from the economic activity of Asia; developed economies have both consumer and producer responsibility higher than direct emissions (Rodrigues et al. 2010, p. 79). The empirical analysis in this chapter shows the complex structure of international carbon flow, where it is difficult to distinguish between the responsibilities of developed and developing countries.

[8]A negative value indicates CO_2 inflows from foreign countries and regions.

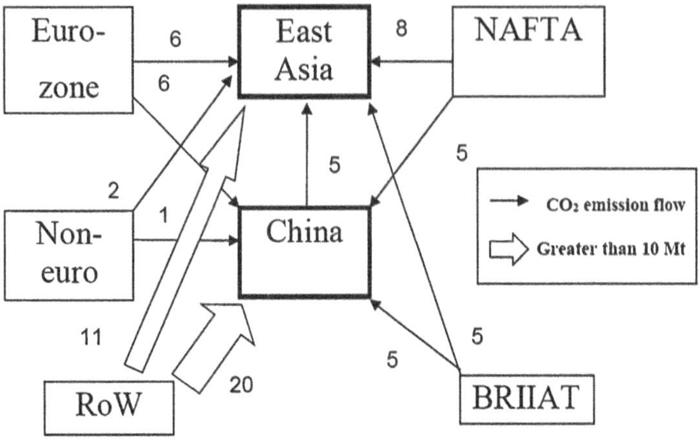

Fig. 16.10 Structure of international carbon transfer in China and East Asia (unit: Mt CO_2)

Nevertheless, the anthropogenic component of climate change has been attributed to developed countries, who should take the initiative to reduce CO_2 (Zhou and Qian 2012, p. 27). In the Framework Convention on Climate Change, the concept of common but differentiated responsibilities was introduced based on the fact that cumulative carbon emissions and per capita emissions vary between developed and developing countries (Zhou et al. 2015, p. 27).

The impact of the IPCC Report and advancement of scientific understanding of climate change has led to recognition that domestic CO_2 reduction policies are mutually beneficial and the development of low-carbon societies should be pursued in both developing and developed countries. In order to deal with differences in obligations for developed and developing countries, the Paris Agreement states that:

> Developed country Parties should continue taking the lead by undertaking economy-wide absolute emission reduction targets. Developing country Parties should continue enhancing their mitigation efforts.

The concept of concentric differentiation is introduced, which encourages a shift over time to economy-wide emission reductions targets in the context of differing national circumstances (MOFA 2015; Takamura 2016). Every country must implement domestic climate change policies and submit nationally determined contributions every 5 years. Realization of a low-carbon society is a priority in Asia, in particular in East Asia.

16.6 Conclusion

This chapter quantifies international carbon flow as a result of economic globalization based on input-output analysis using the World Input-Output Database (WIOD). Trends in CO_2 emissions induced from 7 regions and 35 industries were analyzed as a case study of the production of transportation equipment sector.

Total CO_2 emissions induced in China are larger than in the other six regions. Production-based emissions tend to be larger than consumption-based emissions in China and East Asia, while the inverse is true elsewhere. The ratio of production- to consumption-based CO_2 emissions induced by a region to itself is close to one. CO_2 emissions from manufacturing transportation equipment such as automobiles are larger than emissions from purchasing in this sector in East Asia. Developing countries tend to have higher consumption- than production-based emissions. Furthermore, consumption-based calculations tend to underestimate CO_2 emission in developing countries and emerging countries compared with production-based estimates. It is important to compile further analysis on global carbon flows in consumption-based cases. In considering burdens of responsibility for global CO_2 emissions, ratios must be taken into account, allowing identification of the hidden, larger amount of emissions in RoW (developing countries) and BRIIAT (emerging countries). High volumes of CO_2 emissions flow to RoW from the Euro-zone, NAFTA, RoW, East Asia and BRIIAT; CO_2 emissions must be reduced in both developed and developing countries.

The East Asian area emits more CO_2 from manufacturing transportation equipment than it does from purchasing in this sector, while in developing countries, CO_2 emissions in consumption are higher than those of production. This analysis highlights the importance of assessing trends in emissions from emerging and developing countries when considering global burdens of responsibility for CO_2 emissions.

References

Bastianoni S, Pulseli FM, Tiezzi E (2004) The problem of assigning responsibility for Greenhouse Gas emissions. Ecol Econ 49:253–257

Boitier B (2012) CO_2 emissions production-based accounting consumption: In-sights from the WIOD databases, version, 1–23

Davis SJ, Caldeira K (2010) Consumption-based accounting of CO_2 emissions. Sustain Sci 107 (12):5687–5692

Dietzenbacaher E, Los B, Timmer M (2013) The world Input-Output tables in the WIOD database. J Life Cycle Assess 9(2):91–96. [in Japanese]

EDMC (The Energy Data and Modelling Center. The Institute of Energy Economics) (2012) EDMC handbook of Japan & world energy & economic statistics. The Energy Conservation Center, Tokyo. [in Japanese]

EDMC (The Energy Data and Modelling Center. The Institute of Energy Economics) (2015) EDMC handbook of Japan & world energy & economic statistics. The Energy Conservation Center, Tokyo. [in Japanese]

EDMC (The Energy Data and Modelling Center. The Institute of Energy Economics) (2018) EDMC handbook of Japan & World energy & economic statistics. The Energy Conservation Center, Tokyo. [in Japanese]

EDMC (The Energy Data and Modelling Center. The Institute of Energy Economics) (2019) EDMC handbook of Japan & world energy & economic statistics. The Energy Conservation Center, Tokyo. [in Japanese]

Fujikawa K, Ban H (2016) An empirical study on the interdependency of energy con-sumption and CO2 emissions under the international input-output structure of the Asia-Pacific region. In: Lee S, Pollitt H, Seung-Joon P (eds) Low-carbon sustainable future in East Asia: improving energy systems, taxation and policy cooperation. Routledge, London, pp 270–287

Fujikawa K, Shimoda M, Watanabe T, Zuoyi Y (2015) Interdependence of CO_2 emissions in East Asia. In: Kiyoshi F (ed) Input-output analysis and computable general equilibrium analysis in China's economy. Horitsu-Bunkasha, Kyoto, pp 129–142. [in Japanese]

Guan D, Lin J, Davis SJ, Pan D, He K, Wang C, Wuebbles DJ, Streets DG, Zhang Q (2014) 'Reply to Lopez et al: Consumption-based accounting helps mitigate global air pollution' (PNAS letter)

Hoshino Y, Sugiyama T (2012) Experiment of future estimation of trade Embodied CO_2 emissions in developed countries. SERC Discussion Paper, SERC12002, 1–11. http://criepi.denken.or.jp/jp/serc/discussion/download/12002dp.pdf

Lee S, Pollitt H, Seung-Joon P (eds) (2016) Low-carbon sustainable future in East Asia: improving energy systems, taxation and policy cooperation. Routledge, London

Leontief W (1970) Environmental repercussions and the economic structure: an input-output approach. Rev Econ Stat 52(3):262–270

Lin J, Pan D, Davis SJ, Zhang Q, He K, Wang C, Streets DG, Wuebbles DJ, Guan D (2014) China's international trade and air pollution in the United States (PNAS Article)

MOFA (Ministry of Foreign Affair of Japan) (2015) Paris Agreement

Mózner ZV (2013) A consumption-based approach to carbon emission accounting-sectoral differences and environmental benefits. J Clean Prod 42:83–95

Na S (2006) Climate change and international cooperation: efficiency, equity, sustainability. Yuhikaku, Tokyo. [in Japanese]

Pang J, Alexandri E, Lin SM, Lee S (2016) Measuring both production-based and consumption-based CO2 emissions of different countries based on the multi-region input-output model. In: Lee S, Pollitt H, Park SJ (eds) Low-carbon sustainable future in East Asia: improving energy systems, taxation and policy cooperation. Routledge, London, pp 247–269

Peters GP, Hertwich EG (2008) CO2 embodied in international trade with implications for global climate policy. Environ Sci Technol 42(5):1401–1407

Peters GP, Davis SJ, Andrew R (2012) A synthesis of carbon in international trade. Biogeosciences 9:3247–3276

Rodrigues J, Marques A, Domingos T (2010) Carbon responsibility and embodied emissions theory and measurement. Routeledge studies in ecological economics. Routledge, London

Takamura Y (2016) Evolving international climate change regime: The Paris Agreement and its prospects and challenges. Environ Res Q 181:11–21. [in Japanese]

Timmer M (ed) (2012) The World Input-Output Database (WIOD): contents, sources and methods.

Ueta K (2010) Environmental governance for sustainable development in Asia. Res Environ Disrupt 39(4):2–8. [in Japanese]

Wyckoff AW, Roop JM (1994) The embodiment of carbon in imports of manufactured products: Implications for international agreements on greenhouse gas emissions. Energy Policy 22 (3):187–194

Yamazaki M, Su X, Sun F, Zhou W (2012) Estimating international car-bon flow using I-O model. In: 17th Annual meeting. The Society for Environmental Economics and Policy Studies (SEEPS), Tohoku University, Sendai. [in Japanese]

Yoshinaga K (2004) Factor decomposition analysis for reduction of CO_2 emission by Germany: input-output analysis of German gasification. Keizai Ronshu 54(3/4):557–580. [in Japanese]
Yoshinaga K (2013) An analysis of CO_2 emissions by EU27 using world input-output database. Keizai Ronshu 63(2):135–163. [in Japanese]
Zhou W, Qian X (2012) Study on the historical change China's climate change policies-from local to global environmental issues. Policy Sci 19(2):15–28. [in Japanese]
Zhou W, Qian X, Su X, Li F (2015) Strategic choices for low-carbon china and proportion of East Asia low-carbon community. J Policy Sci 9:25–49

Chapter 17
Global Recycling System for an East Asian Low-Carbon Society

Weisheng Zhou, Kyungah Cheon, and Xuepeng Qian

Abstract In order to realize the global sustainability of the East Asian Low-Carbon Community centered on China, Japan, and Korea, it is essential to review the modern civilization of mass production, mass consumption, and mass disposal and to realize a recycling-oriented society aimed at maximizing resource use and minimizing environmental impact. A recycling society is one in which economic, resource, and environmental policy are integrated, promoting a circular economy from the three aspects of small circulation (local recycling), middle circulation (glocal recycling), and general circulation (global recycling). Small and middle circulations are domestic recycling methods, whereas general circulation is an international recycling method. The global recycling system (GRS) includes these three types of recycling. Therefore, general circulation aims to realize a wide-area recycling society that transcends national borders, such as in the China–Japan–Korea (CJK) circular economy model. In June 2015, the Chinese government, in cooperation with Japan and South Korea, selected Dalian Zhuanghe City as a CJK circular economy model in China and a practical example of a GRS in the CJK partnership. It has encouraged economic cooperation in Northeast Asia, strengthening exchanges among China, Japan, and Korea and promoting the development of a regional circular economy. It will contribute to achieving the ultimate goal of sustainable development or minimizing the environmental load.

W. Zhou (✉)
College of Policy Science, Ritsumeikan University, Ibaraki, Osaka, Japan
e-mail: zhou@sps.ritsumei.ac.jp

K. Cheon
Research Center for Sustainability Science, Ritsumeikan University, Ibaraki, Osaka, Japan

X. Qian
College of Asia Pacific Studies, Ritsumeikan Asia Pacific University, Beppu, Oita, Japan
e-mail: qianxp@apu.ac.jp

W. Zhou et al. (eds.), *East Asian Low-Carbon Community*,
https://doi.org/10.1007/978-981-33-4339-9_17

17.1 Addressing Global Climate Change Through a Recycling-Oriented Society

A low-carbon society is desirable to both reduce CO_2 emissions and achieve sustainability through maximizing resource use and minimizing environmental load and emissions. Development of a recycling-based society is especially important as a measure to tackle global climate change.

The formation of a recycling-based society contributes to global climate change countermeasures through the principle of the 3Rs: reduce, reuse, and recycle. First, emissions of waste are reduced as much as possible, and then items are reused as much as possible before being recycled. One example of this is the collection of heat through burning waste to generate power, contributing to a reduction in greenhouse gases. Modern civilization has evolved with the aid of mass production, mass consumption, mass disposal, and mass pollution, all of which continue to increase. However, environmental limits are becoming evident on a global scale. In order for humankind to thrive, development must satisfy the needs of future generations while also satisfying those of the current generation. This can be achieved by consideration of the environmental waste purification function and the natural resource supply function. As an alternative to mass production, mass consumption, and mass disposal, a resource-saving and recycling-based social system should be constructed, in which recycling activities are incorporated into the market. It is important to foster a long-term perspective of the environment and resource conservation for future generations and provide incentives to promote industrial structures, lifestyles, and technological innovations that produce minimal waste. In particular, appropriate social regulation is considered essential for building a resource-saving and recycling-based society.

The ultimate goal of the East Asian Low-Carbon Community is global sustainability, and for that purpose it is essential to realize a recycling-oriented society aimed at maximizing resource use and minimizing environmental impact. To prevent environmental destruction by collecting mineral resources, recycling exhaustible resources, and recycling environmental pollutants, both local and global recycling can be used.

17.2 Concept of a Global Recycling System

17.2.1 Basic Concept of a Global Recycling System

The global recycling system is such that:

> waste trade for the purpose of recycling must have been established along with arterial flow (the arterial industry: production, distribution, and consumption). However, due to delays in the development of venous flow (disposal, resource recycling, and final disposal) in developed countries, there are problems such as illegal export/import and the risk of

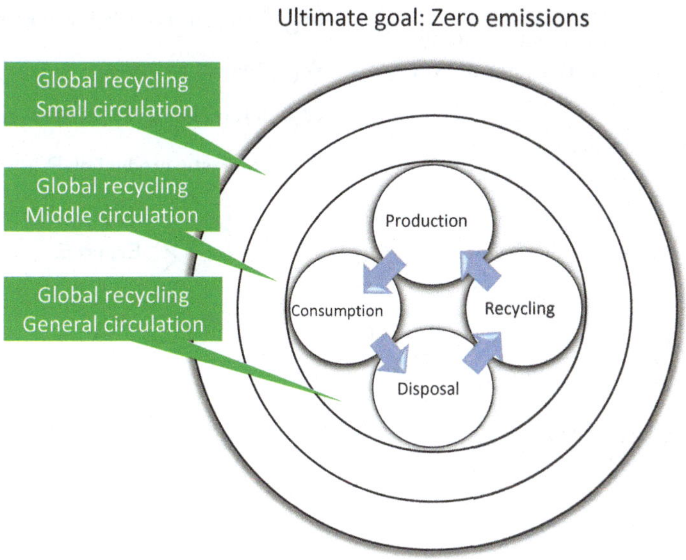

Fig. 17.1 Conceptual diagram A of global recycling system (spatial distribution)

> environmental pollution. By constructing an international resource recycling system and system operation rules, an international resource recycling social system that maximizes resource productivity and minimizes environmental impact can be realized. (Koizumi and Zhou 2005, Koizumi 2006)

A recycling society is one in which economic, resource, and environmental policies are integrated, promoting a circular economy from the three aspects of small cycle, middle cycle, and general circulation. The whole production consumption pattern, including clean production technology, material circulation between urban and rural areas, utilization of local resources, and utilization of natural energy, is considered, in addition to waste. This circulation system is shown in Fig. 17.1 (Koizumi and Zhou 2005).

Small circulation (local recycling) refers to efforts at a company or community level to minimize the generation of pollutants, using clean production to reduce the amount of substances and energy used in products and services. Recycling activities that recycle food waste generated by local governments into fertilizers and feeds for use in local governments or that reuse waste materials generated within a factory correspond to small circulation (local recycling).

Middle circulation (glocal recycling) refers to efforts at regional and national levels, promoting development of eco-industries in companies, industrial parks, and economic development zones. Byproducts or wastes from upstream production processes are used as raw materials for downstream production processes, thereby constructing an ecological industrial complex that forms an ecological industry chain of metabolic and symbiotic relationships between companies. The ultimate aim is to achieve zero emissions at the local level. In recent years, due to the increasing

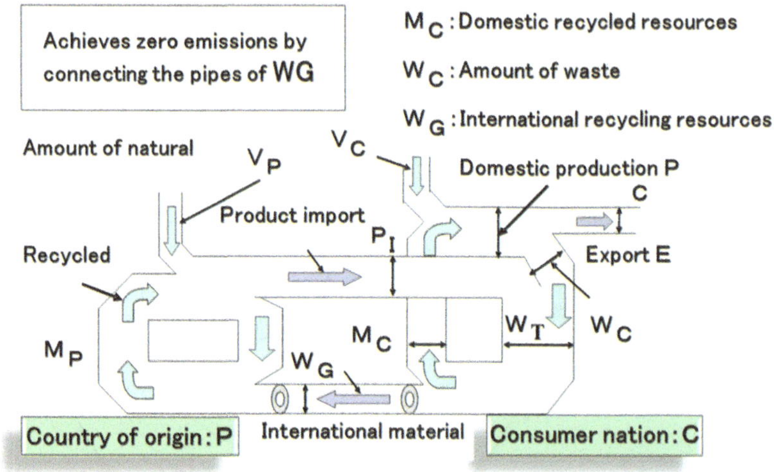

Fig. 17.2 Conceptual diagram B of global recycling system (material flow) (Koizumi and Zhou 2005)

sophistication of waste treatment facilities and rising construction costs, wide-area waste treatment has become popular. Similarly, wide-area recycling is also spreading, as seen in the Eco Town Plan (Ministry of the Environment 2020).

General circulation (global recycling) refers to social and international efforts in which green consumption is promoted and separate collection systems are established for waste. In addition, international industries for recycling and disposal of waste and discarded resources are established to maximize resource utilization and minimize environmental impact. Ultimately, general circulation aims to realize a wide-area recycling society that transcends national borders, such as in the China–Japan–Korea (CJK) circular economy model.

Small and middle circulations are domestic recycling methods, whereas general circulation is an international recycling method. The global recycling system (GRS) includes these three types of recycling (Koizumi and Zhou 2005, Koizumi 2006). As trade becomes increasingly globalized, it is important to build a GRS with international waste movement, in order to maximize resource utilization and to effectively reduce global emissions.

Figure 17.2 shows the concept of a GRS as a material flow diagram (Koizumi and Zhou 2005).

The global recycling system (GRS) is designed such that:

> waste trade for the purpose of recycling must have been established along with arterial flow (the arterial industry: production, distribution, and consumption). However, due to delay in the development of venous flow (disposal, resource recycling, and final disposal) in developed countries, there are problems such as illegal export/import and the risk of environmental pollution. By constructing an international resource recycling system and system operation rules, an international resource recycling social system that maximizes resource

productivity and minimizes environmental impact can be realized. (Koizumi and Zhou 2005, Koizumi 2006)

If ordered international trade is dubbed a global trade system, then ordered international resource waste trade can be called a global recycling system. However, global recycling is not effective for all materials. Effectiveness depends on economic conditions (resource demand, wages), technological progress, scarcity of resources, and changes in the surrounding environment, such as laws and systems. Even if ineffective at a given point, the system is dynamic, thus can become effective under favorable conditions.

17.2.2 Appropriate Judgment of a Global Recycling System (GRS)

In addition to the benefits of effective resource utilization, a GRS is characterized by environmental pollution from waste products that has unequal effects on the countries exporting and importing waste. Importing countries are not unilaterally damaged by environmental pollution and can enjoy the benefits of securing low-cost resources and expanding employment. It may also be advantageous from a global environment point of view, through, for example, a reduction in air pollution load due to avoidance of conversion from mining to materials. Nevertheless, the environmental benefits are externalized because they are not accounted for within the market. Importing countries risk introducing illegal waste and contamination. If technology for preventing environmental pollution is unavailable and environmental management is insufficient, adverse health effects also become a concern.

Exporting countries enjoy the benefits of reducing waste disposal costs, extending the lifetime of disposal plants, and avoiding pollution. However, if the venous industry is hollowed out and export waste is leaked via illegal routes, there are costs associated with rectifying the situation and a risk of bad publicity for environment-polluting exporters or export companies. The impact of the import and export of resource waste on society is classified into three components: economy, environment, and society. The costs, benefits, and sustainability of domestic and international recycling methods are summarized in Fig. 17.3.

1. Economy refers to the benefits and costs internalized in the market economy. In the recycling process and during waste disposal, the lower price of recycled resources is a benefit, and costs are those associated with recycling and landfill.
2. Environment is a factor that encompasses influences on noise, vibration, landscape, soil, water quality, and air quality (CO_2, SO_x, NO_x, etc.) that change individual utility levels.
3. Society includes the social impact of prolonged operation of the disposal site, including employment.

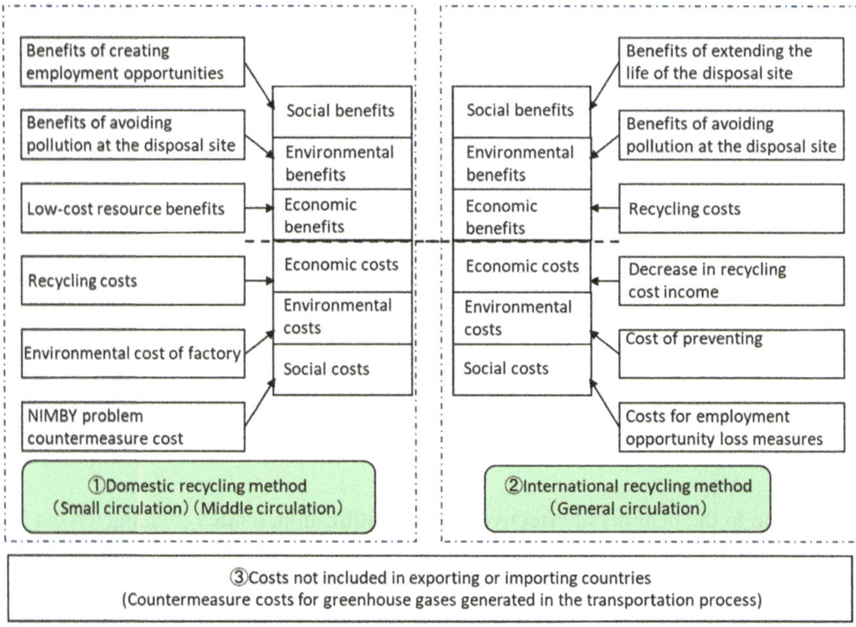

Fig. 17.3 Expected costs and benefits of domestic and international recycling methods (Suzuki et al. 2008)

In tandem with economic globalization and diversification of trade, industrial transfer from developed to developing countries has been promoted, contributing to international development. The flow of these products is called arterial flow. International arterial flow has steadily occurred based on rules from the World Trade Organization (WTO) and other bodies regarding trade, contributing to development of the global economy. When products created by resource-consuming manufacturing industries reach the end of their life and become waste, some are recycled, but residues are often landfilled and pollutants leech from landfill residues into the soil. As a result, environmental and health problems occur due to drainage of pollutants into rivers, overcrowded landfill sites, and limited exhaustible resources; countermeasures are essential. In order to solve these problems, the OECD has advocated an extended producer responsibility (EPR) policy, and OECD countries, such as Europe and Japan, have implemented various recycling laws. The GRS must evaluate whether waste should be treated in small, medium, or general cycles. In the general cycle, there are many advantages and disadvantages for both exporting and importing countries, and waste trade may or may not be appropriate.

In order to judge suitability, the environmental, economic, social, and legal interests of both countries must be considered and a stakeholder analysis conducted. A qualitative evaluation has been attempted considering waste electrical appliances and waste PET bottles between Japan and China (Koizumi et al. 2004). There are many advantages for both countries; if there are few disadvantages, it will be a proper GRS. If the opposite is true, it will be an inappropriate GRS. The impact of the GRS on global climate change must be quantitatively evaluated, and its wider environmental impact can be elucidated using life cycle assessment (LCA). A systematic GRS evaluation method has been established, using environmental impact evaluation methods such as stakeholder analysis, questionnaire analysis, simulation analysis by LCA, and resource productivity analysis (Koizumi et al. 2004). In order to prevent environmental problems caused by the international movement of resource waste without technology transfer to developing countries and to promote the formation of an international recycling-based society, specific policies are required. Under multilateral trade, the "expanded producer responsibility" policy, which targets traditional enterprises, has changed to the "extended producer responsibility" policy, which targets countries, especially where developed countries are trading partners with developing countries. Producers have a moral responsibility to support the construction of recycling-based social systems and also have social responsibility (CSR), which includes aspects such as technology transfer.

The requirements for establishing a GRS can be broadly summarized into three categories:

1. Improving resource productivity and controlling natural resource consumption (maximizing resources and energy efficiency).
2. Reducing the global environmental load (minimizing inputs and emissions).
3. Avoiding the risks of environmental pollution caused by trade and the collapse of recycling-based social systems; the safety of countries involved in trade can be ensured (maximizing benefit to society as a whole).

As globalization of trade progresses, it is important to establish a GRS with the ultimate goal of maximum resource utilization and effective zero global emissions. The Japan, China, and South Korea (CJK) circular economy model, which is one of general circulation, is introduced below as an example of GRS.

17.3 Circular Economy Model in Japan, China, and South Korea

17.3.1 Proposal for a Circular Economy Model for Japan, China, and South Korea

In order to realize a low-carbon community in East Asia, a new strategic, mutually beneficial cooperative framework for resource circulation and environmental load reduction must be established. With simultaneous globalization and regionalization, East Asia faces two challenges: resource constraints, which represent resource depletion, and environmental constraints, which include waste disposal sites reaching capacity, secondary and transboundary pollution, and global climate change. To overcome these challenges, energy efficiency must be improved and cross-border resource cycles developed. This will lower the environmental burden of society through the development and transfer of cutting-edge technologies across a wide area.

The second China–Japan–Korea Summit to maximize resource waste utilization and minimize environmental pollution took place amid rapid changes in the global economy and an increasing need to respond to climate change and calls for a green economy. In October 2009, the Joint Statement on Sustainable Development (CJK Summit 2009) was announced in Beijing. It declared that in the context of reduce, reuse, and recycle, the establishment of a CJK circular economy model would be explored, promoting joint efforts to achieve resource conservation and environmentally friendly industrial structures, growth patterns, and consumption patterns. The importance of the CJK circular economy model was further emphasized in the fourth (2011, Japan) and sixth (2015, South Korea) summits; expectations for project selection and model implementation were announced at the latter (CJK Summit 2015).

In June 2015, the Chinese government, in cooperation with Japan and South Korea, selected Dalian Zhuanghe City as a CJK circular economy model in China and a practical example of a GRS in the CJK partnership. It has encouraged economic cooperation in Northeast Asia, strengthening exchanges among China, Japan, and Korea and promoting the development of a regional circular economy. The creation of an industrial cluster of renewable resources at the East Asian level and the introduction of a wide-ranging circular economic zone across national borders, from small to middle to general circulation, will allow resource utilization to be maximized. It will contribute to achieving the ultimate goal of sustainable development or minimizing the environmental load.

The Home Appliance Recycling Law of Japan establishes a mechanism for addressing environmental pollution in the waste refrigerator supply chain at the shipment stage at a Japanese recycling factory, so a GRS would be expensive. After CFCs are collected from compressors and heat exchangers through a limited number of controllable recycling factories, the lubricating oil of each compressor is drained before the compressor is exported to China (Fig. 17.4).

17.3.2 Outline of the CJK Circular Economy Model Concept

The CJK circular economy model base is located in the Dalian Circular Industrial Economy Zone (Zhuanghe City, Dalian) in an area of 97.8 km^2 (Fig. 17.5). The economic zone is located on the eastern side of the Liaodong Peninsula, on the northern coast of the Yellow Sea. It is a national strategic priority park for the Liaoning Coastal Economic Zone. On the east side of the park, Zhuanghe Port is the closest port in the North Yellow Sea to Japan and South Korea and receives renewable resource imports. The Dalian National Eco-Industrial Model Park is located in the core area of the Dalian Recycling-Type Industrial Economic Zone.

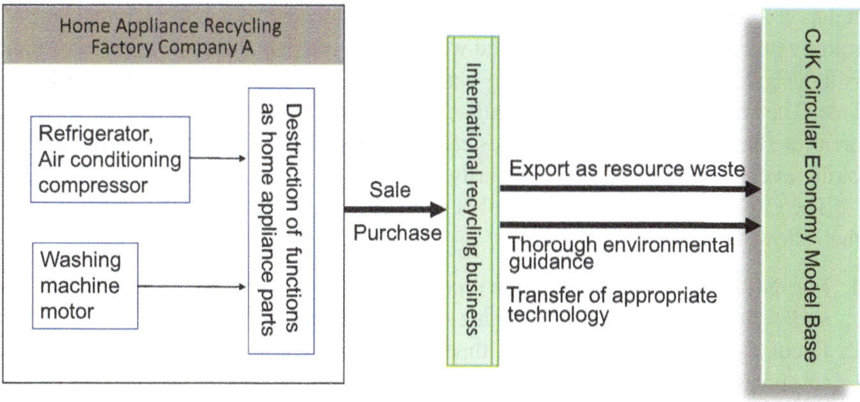

Fig. 17.4 Main flow of China–Japan–Korea (CJK) circular economy model

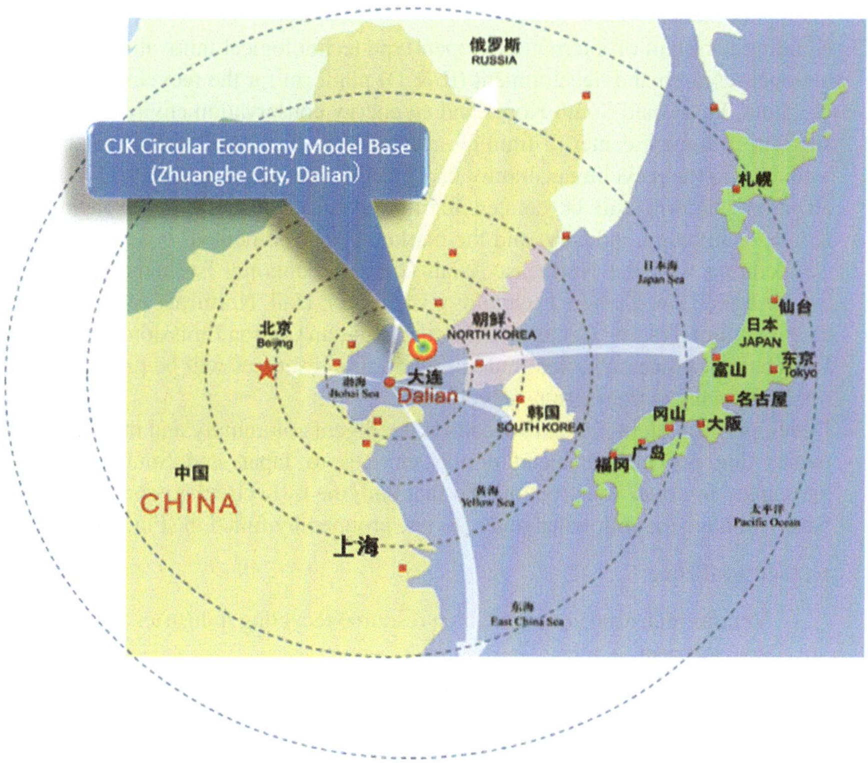

Fig. 17.5 Location of China–Japan–Korea (CJK) circular economy model base

It has a standard screen area of 12 km^2 and is located alongside China's urban mining model and a national imported waste park-type management model. The area is a designated park for dismantling waste electronic equipment in Liaoning Province. The business office, customs office, and inspection and quarantine supervision areas are already operational. World-leading technology from Japan will be used to build a venous labor safety and security system (Dalian City Government 2014).

The CJK circular economy model is a practical example of a GRS and promotes the following projects:

1. Forming cycles to circulate three renewable resources within parks in the Dalian and Tohoku areas, across Northeast Asia, and globally.
2. Focusing on development of three major industries—recycling renewable resources, energy-saving environmental protection facilities, and manufacturing industries—to achieve the organic combination of arterial and venous industries. Technologies and products from Japan and South Korea will enter the Chinese market, developing arterial and venous industries that mainly focus on high-end intelligent plants.
3. Building a system of international open-type technological innovation, an international research and development (R & D) platform for the recycling economy of China, Japan, and South Korea and an energy conservation environment. The site will become a center of training and technological development for human resources in the recycling economy in all three countries. A market and industrialization platform will be created for the advanced technology of Japan and Korea; technology, industry, and the market will be combined.
4. Cooperating with the Northeast Asian Circular Economy Exchange Center to establish the Low-Carbon Technology Exhibition Hall, Northeast Asian Renewable Resource Trading Center, and Northeast Asian Carbon Emissions Exchange. Distribution, processing, and trade of renewable resources will be promoted and offshore financial transactions developed.
5. Developing an international low-carbon intelligent community and inviting companies that provide advanced public services to Japan and South Korea to participate in urban construction. To that end, the CJK circular economy model base will have the following seven zones, grouped by function (Fig. 17.6).

    - Industrial zone:

        - Venous industrial zone (in which resource recycling industries are the main development target)
        - Arterial industrial zone (mainly related to emerging industries such as equipment manufacturing, ocean and wind power generation, new materials, and liquefied natural gas (LNG) comprehensive use)
        - Eco-recycling agricultural and fishery zone (organic farming, industrialized aquaculture, and multifunctional gathering, with a focus on developing fishing port services that integrate ports and towns)

    - Functional zone:

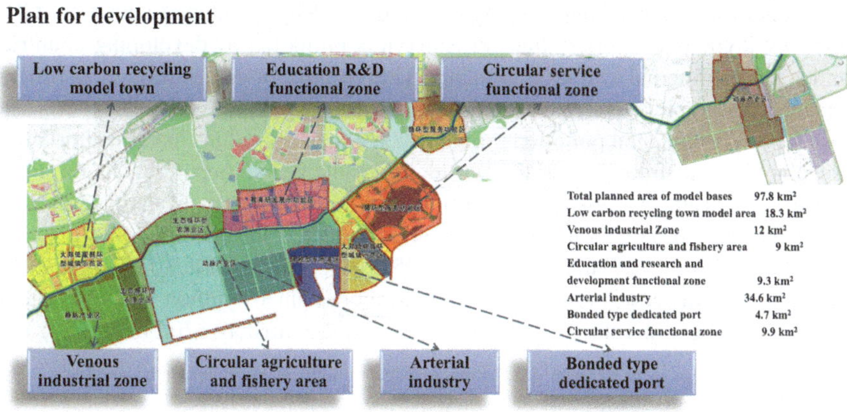

Fig. 17.6 Functional plan of China–Japan–Korea (CJK) circular economy model base

- Circular service functional area (developing industries such as business finance, cultural tourism, health, and medical care)
- Education and R & D functional zone (for exhibitions, research and development, and educational technology service industry)
- Low-carbon recycling model town (developing low-carbon technology and smart, green, sustainable new towns)

• Port zone: Bonded-type dedicated port (dedicated to modernization and smartness, centering on imported recycled resources and pelagic fisheries)

17.4 Challenges of and Prospects for Global Recycling Systems

China requires resources for both export and domestic demand. It has been importing scraps of iron, copper, and aluminum and waste plastics and paper at an annual rate of 30% since 1998. Imported resource waste uses low-wage, abundant local migrant workers and attains a higher recycling rate than can be achieved in developed countries, through detailed manual disassembly and manual selection. This makes an unexpected contribution in terms of maximum utilization; resources are secured, and employment is provided for migrant workers in rural areas. For this reason, the Chinese government has encouraged the import of resource waste. However, the residue from dismantled electronic waste is illegally dumped into rivers in rural areas, and cathode ray tubes containing lead are abandoned. The poor quality of the working environment also has implications for the health of workers.

In contrast to the global trade system promoted by developed countries, GRS does not involve technology transfer, since it is promoted by developing countries under strong internal resource demand. The volume of trade through unofficial channels has rapidly increased, and negative aspects such as illegal exports and resultant environmental pollution problems have surfaced. China has been importing recyclable waste from across the world for many years. According to the latest Chinese government statistics, China bought 49.6 million tons of garbage in 2015 alone. The European Union exports half of its collected and classified plastic waste, 85% of which goes to China. Ireland sent 95% of its plastic waste to China in 2016. In the same year, the United States exported more than 16 million tons of waste to China, equivalent in value to 5.2 billion USD (approximately 570 billion CNY):

> There is a lot of dirty or even dangerous garbage in solid waste that can be used as raw materials,

China's Environment Ministry said in a notification to the WTO. However, from January 2018, imports of some of the 24 types of solid waste were banned (State Council 2017). Imports of waste products have thus been falling, also as a result of slowing economic growth, declining resource demand, and falling resource prices, although illegal imports and secondary pollution are the most serious causes. In future, imports and exports of resource waste are expected to resume due to economic growth and resource depletion. However, if waste from recycled resources that are no longer in demand is incinerated or dumped in landfill with other waste, it could lead to an environmental catastrophe. This phenomenon is a challenge for a GRS. In order to overcome this, appropriate treatment technologies must be applied to the import and export of resource waste, to ensure thorough environmental guidance and compliance with the Basel Convention, and to reduce the amount and hazardous nature of resource waste. A mechanism must be established to eliminate the risk of environmental pollution across all supply chains, including physical properties, shipping conditions, land transportation, port storage, port handling, maritime transportation, and recycling plants of exported circulated resources. The CJK circular economy model will play an important role in this. In environmental cooperation in the East Asian region, regional facilities that can serve as bases should be established, where resources can be consolidated, allowing eventual expansion. Such a process can contribute to achievement of the United Nations Sustainable Development Goals.

References

2nd Japan, China and Korea Summit (2009) Joint Statement on Sustainable Development among the People's Republic of China, Japan and the Republic of Korea, Beijing, China,10 Oct 2009. https://www.mofa.go.jp/region/asia-paci/jck/meet0910/joint-2.pdf. Accessed 5 June 2020

6th Japan, China and Korea Summit (2015) Joint declaration for peace and cooperation in Northeast Asia, Seoul, Republic of Korea. https://www.mofa.go.jp/mofaj/a_o/rp/page3_001450.html. Accessed 5 June 2020

Dalian City Government (2014) China-Japan-Korea circular economy model base-basic concept and implementation plan

Koizumi K (2006) A study on the construction of global recycling system for resource waste. Doctoral dissertation, Ritsumeikan University (in Japanese)

Koizumi K, Zhou W (2005) Evaluation of global waste recycling system by AHP (analytic hierarchy process). J Policy Sci 12(2):23–30. (in Japanese)

Koizumi K, Zhou W, Obata N (2004) Global recycling system for waste -A case study of a recycling system for discarded home appliances in Asia. J Policy Sci 11(1):43–49. (in Japanese)

Ministry of the Environment (2020). http://www.env.go.jp/recycle/ecotown/. Accessed 5 June 2020

Suzuki Y, Koizumi K, Zhou W (2008) A study on export resistance of resource waste for recycling--evaluation of recycling behavior and psychological impact by selection experiment. Environ Technol Soc Environ Conserv Eng 37:340–346. (in Japanese)

Chapter 18
Building a Recycling-Oriented Society Through Collaboration Between Urban and Rural Areas: Sustainable Domestic Waste Treatment "Pujiang Model"

Kyungah Cheon, Chong Zhang, Xuepeng Qian, and Weisheng Zhou

Abstract With the rapid development of Chinese rural economy, the consumption has also achieved an incredible increase in rural areas. As a result, the waste problem has become a topic of concern for all. However, the existing approach to waste disposal remains problems including a high transfer cost and underutilized resources. Therefore, we focused on and analyze "the household waste disposal system model in Pujiang," which is an approach to dispose waste in a rural area and has achieved certain results. The result shows this approach is a usage of new public-private partnership (PPP) model which solved the disadvantage of the conventional PPP model. The new PPP model means the government purchases public service from the enterprise and could built a partnership between public and private involving more participation of local people. In this research, we evaluate the cooperation among the administration, the enterprises, and residents in Pujiang model from the perspectives of economy, environment, and society. Depended on the evaluation, we summarize the features of the model as following and make recommendations for promoting the model to other areas in China. The features of Pujiang model could be concluded in following six aspects: (1) waste reduction and recycling (organic fertilizer mainly); (2) waste collection and processing cost reduction; (3) soil restoration and improvement of agricultural production environment; (4) areas such as new tourist spots and places to visit economic promotion; (5) improvement of

K. Cheon
Research Center for Sustainability Science, Ritsumeikan University, Ibaraki, Osaka, Japan

C. Zhang
Zhejiang Gabriel Biotechnology Co., Ltd., Zhejiang, China

X. Qian
College of Asia Pacific Studies, Ritsumeikan Asia Pacific University, Beppu, Oita, Japan
e-mail: qianxp@apu.ac.jp

W. Zhou (✉)
College of Policy Science, Ritsumeikan University, Ibaraki, Osaka, Japan
e-mail: zhou@sps.ritsumei.ac.jp

W. Zhou et al. (eds.), *East Asian Low-Carbon Community*,
https://doi.org/10.1007/978-981-33-4339-9_18

environmental awareness of residents and (6) achievements such as presentation of new waste disposal system of public-private partnership.

18.1 Introduction

As described in Chap. 17, in order to realize a recycling-oriented society covering a large area, circulation must be divided into medium circulation and global circulation. In this chapter, we analyze the Chinese Pujiang model of rural and urban mixed-type household waste disposal through public-private partnerships and discuss its role in the establishment of an East Asian Community.

Currently, the Chinese government aims to move toward a "recycling-oriented economic society" that utilizes regional characteristics to properly dispose of household waste according to national policies on improvement of living standards. The government is paying particular attention to Pujiang County as an example of a model of a household waste disposal system on the rural-urban fringe. The system has been in operation since 2014 and is being considered for application to other similar areas.

This chapter summarizes household waste issues currently faced in rural China, with respect to the desire for a sustainable recycling-oriented society. Based on interviews and field research in Pujiang County, we analyze the sustainability of the model from the perspectives of economy, the environment, and society. Although this solution to waste is unique to rural areas, we explore applicability of solutions to waste problems in other areas in China.

18.2 Household Waste Disposal in Rural China

One of the major characteristics of the household waste disposal problem in China is its severity in rural-urban fringe areas. According to the Sixth National Population Census (2010), 70.86% of the population in China is in rural areas. The economy in rural areas has rapidly developed due to the cooperative promotion of industrialization, information technology, urbanization, and modernization of agriculture. With this rapid economic development, waste disposal in rural areas has become an urgent problem. The total amount of household waste generated in China is now the largest globally, and the amount of waste per capita is rapidly increasing.

The amount of urban household waste collected and transported in 2017 exceeded 150 kg per capita, which is approximately one-third of the average per capita waste from 36 OECD countries (household waste and bulky waste collected by local governments) and approximately half of Japan's daily waste per capita. The volume of waste is expected to continue increasing due to population growth and economic development. Each region of rural China now actively promotes management of household waste, through the uptake of a waste treatment system called "collection in village, transfer from town, disposal in prefecture." However, there are three major problems with this household waste disposal system.

First, all the waste discharged in an area is processed at the prefecture's landfill or incinerator. There is a shortage of final disposal sites, creating a heavy burden on prefectures, which must construct new processing facilities with large processing capacity. For example, the household waste collection and disposal system in Guangxi Zhuang Autonomous Region covers more than 90% of villages, but the amount of waste collected has already exceeded the capacity of the landfill disposal site in the prefecture. By 2020, 138 new incineration sites will be constructed. In Qian'an County, Hebei Province, nine landfill sites were constructed in 2009, but four were closed in 2017 after their disposal capacity was reached. Opposition to waste incineration facilities is common, due to problems such as air pollution and odors emanating from the facilities.

Second, such disposal systems place a heavy financial burden on municipalities in rural areas, which are not economically wealthy, due to high transportation costs. In Tangshan City, Hebei Province, the total area of the city is 1439 km^2, the rural population is approximately 200,000 (population density is low), and the financial income in 2016 was only 1.01 billion CNY. Due to lack of funds, the salaries of waste collection workers are low, and the operation and maintenance status of facilities, cleaning, and transportation are insufficient.

Third, in some areas, non-standardized landfill disposal and incineration facilities are in operation, causing environmental pollution. In order to solve the lack of funds for the disposal of household waste in rural areas, small-scale waste incineration facilities and simple landfills are used to dispose of household waste. However, a lack of environmental technology related to waste disposal, improper management and operation, and inappropriate supervision systems for waste disposal are causing pollution.

The processing system of "collection in village, transfer from town, disposal in prefecture" does not exploit the recycling value of household waste in rural areas nor the socio-economic characteristics of such areas. The current situation is that the waste is simply collected in one place and then discharged or landfilled, which does not contribute to waste reduction and increased recycling.

18.3 Pujiang Rural Household Waste Disposal System

Pujiang County is in the southern part of China and belongs to Jinhua City in Zhejiang Province. It has an area of 920 km^2, a population of approximately 390,000 people (490,000 including the floating population). In Pujiang, there are seven towns, five villages, and three subdistricts. It is divided into 409 administrative villages and 20 communities (housing complexes). In 2016, nominal GDP reached 21.1 billion CNY, and nominal GDP per capita was approximately 50,000 CNY (~8300 USD). The urbanization rate was 60%, which is the average level in China; Pujiang is representative of rural-urban fringe areas.

There are three main reasons for taking the Pujiang Prefectural (County) Waste Disposal Model as representative of the rural-urban fringe. Pujiang County is located in Zhejiang Province in the eastern coastal area and has a nationally average rate of urbanization. It is in the rural-urban fringe in China with a high raw garbage ratio and relatively high levels of economic development. The waste separation compliance rate is approximately 90%, and the waste reduction rate is 50%, representing a great achievement. Thirdly, a public-private partnership (PPP) model has been utilized to create a waste collection and disposal system centered on naturalization of waste in cooperation with the government and private companies. The potential of this policy to contribute to proper disposal of household waste in rural China from the perspectives of the economy, environment, society, and sustainability is analyzed.

18.3.1 Household Waste Disposal in Rural Areas of Pujiang

In Pujiang, the per capita GDP and the amount of cleaning and transportation of household waste per capita are increasing significantly. The Pujiang government noticed the rapid increase in the amount of waste generated and officially incorporated waste separation into government projects in 2010 by announcing the Pujiang Municipal Public Waste Collection and Comprehensive Use Plan.

In 2014, in order to reduce waste disposal costs and utilize private expertise, waste separation collection and household waste disposal services were purchased from the Zhejiang Gabriel Biotechnology Co., Ltd., and a system was introduced for separating and recycling waste. By 2017, 16 waste disposal centers (composts) had been constructed in all prefectures, and 18 natural composting facilities were created. Approximately 200 tons of waste was disposed of every day. Although the amount of cleaning and transportation of waste in Pujiang had remained constant until 2014, this new model of waste disposal has greatly improved the environment of the town. In the same year, Pujiang was selected by the Ministry of Housing and Urban Rural Development of the People's Republic of China (MOHURD) as "the first model prefecture of China's rural waste separation and resource utilization."

18.3.2 Characteristics of the Pujiang Model

18.3.2.1 Separation and Disposal of Household Waste in Rural Pujiang

The Pujiang government conducted a pilot project and fully disseminated the results to the prefecture. After that, Communist Party members oversaw instruction and supervision of waste separation, setting goals to maximize rates of separation, collection, and recycling. Waste is cleaned daily. Each trash bin has a resident number and contact number for the party member in charge of its supervision. If waste is not properly segregated, there is a mechanism to follow up with the party

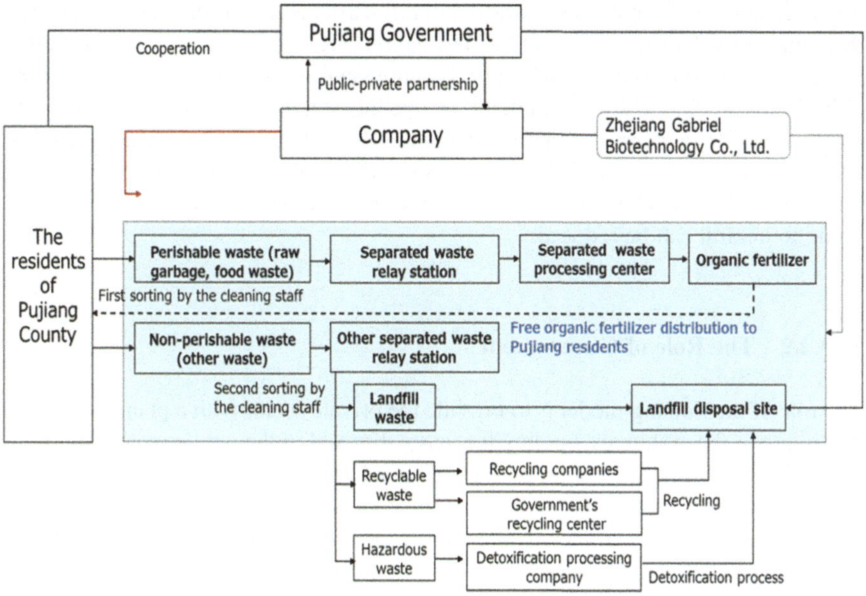

Fig. 18.1 Pujiang Prefecture household waste collection processing flow

responsible for dumping and responsible for supervision by tracking by number (with the "party-plus" system). With this system at its center, in collaboration with a private company, a model for waste separation, collection, and processing centered on the recycling of food waste was constructed (Fig. 18.1). This model takes advantage of the fact that the ratio of food waste to total waste is high in household waste in rural areas.

In urban areas in China and in other developed countries, household waste is generally separated into four types: harmful, perishable, recyclable, and other waste. It is difficult to directly introduce waste separation into rural areas where there is no prior experience of it. Thus, the system was implemented by considering simple and easy-to-understand ways of separating and collecting waste, called the "four kinds of waste/two kinds of waste" concept in Pujiang. The four types are (1) perishable waste (raw garbage, food waste), (2) nonperishable waste (other waste) (3) salable waste (recyclable waste), and (4) hazardous waste (landfill waste). The first sorting is performed by residents and the second by waste cleaning, collection, and transportation staff. First, residents sort household waste into raw garbage and other waste, and then the cleaning staff separates the waste further, checking whether collected waste contains other waste and reporting the rate of waste separation compliance to the village management committee. After that, the separated perishable waste is transported to a waste relay station, collected, and transferred to a waste-processing center where it is made into organic fertilizer. The organic fertilizer produced is provided free of charge to farmers in Pujiang to contribute to soil restoration. Other waste is carried to an ordinary waste relay station where it is divided into three types by the cleaning staff: recyclable waste, landfill waste, and harmful waste, which are then transported to their respective destinations. Recyclables are sold to market recycling companies; where there is no commercial interest due to high recycling costs; items are sent to the government's recycling center. Hazardous waste is sent to a detoxification processing company, and landfill waste is sent to a landfill disposal site. Consequently, waste that can be recycled is collected as much as possible through the waste separation and collection process, and the amount of garbage going to landfill can be reduced.

18.3.2.2 The Role of Government

The role of the Pujiang model is to provide the private sector with a plan to establish and improve the system as a policy for waste disposal in the public works sector, to deal with the proposed public-private partnership, and to pay the full operating expenses to the private sector, which is contracted for waste disposal to be implemented as a PPP project. The government plans to raise the waste separation compliance rate through cooperation between Communist Party members (hereinafter party members) and residents of Pujiang County under the waste separation management responsibility system (a system with Chinese characteristics). The government further increases the waste separation compliance rate through waste separation guidance, public relations, and education. The government aims to improve reliability and transparency through disclosure of information (via Pujiang Prefecture Statistical Yearbook) on the above process.

18.3.2.3 The Role of Companies

As a public service to the government, the company separates, collects, and disposes of household waste discharged by residents of Pujiang County. Private companies recycle the waste generated by residents and transport it to final disposal sites. Planning and proposing a waste treatment system suitable for Pujiang County, developing related technologies, and applying a business model are suitable for management by private companies.

Using the experience and data gained through trial and error in design of the waste disposal service, the government will propose new policies on waste disposal and the use of related technologies. The company reports data on waste disposal management and discloses information (through its homepage) to build trust between the government and residents.

18.3.2.4 The Role of Residents

The residents of Pujiang County separate and dispose of household waste under the guidance of party members. Farmers use organic fertilizers created from recycled waste to grow organic products and consume agricultural products.

18.4 Evaluating the Sustainability of the Pujiang Model

The method of collecting waste has changed from conventional waste disposal (before separation) to a new waste disposal format (after separation). Compared with the cost of mixed collection and separation processing in the conventional system, separated collection and resource recovery processing can significantly reduce costs including those of waste collection and transportation and labor. According to data from the Pujiang government, daily waste emissions were about 514 tons in 2016. Considering the change in the cost of collecting 514 tons of waste before and after introduction of the new system, the collection of resource waste, and the value of the organic fertilizer produced, the economic benefits of waste separation (currently all organic fertilizers are distributed free of charge to residents of Pujiang County) can be estimated at approximately 39 million CNY per year.

18.4.1 Reduction in Waste Collection Processing Costs

The daily cost of waste disposal in this area is approximately 10% lower than the costs of mixed collection and mixed processing in the conventional system. The main reasons for the reduction in costs after separation are reduction in collection cost, reduction in transfer cost, and reduction in disposal cost. Although there is a cost to separating waste, this is less than the reduction in other associated costs. In 2016, the daily cost of waste disposal was 26,400 CNY (Fig. 18.2), equivalent to an annual reduction in costs of approximately 9.636 million CNY.

18.4.1.1 Collecting and Processing Costs Before Waste Separation

Before waste separation began in 2014, there was one garbage bin for every eight households in the town, and cleaners collected waste from each garbage bin and public places and carried it to the waste relay station. After that, it was transported to "town" by a waste transport vehicle, transferred from "town" to "prefecture," and was disposed of at the "prefecture" landfill site. In this centralized collection process, all waste was collected and disposed of at the landfill. We calculated the processing costs had this centralized system been in place in 2016 to be 174,000 CNY for

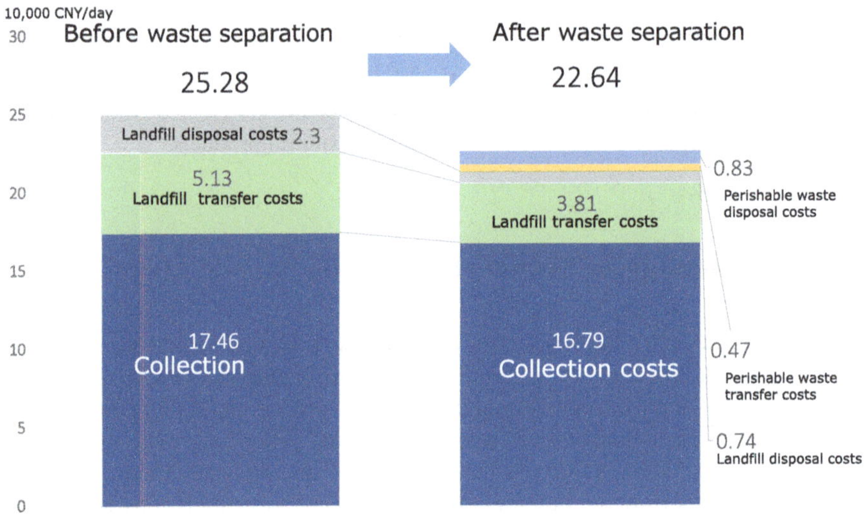

Fig. 18.2 Daily processing costs before and after introduction of waste separation processing in Pujiang County

collecting from the village, 513,000 CNY for transferring to the prefecture and 23,000 CNY for landfilling at the treatment plant. Therefore, total daily waste collection and treatment costs would have been 252,800 CNY.

18.4.1.2 Reduction in Collection Costs

Regarding waste collection costs, the standardization of waste disposal by the separation treatment has significantly reduced illegal dumping of waste in public facilities, rivers, and roads in Pujiang, thus reducing the workload and increasing the efficiency of cleaning staff. Although the number of households is increasing, the number of cleaners has been reduced from 150 cleaners per 10,000 households to 146 per 10,000 households, resulting in a reduction in waste collection cost from 174,000 CNY to 167,900 CNY.

18.4.1.3 Reduction in Transportation Costs Through Decentralization

Following implementation of the new system, 514 tons of waste are transported daily to the waste relay station after household waste has been separated and discharged in Pujiang (approximately 158 tons of raw garbage and 356 tons of other waste separated by cleaning staff). Approximately 1 ton of recyclable waste is collected, and 355 tons of remaining waste is transported. The volume of hazardous waste is negligible, so is excluded from the cost calculation.

Before introduction of waste separation, discharged household waste was transported to the single landfill in Pujiang Prefecture and disposed of through "dispersed" collection. In terms of processing, in keeping with the concept of local production for local consumption, waste that can be recycled into resources is instead preferentially transported to a nearby waste disposal center for processing, and only the remaining waste is transported to the landfill. According to 2016 data, the amount of waste transported to the waste disposal plant was reduced from 514 tons to 355 tons, and the transferring cost reduced from 55,100 to 38,100 CNY.

18.4.1.4 Reduction in Landfill Disposal Costs

If the proportion of raw garbage in total waste is large, the amount of organic substances that decompose in the filtrate will increase, and the processing cost of the filtrate will increase. Conversely, if raw garbage is removed before landfill disposal, processing costs decrease. According to data from the Pujiang Prefectural Health Bureau, the disposal cost after separation is about 7040 CNY.

18.4.1.5 Increased Cost of Waste Collection Processing

Raw waste is transported to a nearby processing center where it is recycled into organic fertilizer by a waste-processing machine. The cost of transporting 158 tons of waste a day is approximately 4700 CNY, and the cost of producing organic fertilizer is approximately 8300 CNY. In total, the transfer processing cost of raw garbage will be approximately 13,000 CNY.

As described above, even if the predicted costs for transferring and processing raw garbage are incurred, the overall waste disposal costs are reduced through reducing waste collection costs, landfill waste transfer costs, and landfill waste costs.

18.4.2 Economic Effects of Recycling

As shown in the flow chart for processing waste in Pujiang Prefecture (Fig. 18.3), raw (rotting) garbage is fermented and decomposed in a short time using a microbial fungus agent in waste mechanical fermentation equipment. In 10 days, approximately 511 kg of organic fertilizer can be produced from 2 tons of waste. Currently, this is provided free of charge to farmers in Pujiang County in order to incentivize them to separate their waste, but in the future it will be sold. According to a report from August 2018 (Pujiang County, Zhejiang), approximately 200 tons of raw garbage was processed per day in 2017, producing organic fertilizer worth three million CNY by the end of the year. Recyclable waste, which was originally disposed of by villagers under the two-time sorting system and mixed with general waste, is now sorted and collected by cleaning staff. Approximately 1 ton of

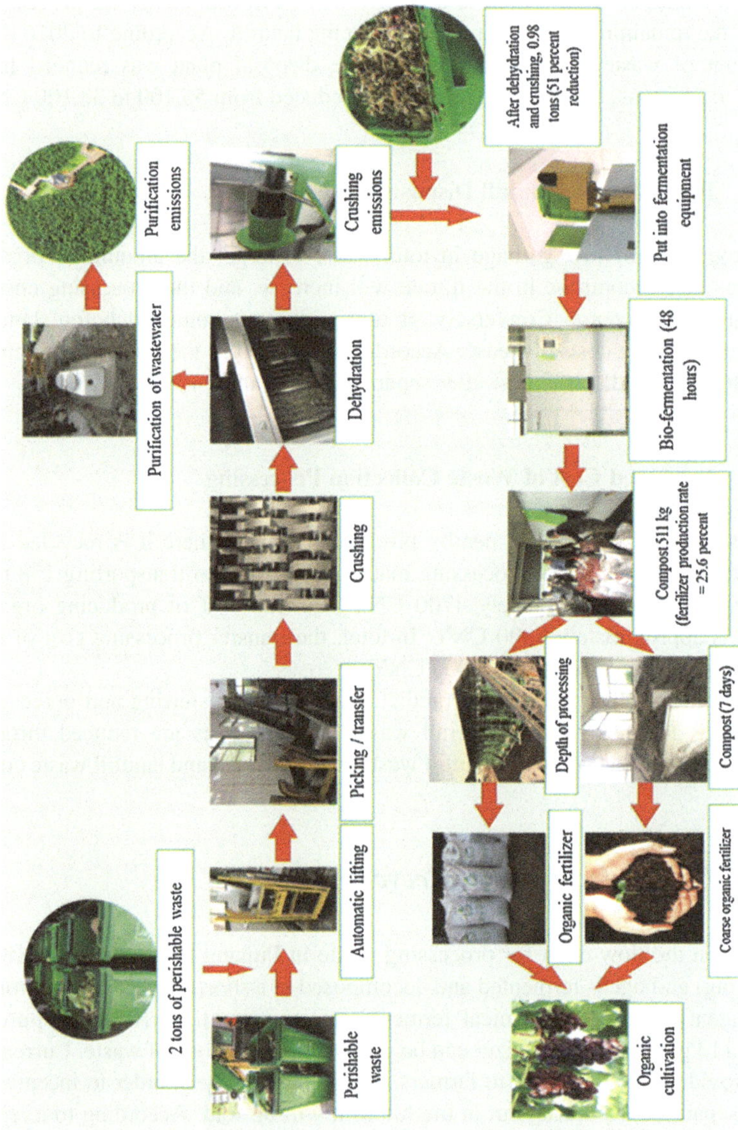

Fig. 18.3 Raw garbage disposal flow in Pujiang Prefecture

recyclable waste is collected each day, worth approximately 400 CNY or 146,000 CNY per year.

The collection and disposal costs before and after separating garbage can be reduced by 9.636 million CNY per year. In addition, an annual profit of 146,000 CNY can be generated from the sale of recyclable waste, enabling the production of organic fertilizer with a total value of three million CNY. Thus, the annual profit obtained by waste separation is estimated to be approximately 12.8 million CNY.

18.4.2.1 Reduction of Financial Burden in Waste Separation Processing

The cost of government waste disposal in rural areas of China has increased sharply since 2013, becoming a heavy burden on local government finance. Until 2013, waste disposal costs were not included in hygiene management costs and were not separately disclosed. The release of separate statistics on these costs since 2013 is an indication of increasing government focus on waste disposal problems. It became clear that the financial burden of waste disposal could be reduced by implementing a new waste separation disposal system in Pujiang Prefecture. Inhabitants could also obtain economic and social benefits by collecting resource waste and receiving organic fertilizer free of charge.

In addition to reducing the financial burden on the government, waste disposal can also benefit private companies and residents. As waste disposal costs are expected to continue to increase in the future, the Pujiang model is expected to be applied to other regions in China.

18.4.3 Environmental Effects of the Pujiang Model

18.4.3.1 Waste Reduction by Intermediate Treatment

In 2016, the total amount of general waste, excluding industrial waste, collected in Pujiang County was approximately 900 tons per day, of which approximately 514 tons was domestic waste. As a result of the separation process, 158 tons of raw garbage is made into organic fertilizer by intermediate disposal of household waste, and approximately 1 ton of resource waste is collected (Fig. 18.4). The amount of waste finally transported to landfill is approximately 355 tons. The reduction of process waste by intermediate treatment is about 31 percent. The amount of industrial waste finally transported to landfill is 95 tons, due to the separate collection of raw garbage and recyclable waste, and the amount of waste finally entering landfill is 450 tons. The rate of waste reduction due to intermediate disposal has reached 50%. This reduction in weight doubles the useful life of the landfill. Since all waste that cannot be recycled is landfilled directly as landfill waste, problems such as soil pollution and wasted land resources remain; therefore an incineration facility was not constructed. It is expected that the rate of waste

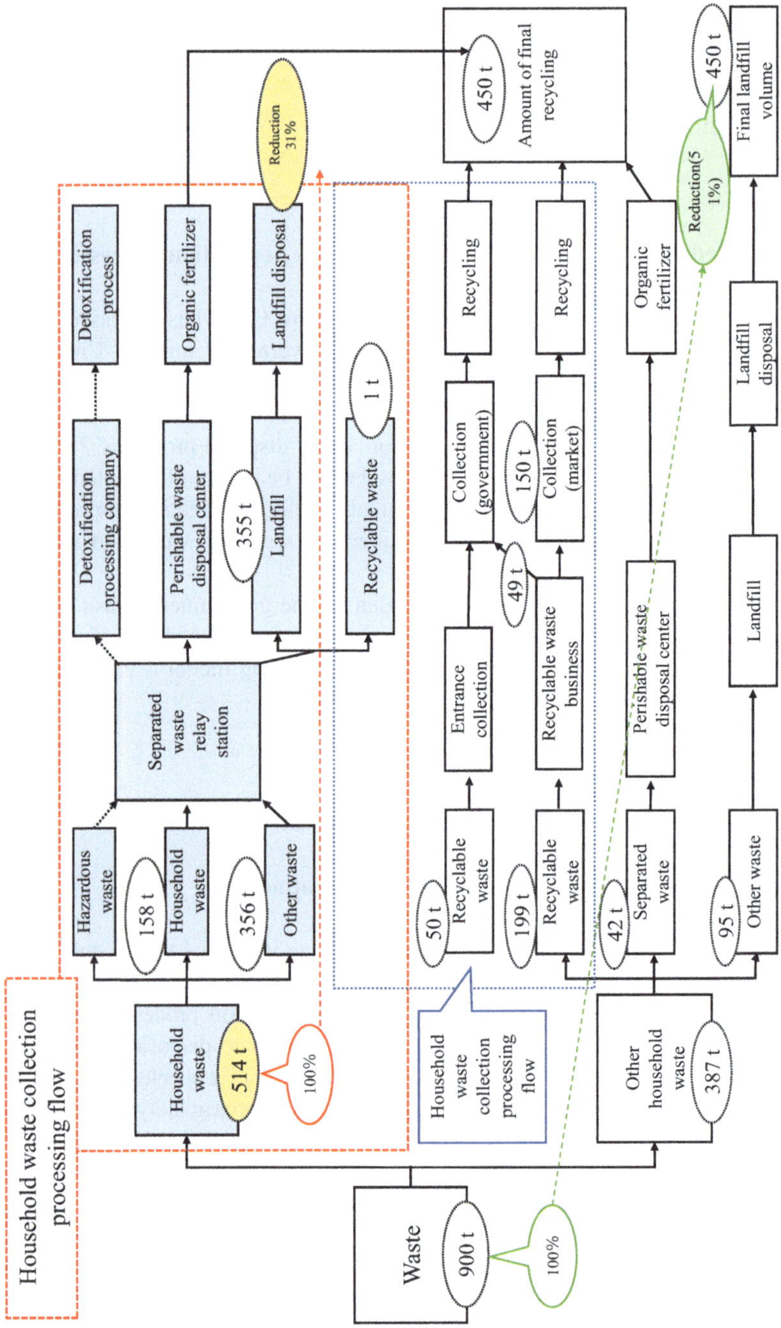

Fig. 18.4 Pujiang Prefecture waste collection processing flow

reduction resulting from intermediate treatment of waste will be further increased by operating the waste incineration facility and using incineration ash and residual heat. Further, a methane gasification facility has been installed, and when the market needs of organic fertilizer are satisfied in future, residual waste can be subjected to methane fermentation, producing biogas for renewable energy.

18.4.3.2 Promotion of Organic Agriculture Through Provision of Organic Fertilizers

Between October 2014 and June 2015, the Gabriel Biotechnology Company tested the effects of organic fertilizers obtained from household waste. Use of the organic fertilizer increased grape production by 17%. Sweetness increased by 2.9 points, there was a clear reduction in the number of cracked grapes, and insect damage was less common. Waste that had previously been illegally disposed of or landfilled plays a major role in organic cultivation of agricultural products by utilizing microbial fermentation technology. In this way, the system is a small-circulation system in which organic fertilizer is produced from raw garbage and farm products are supplied to residents by utilizing the fertilizer.

18.4.4 Societal Effects of the Pujiang Model

18.4.4.1 Toward a Clean and Beautiful Rural Society

Waste separation has already played an important role in improving the rural environment in Pujiang. It also helps improve the environment and landscape of rural areas through separate waste collection. Pujiang was able to greatly improve the living environment of residents by promoting proper waste separation. It has been dubbed "the first prefecture in China to build a national ecological civilization" and "the first prefecture in the country to sort rural household waste." In addition, organic fertilizer, which is recycled from raw garbage, is distributed to local farmers free of charge, contributing to soil restoration and improvement by organic farming. The government hopes to increase the economic power of farmers, create jobs in rural areas, and improve the expertise of farmers to complement development of rural tourism in China based on environmental improvement and organic farming. In addition, the government expects to maintain traditional culture in rural areas, establish infrastructure, and exchange visitors with urban areas.

18.4.4.2 Establishing Trust Between the Government and Residents

Under the party-plus system, party members were able to actively demonstrate exemplary behavior, guiding community residents face-to-face, narrowing the distance between party members and residents, and overseeing thorough separation of waste. Under its policy of operating a market and dealing with waste, the government is trying to return maximum profits to the people by improving the rate of waste separation by residents. The profits obtained from waste collected by the cleaners in two separate collections provide personal income for the cleaners. Companies also recycle and use collected food waste and give free organic fertilizer to farmers. A street survey of 200 residents showed that 96.2% of respondents agreed to separate waste collection and 81% were satisfied with the clean living environment. Proper waste disposal was found to have improved residents' opinions of the government and promoted trust in the local community.

18.4.4.3 Improvement of Environmental Consciousness

The rate of waste collection compliance (accuracy rate) of residents in Pujiang is very high, at more than 90%. Residents have become more concerned about environmental hygiene. The improved efficiency of waste separation has promoted the use of resources and improved awareness of environmental protection. Since implementation of the Pujiang County and the City Waste Separation and Comprehensive Utilization Plan, the Pujiang government has actively promoted separate waste collection, by fully utilizing media and network platforms and disseminating knowledge about separate waste collection. In addition, a waste segregation duty supervision team was established in 429 villages in all prefectures, allowing women and junior high school students to play a central role. Organizations such as the Communist Youth League and the Senior Citizens' Association also actively participate. A separate promotion group has been created for floating waste. With the production of an animation of waste separation, an atmosphere has been created in which everyone is responsible for separating waste. There are more than 2000 promotional drawings and 120,000 approval signs, which also explain to residents how to separate household waste. Finally, the government has been actively engaged in school education and field trips to waste disposal facilities to further understanding of the segregated collection of waste.

Although separate waste collection has been helpful for improving awareness of environmental protection and rural living conditions, some claim that it has increased the number of public officials and residents. After official implementation of the Rural Waste Separation Management Ordinance in Jinhua, Zhejiang Province, in 2018, the separation of waste was introduced in the evaluation of executives, and fines were imposed on retailers and individuals who did not adequately recycle, causing antipathy among some executives and residents.

18.4.4.4 Promotion of Public-Private Partnerships

In December 2014, the Ministry of Finance in China established the Guidelines for Operating a Model of Government and Social Capital Cooperation Models and presented concrete public-private partnership proposals in the field of public works. Currently, China has the world's largest PPP market and is known for its large number of state-owned enterprises, including local government-related projects such as sewage treatment, flooding, waste disposal, and infrastructure maintenance, as well as public services. Much attention is focused on waste treatment businesses in rural areas, which lack funds, experience delays in the application of related technologies, and lack human resources. PPP is one way to solve these problems. Developed countries have already introduced the PPP model to the practice of rural environmental governance, such as in agricultural ecosystem management and rural environmental services. In Japan, third-party companies are entrusted with the construction, operation, and maintenance of sewage facilities.

As shown in Fig. 18.5, the new PPP pilot project of Pujiang waste disposal, which started in 2014, is a management contract in which ownership is held by the government, but private companies are responsible for the operation, protection, and customer service of public facilities under a 3-year contract. In the budget set by the Pujiang Prefecture Government for waste disposal, the amount contracted with the enterprise is paid in advance, and the entrusted enterprise manages and operates waste disposal appropriately (collection, transportation of household waste) within the funds paid. Any expenses saved by the company can be retained as profits. In order to ensure reliability, companies are proposing data reporting and management of waste collection processing systems, as well as new policies on waste disposal and the use of related technologies. The advantages of the new PPP model for waste disposal in Pujiang County include improvements in management, such as curbing fiscal spending and using technology related to waste disposal by private companies maintaining management and operation. As a private company, the procedure is less complicated than that of traditional PPP model businesses in China; the procedure to become involved is easy. Start-up companies who may have insufficient funds to launch themselves often have new ideas and new technologies. Business opportunities can be provided, and efficiency can be improved through management by private companies. For residents, it is possible to secure high-quality public services through monitoring, with the expectation of information disclosure by the private companies.

With implementation of the new PPP model, government funds can be used more efficiently, related technology can be introduced, and the quality of public services in rural waste disposal can be improved through the reduction of waste disposal costs to the Pujiang government and the use of private technology. However, the government still has a heavy financial burden, and there are disadvantages, such as increasing the workload of party members by instructing residents to separate waste and the requirement for frequent renewal of short-term (3-year) contracts. Moreover, there are a limited number of private companies that can respond to these requirements, and policies introduced to improve people's awareness of waste problems take considerable time to have an effect, whereas companies need to solve problems and demonstrate their managerial skills over a short period of time. By changing its role from an existing implementer to a supervisor and service buyer, the government may weaken its role of guidance and management in dealing with waste.

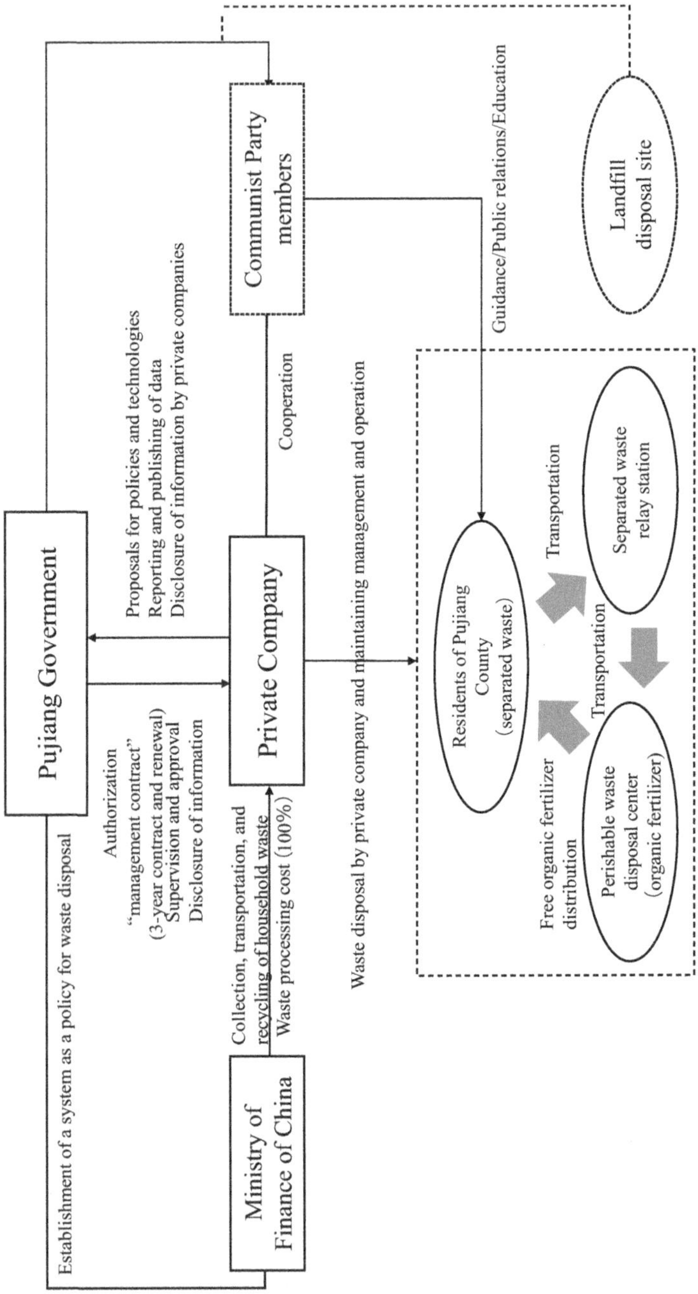

Fig. 18.5 Pujiang as a model of the new public-private partnership (PPP) system for waste disposal

18.4.5 Holistic Construction of the Pujiang Model Household Waste Disposal System

There are three main reasons why waste separation in Pujiang has achieved great results. First, in Pujiang, the PPP model was used to build a public-private partnership system in which the government purchases services from private companies. The government established a recycling recovery system focusing on food waste by securing high levels of waste discrimination and managing private companies. By discriminating and recycling food waste from household waste, the effect of reducing waste by interim treatment was significantly increased, reducing the burden on waste disposal plants, reducing the cost of transportation and disposal of waste by collecting food waste, and reducing the financial burden. In 2014, the Pujiang government set Zhangjiajie as a pilot area for separate waste collection and disposal, distributing to each community (housing complexes in Zhenjiang) in 2016. Coverage of all rural areas reached 100% in 2017. In the same year, Pujiang was selected by the Housing Department of China as a model for the country's first separate collection of rural waste and use of resources. Since 2018, Pujiang has been building a waste collection and processing information system that will allow a move toward a sustainable society of waste reduction and recycling (Fig. 18.6).

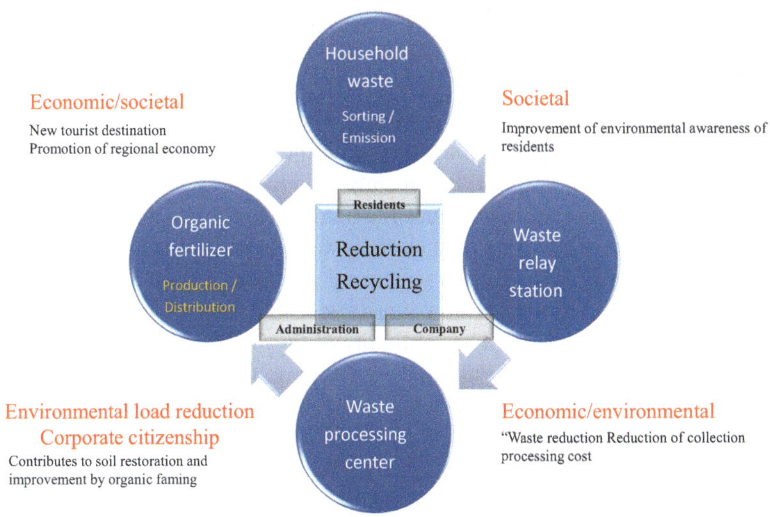

Fig. 18.6 Model of Pujiang household waste disposal circulation system

18.5 Conclusion

The rural economy of China has developed rapidly, consumption has spread to rural areas, and waste is becoming an apparent problem. However, the conventional centralized processing method of "collecting in village, transferring from town, processing in prefecture" has a high transfer cost and does not allow local resource recycling. The current Chinese PPP method is expected to provide a solution to the household waste problem in rural areas because it can utilize the capital of private companies and their technologies. However, the investment required is huge and the cost recovery period is long. Therefore, there are limits to the possible responses from private companies. The Pujiang model utilizes the conventional PPP model and chooses the way the government buys public services from companies to solve the shortcomings of the conventional PPP model. This establishes a government-civilian partnership and a household waste disposal system in which all residents can participate. In this model, the government implements a party-plus system, thoroughly instructing residents on how to separate, supervise, and manage waste, increasing the waste separation rate with a simple and easy-to-understand waste separation method. The company has succeeded in reducing the amount of raw garbage going to landfill. Instead, garbage is recycled; as high proportions of raw garbage characterize household waste in rural areas, the waste collection process centers on raw garbage. The waste disposal method changed from centralized to decentralized due to successful resource recovery from food waste. The life span of the landfill was doubled, and transportation and disposal costs drastically reduced. By providing organic fertilizer produced from raw garbage to residents free of charge, local soil restoration, organic agriculture, and development of the tourism industry have been promoted. (Annual tourism revenue in 2015 was 49.61 billion CNY, an increase of 75.2% on the previous year, and Zhejiang is now in the top ten provinces for tourism.) The Pujiang model has shifted toward recycling of waste by constructing a regional recycling system in cooperation with the government, private companies, and residents.

In the future, this model will be useful as a waste disposal method in both other rural areas of China and other regions. However, the reduction of waste from source is still insufficient in Pujiang, and it is mainly government finance that funds waste disposal. In the future, in order to reduce the financial burden to the government and the environmental load, a volume-rate waste system should be introduced, which can curb waste emissions, overhaul the system of related agencies operating enacted laws, and promote effective utilization of organic resources from food waste (such as using livestock feed or biogas). The continuity of this model from the perspective of economic, environmental, and societal aspects was evaluated by looking at the cooperation of the administration, private enterprises, and residents. Key characteristics of this model, which have implications for policy development in other regions, are (1) waste reduction and recycling (mainly organic fertilizer); (2) reduction of waste collection and processing costs; (3) soil restoration and improvement of agricultural production environment; (4) creation of areas such as new tourist destinations; (5) improvement of environmental awareness of residents; and (6) presentation of a new waste disposal system based on public-private partnership.

Acknowledgment This study was financially supported by National Key R&D Program of China (grant number 2018YFD1100601-07).

Bibliography

Bai Y, Wu W, Wu D (2016) Loading of non-point source pollution of rural domestic waste in Paihe River Valley. J Ecol Rural Environ 32(4):582–587. (in Chinese)

Choi J, Lee S (2010) Key risks and success factors on the China's public-private partnerships water project. Kor J Constr Eng Manage 11(3):134–144. (in Korean)

Contemporary Green Economy Research Center (2016) Research on rural domestic waste treatment, p 64. (in Chinese)

Cun M (2005) A study on the construction of the organic waste recycling system in Kunming City, China. Master Thesis, Ritsumeikan University Graduate School of Policy Science. (in Japanese)

Ehara N (2016) Current status and expectations of the PPP model in China. Int Trade Invest Q J 2016(106):95. (in Japanese)

Interview with President of Zhejiang Gabriel Biotechnology Co., Ltd. http://www.gabriel-gp.com/qyjj.html

Jia X, Chen Y, Zhao Y, Dong X (2018) The dilemma and mechanism model innovation of rural domestic waste Management in China: based on the investigation of rural domestic waste Management in Hebei, Zhejiang, Guangxi and other places. Environ Sustain Dev 3:66–68. (in Chinese)

Kitagawa H (2018) Progress and issues of city waste policy in China: comparative study with Japan and Taiwan. Ryukoku J Policy Sci 7:55–70. (in Japanese)

Lee J (2016) China's rapidly changing public-private partnership project. World City 13:48–55. (in Korean)

Lu S, Li X, Du H (2018) Comparative analysis and suggestions on typical models of rural domestic waste treatment. World Agric 2016(2):4–210. (in Chinese)

Ministry of Finance of the People's Republic of China (2016) Construction statistics yearbook (2016). http://www.mohurd.gov.cn/xytj/tjzljsxytjgb/jstjnj/index.html. Accessed 5 March 2019

Peng Z, Li S, Liu Z (2016) Characteristics of transboundary non-point source agricultural pollution in the Taihu Valley. J Ecol Rural Environ 32(3):458–465. (in Chinese)

Wu D, Wang S (2014) The research and development of PPP model in China. J Eng Manage 6. (in Chinese)

Yu H (2019) A study on policy on the collection and disposal household waste in urban and rural areas in China. Master Thesis, Ritsumeikan University Graduate School of Policy Science. (in Japanese)

Zhang C (2017) A case study of recycling-oriented agriculture centered on soil restoration. J Policy Sci Res Notes 24:353–363. (in Japanese)

Chapter 19
Potential for Cooperation Among China, Japan, and South Korea in Renewable Energy Generation

Yishu Ling, Weisheng Zhou, and Xuepeng Qian

Abstract As an important countermeasure for climate change, the promotion of renewable energy plays an important role in reducing emissions of greenhouse gas (GHG) and facilitating sustainable development. Renewable energy cooperation has also become an important issue among China, Japan, and South Korea. In recent years, economic potential of environmental industry has been well recognized, besides the requirement of both ensuring energy security and environmental protection. After clarifying the status and issues of renewable energy from the perspectives of economy, environment, and technology in the three countries, they can cooperate in multiple ways in governmental and private projects, and cooperation has developed from government-oriented stage to business-oriented stage. Under the framework of the "East Asian Low-Carbon Community," we find a pathway to realize the tripartite cooperation among China, Japan, and South Korea in the renewable energy field. A systematic scheme was concluded to considering the roles of factors and dynamic relation for cooperation. Business power is expected to be more active in promoting cooperation.

19.1 Introduction

As a vast consumer of energy, East Asia was responsible for nearly 49.10% of global CO_2 emissions in 2018. Notably, rapid economic growth in China, Japan, and South Korea has been driven by massive consumption of fossil fuel. In 2018, China, Japan,

---

Y. Ling
Graduate School of Policy Science, Ritsumeikan University, Ibaraki, Osaka, Japan
e-mail: ps0290ih@ed.ritsumei.ac.jp

W. Zhou (✉)
College of Policy Science, Ritsumeikan University, Ibaraki, Osaka, Japan
e-mail: zhou@sps.ritsumei.ac.jp

X. Qian
College of Asia Pacific Studies, Ritsumeikan Asia Pacific University, Beppu, Oita, Japan
e-mail: qianxp@apu.ac.jp

W. Zhou et al. (eds.), *East Asian Low-Carbon Community*,
https://doi.org/10.1007/978-981-33-4339-9_19

and South Korea created 23.62% of global GDP, while emitting 33.44% of global CO_2 emissions (IEA 2019).

Geographic challenges such as the logistics of land transportation and historically rooted issues have tended to limit cooperation among China, Japan, and South Korea (Calder and Ye 2008, Kim et al. 2014). However, building an energy cooperation scheme within the region could achieve the goals of sustainability, stability, and security (3S), including mitigating insecurities about access to energy resources, developing energy sustainability, and maintaining stability of the energy market (Kim et al. 2014; Freeman 2015). Energy cooperation should promote the enhancement of energy efficiency, increase the proportion of renewable energy in the energy mix, and improve energy access and connectivity across the region (ESCAP 2017). In addition to environmental conservation responsibilities, the economic potential of environmental issues has been well recognized. China, Japan, and South Korea have incorporated low-carbon development into their dominant national strategies and have prioritized environment and energy-related industries, which provides a backdrop for mutually beneficial cooperation. Considering the different stages of industrial technology development, the benefits of environmental cooperation have been qualitatively and quantitatively analyzed in many scenarios, providing evidence of economic and environmental gains, in particular in the field of renewable energy generation to reduce emissions. As well as developing national countermeasures, regional cooperation in East Asia is needed in order to successfully achieve reduction targets (Zhang 2018; Meng and Serafettin 2020). Thus, establishing an East Asian Low-Carbon Community including Japan, China, and Korea should be considered as a priority measure (Zhou 2019).

Considering the immense potential in renewable energy and clean technology, this chapter argues that China–Japan–South Korea cooperation would widen economic relations among the countries. Challenges and potential for international environmental cooperation among China, Japan, and South Korea are discussed. The mechanism of international environmental cooperation is explored from government level to industrial level. We conclude that the three countries should cooperate in the framework of the East Asian Low-Carbon Community.

19.2 Existing Cooperation Among China, Japan, and South Korea in Renewable Energy

As discussed in Chap. 2, after 2008, cooperation among China, Japan, and South Korea in energy and the environment changed to a mutually beneficial arrangement and expanded to include renewable energy. The China–Japan–Korea Renewable Energy and New Energy Technology Cooperation Forum was held in Beijing in 2008. The meeting discussed and adopted the Common Initiative on Strengthening China–Japan–Korea Renewable Energy and New Energy Science and Technology Cooperation. This initiative reached consensus on the principles of mutual benefit, protection of intellectual property rights, prioritization of renewable energy cooperation, policy support, and promoting movement of scientists and engineers among the three countries (The Ministry of Technology 2008). In 2011, the fourth China–Japan–South Korea trilateral summit called for "sustainable growth first by enhancing renewable energy and energy efficiency collaboration" and approved development of green energy and energy efficiency to promote sustainable growth.

The Tripartite Environment Ministers Meeting (TEMM) is held annually to offer a platform for dialogue and to encourage cooperation. The 21st TEMM held in Japan in 2019 reviewed implementation of the China–Japan–Korea Environmental Cooperation Joint Action Plan (2015–2019). The plan had deepened exchanges and cooperation around ecological and environmental issues, and China had advocated a model of "China, Japan, and South Korea + X" for cooperation to complement the Belt and Road Green Development International Alliance (Trilateral Cooperation Secretariat 2019).

Cooperation among the three countries began with specific industry cooperation. At the Gansu Natural Energy Research Center (GNERI) in Lanzhou, a Sino-Japanese cooperative project was established to demonstrate a hurricane-type wind power irrigation system. China and South Korea have begun to cooperate in the development of tidal energy (China-Korea Joint Ocean Research Center (CKJORC) 2020).

In summary, cooperation among the three countries has gradually transitioned from consultation and dialogue to the launching of business projects. However, political factors continue to influence bilateral relations, often causing fluctuations in cooperation. Furthermore, China–Japan–Korea cooperation is grappling with the issue of how to balance dominance within the cooperative relationship. Development of a framework for cooperation that is acceptable to all three countries would contribute to deepening their cooperation.

19.3 Renewable Energy Policy and Technology Development in China, Japan, and South Korea

19.3.1 Renewable Energy Development Targets in China, Japan, and South Korea

Given the potential for energy conservation and CO_2 emission reductions among China, Japan, and South Korea, it is necessary to compare policy instruments and technology development among the three countries and with international experience from the European Union.

Some renewable energy industries are growing much faster than others. China has the world's largest installed capacity of wind energy and hydropower and ranks third in bioenergy (Table 19.1). Japan holds the third position for solar photovoltaic (PV) and is a top ten producer of bioenergy, hydropower, and geothermal. South

Table 19.1 Renewable energy development in China, Japan, and South Korea

Sector	Country	2010	2012	2014	2016	2018	Capacity development targets
Total renewable energy	China	233,260	267,903	302,108	359,519	414,653	15% non-fossil final energy, 2020; 3% non-hydroelectricity, 16% electricity, 2020
	Japan	36,041	37,413	3898	46,072	56,088	10% primary energy, 2020
	South Korea	2819	3322	3687	4330	5716	6.1% primary energy, 2020; 11.5% 2030; 20% 2050
	China	196,290	232,980	380,440	319,530	352,261	300,000 MW (2020)
Hydro-power	Japan	47,736	48,934	49,597	50,117	50,117	150,000 MW (2020)
	South Korea	5525	6447	6466	6485	6494	
	China	29,633	61,597	96,819	148,517	184,696	
Wind energy	Japan	2294	2562	2753	3246	3653	
	South Korea	382	464	612	1067	1311	1570 MW (2022)
	China	1022	6718	28,388	77,788	175,018	20,000 MW (2020, total solar)
Solar photovoltaic	Japan	3618	6632	23,339	42,040	55,500	14,000 MW (2020); 53,000 MW (2030)
	South Korea	650	1024	2481	4502	7862	3500 MW (2030)
	China	3	8	14	14	14	
Concentrated solar power	Japan						
	South Korea		4	229	229	229	
	China	4	4	4	4	4	
Marine energy	Japan						
	South Korea	1	255	255	255	255	
	China	3446	4617	6653	9269	13,235	30,000 MW (2020)

Bioenergy	Japan	1956	1514	1708	1955	2272	
	South Korea	161	197	601	1066	1419	
	China	24	26	26	26	26	
Geothermal energy	Japan	537	512	508	526	536	
	South Korea						

Source: IRENA (2019)

Korea has emerged as a world leader in marine energy and is a critical player in solar PV. Wind and solar powers are experiencing rapid growth in all three countries. Grid-connected solar PV has expanded at an average annual rate of 60% since 2010, while wind energy has had an average annual growth rate of 20% since 2010. Viewed in terms of capacity development targets, China achieved its 2020 target for wind energy in 2018. Japan and South Korea both achieved their respective 2030 targets for solar PV in 2018. Hydropower is a mature sector with a relatively slower rate of growth of approximately 3.2%, similar to that of geothermal and bioenergy. Other renewable energies, including marine energy (wave and tidal), remain small-scale.

19.3.2 Renewable Energy Policy in China, Japan, and South Korea

We compared national policies from the perspectives of legislation, subsidies, tax incentives, tradable certification, and green power certification (Table 19.2).

Japan and South Korea have provided various subsidies for the development of renewable energy. Japan has provided more than 132.2 billion JPY for residential PV power generation in order to promote substantial development of PV installations. South Korea has adopted subsidies for projects with market potential and projects entering the commercialization stage. Renewable energy subsidies in China are more diversified; the central government provides key industries with free subsidies and loan discounts. The renewable energy electricity price system has promoted industrial development and construction of facilities.

In terms of tax incentives, the renewable energy industry receives preferential treatment. The 10% cost of reforms by Japanese companies in energy saving can be deducted from income tax. South Korean renewable energy manufacturers and consumers receive up to 80% of loan amounts. Policies for the acquisition of land for wind power in China include land tax reduction and exemption, half value-added tax, and other collection policies. Furthermore, tradable certification and a green power certification system encourage industrial development through market mechanisms.

19.3.3 Renewable Energy Technology Development in China, Japan, and South Korea

Renewable energy resources have two main disadvantages, fluctuation and low density (IRENA 2019). Therefore, we divided analysis of renewable energy technology development into two parts. The first is the amount of potential renewable energy introduced, which is analyzed in this section from technology and market

Table 19.2 Overview of other renewable energy policy implementation in China, Japan, and South Korea

Category	China	Japan	South Korea
Legislation	– Renewable Energy Law of the People's Republic of China (2006)	– Act on the Promotion of New Energy Usage (1997) – Basic Act on Energy Policy (2002) – Basic Energy Plan (2003) – Act on Special Measures Concerning Procurement of Electricity from Renewable Energy Sources by Electricity Utilities (2011)	– New and Renewable Energy Development and Promotion Law (1987) – The third basic plan for technology development, application and deployment of New and Renewable Energy (2008, published every 5 years)
Subsidy	– Renewable Development Funding (2006) – Allocating Additional Subsidies for Renewable Energy Electricity Prices (2013) – Subsidy for non-hydro renewable energy (2020)	– Renewable energy generation promotion levy (2012) – Subsidy for virtual power plant construction demonstration project utilizing customer-side energy resources – Subsidies for carbon dioxide emission control project expenses (renewable energy electricity/ heat autonomous spread promotion project)	–the 1987 subsidy program for renewable energy demonstration and deployment offered grants to firms for projects for RE technologies – "Local Development Subsidy Program" for RE projects (1996) – "One million green homes" by 2030, absorbed an existing program to install 100,000 solar roofs – "1 GW solar roof" Plan in Seoul (2019)
Tax incentives	– Central government tax, local government tax, and shared tax incentives to develop renewables in renewable industries	– Home energy saving reform tax reduction – Special measures for tax standard for renewable energy power generation facilities – Special measures for the tax standard for biofuel production facilities acquired by biofuel manufacturers	– Loans and tax incentives to develop renewables are both available to both customers and manufactures of thoroughly commercialized renewables systems
Tradable certification or green power certification system	– Renewable energy green power certificate issuance and voluntary trading system (2017)	– Renewable energy green power certification (2001)	– Renewable energy certificates

Sources: National Renewable Energy Center (2018), JEMA (2019), KEMCO (2014)

perspectives. The second is countermeasures against fluctuation, which is assessed in Sect. 19.3.5.

19.3.3.1 Renewable Energy Introduction Potential and Technology Development

China, Japan, and South Korea have abundant renewable energy resources. Due to regional differences in conditions, the focus of renewable energy and the direction of technological development vary by country (Table 19.3). This chapter focuses on solar energy, wind energy, biomass energy, and marine energy and analyzes differences in renewable energy technology development.

Solar Capacity

According to Fig. 19.1, the rate of increase in solar capacity in China, Japan, and South Korea has accelerated since 2011. In the following 3 years, the annual installed capacity of PV in China grew to over 10 GW. However, distributed PV installed capacity is still limited. In 2016, China released its 13th Five-Year Plan for Solar Energy Development with a target to expand capacity to 110 GW, which was achieved in 2018. China has improved the efficiency of monocrystalline, polycrystalline, and thin-film silicon solar cells to 21%, 19%, and 12%, respectively. Also, China is devoted to promoting Chinese brands in solar power generation.

Japan is pacing solar energy technologies by its PV 2030+ roadmap, which sets five stages for developing solar power generation. The second stage ran from 2010 to 2020 with a grid parity of 23 JPY/kWh. During this period, technology development

Table 19.3 The main targets of renewable energy introduction among China, Japan, and South Korea

	China	Japan	South Korea
Solar energy	50 GW by 2020	64 GW capacity by 2030	30.8 GW by 2030
Wind energy	30 GW by 2020	400 MW by 2020 5860 MW by 2030	16.5 GW by 2030
Biomass energy	30 GW by 2020	Biomass energy generation +waste energy generation capacity 4.08 million KL by 2020; 4.94 million KL by 2030	3% of total power generation by 2050
Marine energy	50 MW by2020	51 MW by 2020 554 MW by 2030 7350 MW by 2050	12% of total power generation by 2050
Geothermal energy	527.28 MW by 2020	Max 7.39 MW by 2050	
Small and medium-scale hydropower	80 GW by 2020	14.57 MW by 2050	3% of total power generation by 2050

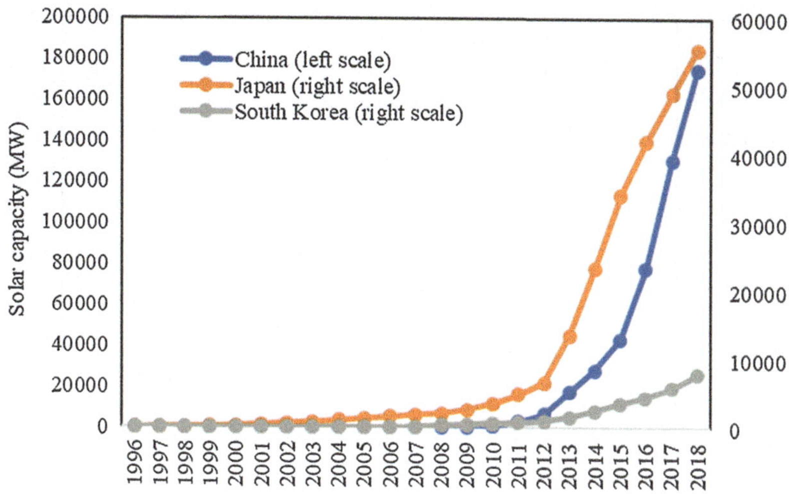

Fig. 19.1 Solar capacity in China, Japan, and South Korea. (Source: BP 2019)

focused on refining production processes to reduce costs. The third and fourth stages will run from 2020 to 2030, with a parity of 14 JPY/kWh. During this period, technology development will focus on low-cost next-generation solar cells and long-life storage batteries. The fifth period from 2030 to 2050 will have a parity of 7 JPY/kWh. During this period, technology development will focus on improving the efficiency of solar panels to 40%.

In South Korea, the Ministry of Trade, Industry, and Energy (MOTIE) planned to help the PV industry reduce the costs of solar module technology to 0.10 USD/cell by 2030 and to increase production efficiency to 24% by the end of the next decade. The government expects PV cell producers to increase the average efficiency of multi-junction devices from 23 to 35%.

Wind Energy Generation

China has been promoting the development of both offshore and onshore wind (Fig. 19.2). The historical focus has been on onshore wind, especially in the northern and western regions of China, where wind resources are abundant. However, insufficient system capacity became apparent. Since offshore wind can be located close to areas with high power demand and is easy to connect to the grid, development of offshore wind has become more important. In 2007, the Donghai Daqiao Offshore Wind Power Generation Project launched. The project installed 34 Sinovel 3 MW units with a total power generation capacity of 102 MW. Several offshore wind farms have been constructed along the coast in water of 5 m depth or less, such as at Rudong (30 MW), Jiangsu Xiangshui (6 MW), and Shandong Rongcheng (6 MW).

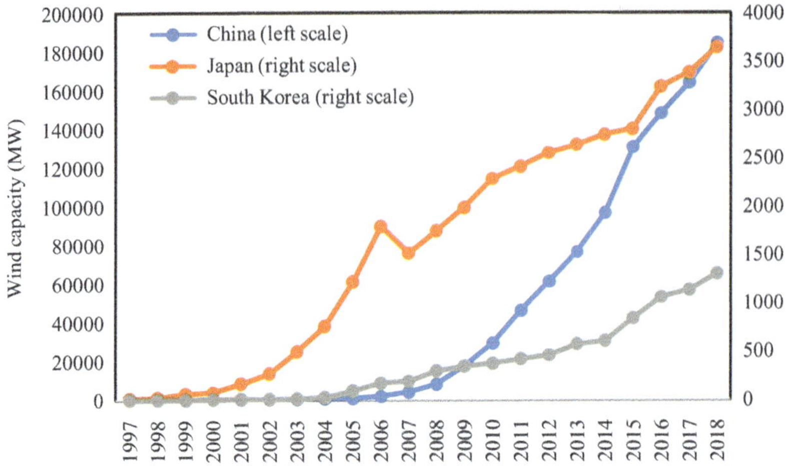

Fig. 19.2 Wind capacity in China, Japan, and South Korea (MW). (Source: BP 2019)

In Japan, installed wind energy capacity has also increased rapidly since 2000. The feed-in tariff (FIT) scheme provides positive incentives for increases in wind energy. Review of the Basic Energy Plan in 2014 led to the release of a more specific plan to develop both onshore and offshore wind. However, compared with China or the United States, uptake in Japan remains slow. Inhibitions to development of wind energy in Japan include ongoing construction of electricity transmission lines and limited availability of land for installation. To overcome these difficulties, Japan is focused on offshore power generation. The Tokyo Electric Power Company and Ørsted plan to build an offshore wind farm near Choshi, to gradually develop offshore wind power markets with huge potential.

Due to similarities in equipment and technology between shipbuilding and wind power generation, the four major South Korean shipbuilders, Hyundai Heavy Industries, Samsung Heavy Industries, Daewoo Shipbuilding, and STX Heavy Industries have invested in the wind power generation industry. In order to accelerate development of wind energy technology, Korean companies focus on mergers and acquisition of global wind energy companies in order to enter the international wind power market. Korea also promotes technology development of offshore wind power generation; projects include a 2.5 GW offshore wind development project in the southwest marine area, development of a large wind turbine, an offshore wind test site (5 MW), and development of a high-performance yaw control system.

Biomass Power Generation

Five kinds of biomass energy utilization technologies exist: direct combustion, thermochemical conversion, biological conversion, liquefaction, and waste treatment. Development of biomass energy utilization technology in China is towards biogas utilization and biomass thermal conversion. Biomass energy utilization in developed countries has mainly focused on conversion of biomass into electricity and transport or as a combustible fuel. In Japan, the Biomass Comprehensive Strategy was released in 2006, combining the aims of mitigating global climate change and creating a recycling-based society. Japan is developing biomass utilization systems considering regional differences in biomass resources (such as lignocellulose, high-humidity biomass, municipal waste, and mixed wet and dry waste) in order to expand the share of biomass in its energy system. The South Korean government has formulated a renewable fuel standard (RFS) system, blending biofuels in gasoline. Energy conversion of domestic waste has also developed rapidly. For instance, Pohang Construction Co., Ltd., has launched a project using solid waste as fuel to generate electricity for more than 40,000 households in Busan.

Marine Power Generation

Technological development of marine energy power generation is promoted mainly in Europe (especially the United Kingdom) and the United States; development in Asia lags. There are five stages of marine energy technology development: concept studies on land (stage 1), design verification and tank test (stage 2), small-scale prototype machine test (stage 3), large-scale prototype machine field test (stage 4), and multiple full-scale machines arranged for an array project (stage 5). These stages are a model case for advancing technology development toward practical application (NEDO 2012). China, Japan, and South Korea are at stages 2–3, which are early stages of development. In order to promote practical use of this resource, essential technologies must be developed at an early stage before proceeding to demonstration tests of small-scale prototypes and full-scale arrays in field conditions.

19.3.3.2 Renewable Energy Industries in China, Japan, and South Korea

Many countries choose to develop renewable energy sectors in a specific narrow range based on analysis of renewable resources, development capacity, and other constraints. China, Japan, and South Korea are all developing renewable energy generation differently; China has advantages in solar energy, Japan in biomass energy, and South Korea in tidal energy. The relative advantages in different sectors provide the potential for technological cooperation.

Regarding manufacturing and production, there has been a marked global shift from Europe and North America to East Asia and especially to China. In 2018, 70% of the world's solar panels, 58% of polycrystalline silicon solar batteries, and 40% of all wind turbines were produced in China. Japanese companies such as Mitsubishi and Sumitomo are also significant producers in wind, solar PV, and biomass sectors. Japanese firms dominate global technology patent registrations. South Korea has a relatively large number of intellectual property rights in the solar, new energy vehicle, and wind energy industries. In terms of wind and solar power, massive companies and large integrated groups in South Korea are the leading patent applicants, such as Interbrand, Daehan Housing Corporation, and Daewoo Construction. In terms of technology development and patent applications for new energy batteries, solar cells, and fuel cells, technology companies in the Korean communication service industry dominate, such as SK Communications, SK Telecom, KT, and SK Broadband.

19.3.4 Renewable Energy Investment in China, Japan, and South Korea

In the decade 2010–2019, global investment in new renewable energy capacity will reach 2.6 trillion USD, and new installed PV capacity will exceed that of any other power generation technology (Fig. 19.3). In East Asia, solar and wind are the two main fields of investment for renewable energy.

China has been the largest investor in renewable energy production capacity in this decade to date. From 2004 to 2018, China committed 865.5 billion USD in investment, followed by the United States which invested 519.3 billion USD, and Japan with 202 billion USD (Fig. 19.4).

Note: The Americas (AMER) comprise the totality of the continents of North and South America. The Antarctic and Southern Ocean Coalition (ASOC) is a global coalition of environmental nongovernmental organizations with more than 150 members in 40 countries worldwide.

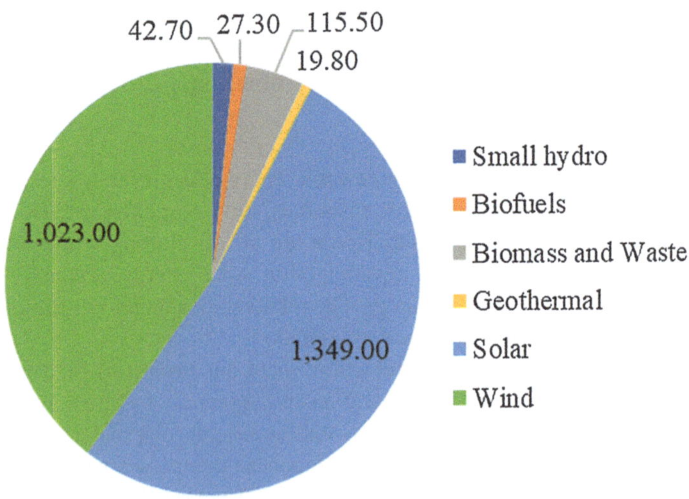

Fig. 19.3 Investment in renewable energy capacity in 2010–2019 (billion USD). (Source: Frankfurt School-UNEP Centre/BNEF 2020)

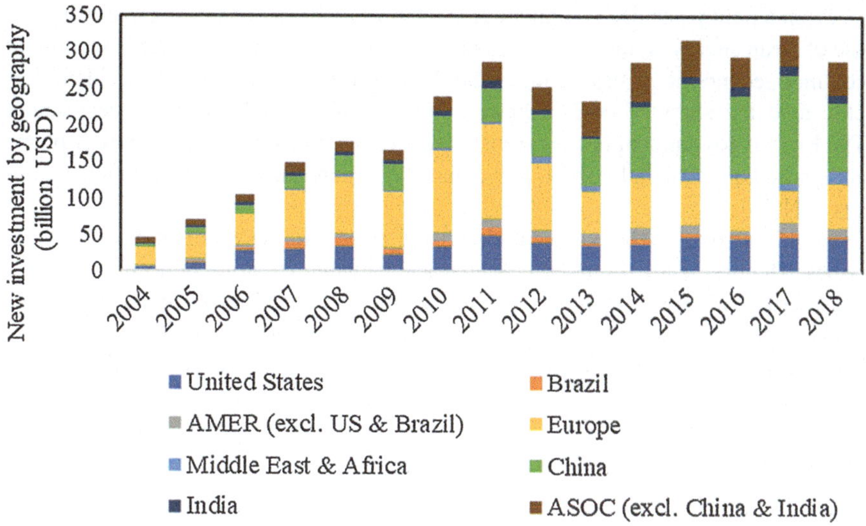

Fig. 19.4 Global investment in renewable energy by region. (Source: Frankfurt School-UNEP Centre/BNEF 2020)

In recent years, a dramatic decline in the cost of wind and solar power has affected policy-making decisions (UNEP Centre/BNEF 2020). These technologies have always been low-carbon and are relatively fast to build. In many countries, wind and solar are currently the cheapest power generation options. China has achieved a relative price advantage in multi-module prices, which has also helped to contribute to expanding PV utilization and reducing costs.

China, Japan, and South Korea are now trying to popularize the use of solar and wind power. China and Japan have some of the highest levels of investment globally. Although South Korea has lower volumes of investment, it is a leader among developing countries. China has made great progress in reducing the cost of PV power generation, providing opportunities for cooperation in renewable energy between the three countries in terms of market sharing and technology transfer.

19.3.5 Challenges to Promotion of Renewable Energy in China, Japan, and South Korea

Renewable energy development and application in the three countries suffers from sizeable capacity gaps, difficult project implementation, low resource utilization, limited grid access, and consumption capacity. Relatively influential policies related to grid connection are the most crucial aspect affecting fluctuation. This section focuses on issues limiting grid connection of renewable energy.

From a policy perspective, although the three countries are striving to expand the use of clean energy, some hindrances remain. The Chinese government is seeking to continue economic restructuring. The Fukushima nuclear power accident has prevented the share of coal in basic power resources in Japan from decreasing. The Korean government has vigorously reduced the number of nuclear power plants and phased out coal and nuclear power but still depends heavily on liquefied natural gas (LNG).

From the perspective of technology and industry, there are three significant issues. The first is the limited physical space available for grid connection, without considering congestion management of the grid system. The second is the abnormally high cost of grid connection, which is linked to the construction of a significant grid system at the expense of power generation companies. The third is the so-called connectable amount of renewable energy, especially of solar and wind power, to the grid. In several regions, utilities set a connectable amount limiting renewable energy, such as solar PV and wind.

19.4 Stakeholders in China–Japan–South Korea Renewable Cooperation

Based on our analysis, feasible cooperation between China, Japan, and South Korea requires an equitable model of profit distribution. Business demands are increasingly influential to international environmental politics, including through lobbying in international negotiations, implementation and technological innovation, and shaping public discourses, private norms, and rule-setting (Falkner 2008). The collective involvement of all stakeholders, including national government, local government, companies, academia, and citizens, in all three countries must be achieved (Fig. 19.5).

Most existing international cooperation focuses on national dialogue and policy advocacy. Cooperation at this level is easily affected by international relations. Another kind of cooperation is exchange and cooperation from academic research institutions, but implementation of specific projects is lacking in strength.

As advocated in the East Asian Low-Carbon Community framework in Chap. 3, the key is to use business power to help build an equitable model for capital investment and profit distribution. Also, as Premier Li Keqiang advocated, the China–Japan–South Korea + X model under the Belt and Road Initiative could provide a new model of cooperation among the three countries (Fig. 19.6).

The model of China–Japan–South Korea + X offers a practice of the principles of negotiation, joint construction, and sharing. It supports cooperation between enterprises from the level of national policy and provides necessary policy support for enterprises in the development of new renewable energy sources. In order to reduce the risk of changing political relations between countries on economic, trade, and

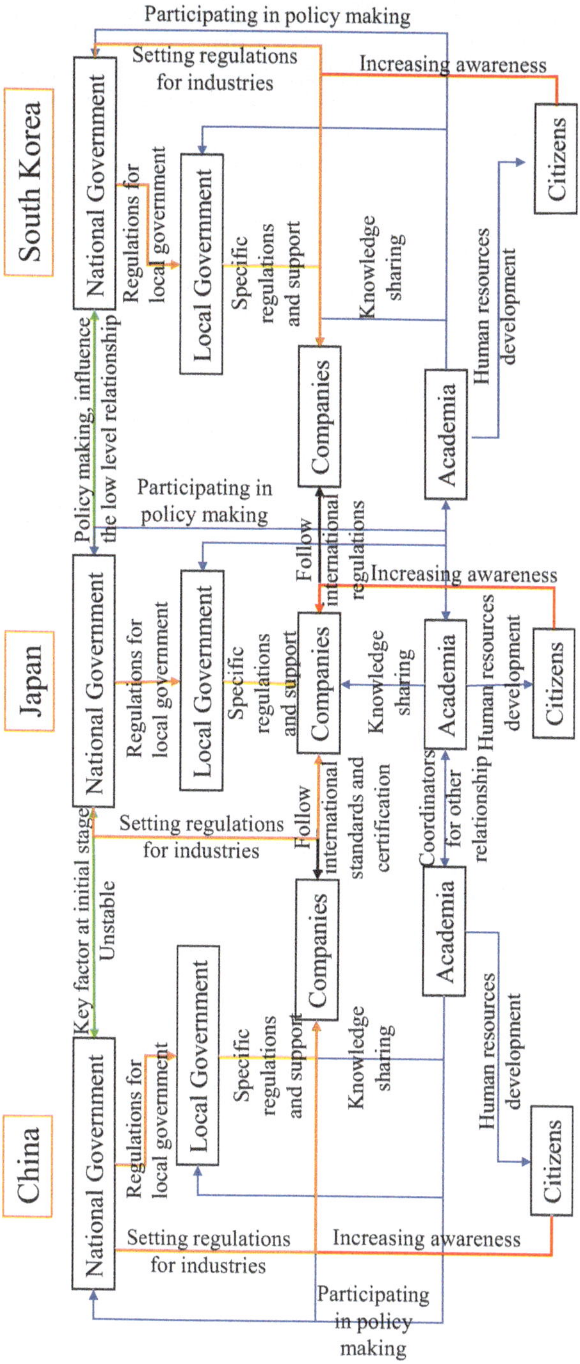

Fig. 19.5 Schematic diagram of international cooperation among China, Japan, and South Korea. (Source: Qian 2015)

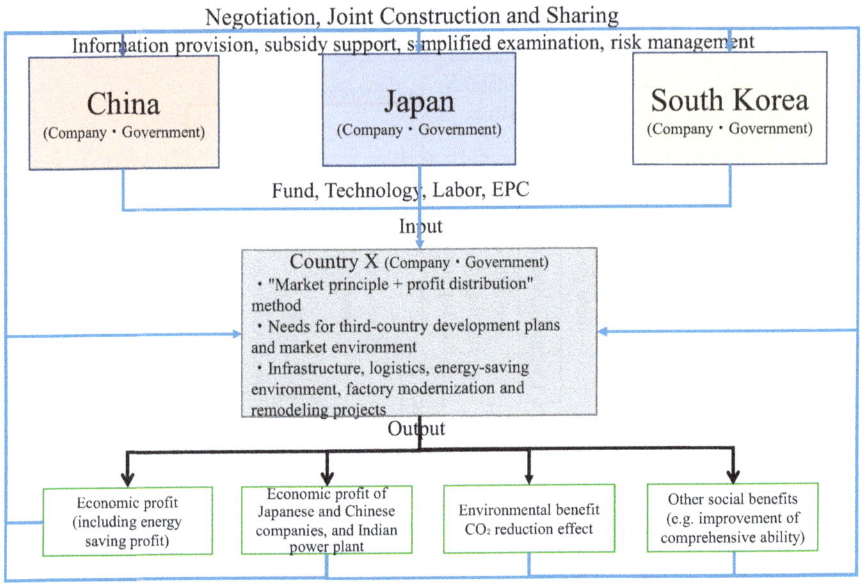

Fig. 19.6 Schematic diagram of China–Japan–South Korea cooperation in country X's market. (Source: Zhou 2019)

technical cooperation, market mechanisms have also been introduced for investment and profit distribution.

19.5 Scheme of Cooperation Among China, Japan, and South Korea in Renewable Energy

By 2050, the proportion of clean energy in primary energy consumption in Northeast Asia will increase to 60% or 4.2 trillion kWh, which is equivalent to reducing annual carbon dioxide emissions by 3.9 billion tons (China Electricity Council 2019). Based on our analysis, we propose a scheme of cooperation among China, Japan, and South Korea in renewable energy generation.

Current cooperation among China, Japan, and South Korea in energy generation is in the form of ministerial meetings or dialogues such as the East Asian Low-Carbon Growth Partnership. The most representative projects of tripartite cooperation in energy are the China, Japan, and South Korea Circular Economy Demonstration Base in Dalian, China, which launched in 2015. In this demonstration base, a town named Dazheng functions as a low-carbon circular economy zone. This project is approved by government dialogue and practiced by enterprises and research institutions, which could offer a model for development elsewhere (Fig. 19.7).

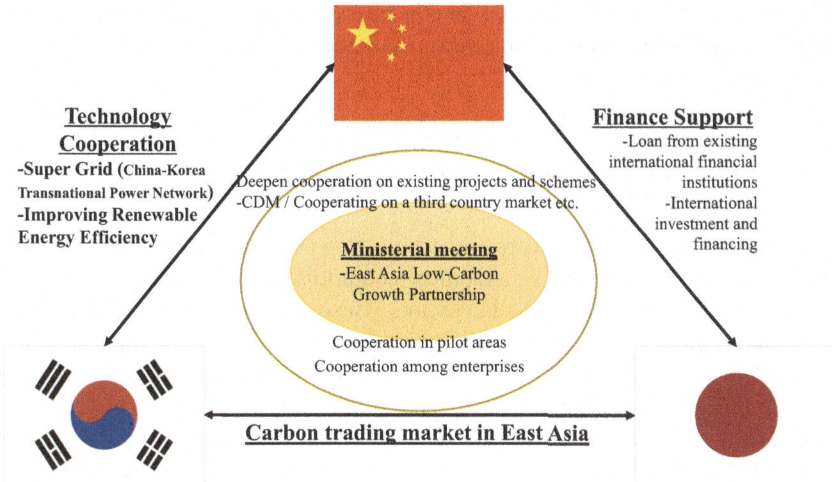

Fig. 19.7 Diagram of cooperation among China, Japan, and South Korea in renewable energy under the East Asian Low-Carbon Community framework

19.5.1 Ministerial Meetings

Although there are no institutions similar to the European Commission among China, Japan, and South Korea, it is still possible to exchange opinions and reach consensus through regular ministerial dialogue. In this way, the three countries can confirm the progress of cooperation and promote projects from the government level.

19.5.2 Deepen Cooperation on Existing Projects and Schemes

Cooperation can be promoted through third-country market projects. At the end of 2017, China and Japan reached a consensus in promoting cooperation projects on a third-country market. This provides a cooperation model that both China and Japan can use to their advantage in terms of technical equipment, capital, human resources, and management modes. Both China and Japan can promote win–win cooperation and jointly develop the market. More than 50 projects have been approved, accounting for 180 billion USD. In renewable energy, ITOCHU Corporation and China CITIC Group will expand investment in German offshore wind power projects. The cooperation model could promote tripartite cooperation among China, Japan, and South Korea. Cooperation can also be promoted through clean development mechanism (CDM) projects, many of which have been implemented in the field of renewable energy. This constitutes another existing scheme that could promote tripartite cooperation among China, Japan, and South Korea.

19.5.3 Technology Cooperation

Development of renewable energy in China, Japan, and South Korea has been constrained by physical and geographical conditions in each country, leading to development focusing on particular sectors. However, solar, wind, and biomass energy have all been developed, which could offer opportunity for technology cooperation among sectors. The fluctuation of renewable energy impedes tripartite cooperation, but the Northeast Asia Grid offers a solution. It is a substantial transnational project, which aims to connect the power grids of South Korea and Japan with China, transforming the abundant renewable energy resources of China and Mongolia into electricity that can be transmitted to Japan and South Korea through ultrahigh-voltage direct current (DC) cables. Part of the construction will be completed by 2022, effectively accelerating opportunities for cooperation.

19.5.4 Financial Support

Finance is another obstacle impeding tripartite cooperation among China, Japan, and South Korea. Therefore, financial support is an integral part of promoting effective cooperation. Firstly, full use should be made of the Asian Infrastructure Investment Bank, Silk Road Fund, and Special Fund for Asian Regional Cooperation. These regional financial institutions should provide an operational model of international investment and financing. Simultaneously, relying on international capital markets will allow continued expansion of funding sources through equity, bonds, and other financing methods. Policy banks can provide preferential loans to support energy companies in technological innovation. Commercial banks can provide various financing channels such as syndicated loans, mergers and acquisitions (M&A) loans, project financing, and issue energy securitization products such as green bonds to improve the efficiency of capital use by energy companies.

19.5.5 Market-Oriented Approach to Cooperation

Instead of building a committee as in the European Union, tripartite cooperation can be promoted through a market-oriented approach. In addition to the model of China–Japan–South Korea + X, existing CDM projects offer a pathway to promote cooperation. Therefore, establishing a carbon trading market among China, Japan, and

South Korea based on carbon trading schemes already existing in each country offers a trading market for using renewable energy to gain profits.

19.6 Conclusion

In order to ensure energy security while solving environmental pollution issues, China, Japan, and South Korea have issued a series of related policies to promote the utilization of renewable energy. In view of geographic conditions and technological limitations, the three countries have developed offshore wind and tidal energy where appropriate, as well as solar, wind, and biomass power generation.

There is a long history of cooperation among these three countries with foreign nations. However, cooperating on specific projects without a cooperation platform is less effective. Policy-oriented cooperation is vulnerable to the political environment. Therefore, a tripartite cooperation model is needed.

Under the framework of the East Asian Low-Carbon Community, we have demonstrated a pathway by which tripartite cooperation among China, Japan, and South Korea can be achieved in the field of renewable energy. The model consists of confirming cooperation intentions via ministerial meetings, deepening cooperation on existing projects and schemes, and technological cooperation in solar, wind, and biomass industries. A superregional grid could improve safeguard mechanisms for renewable electricity consumption, beginning in pilot areas. This is a market-oriented approach that should be financed by regional international financial institutions. A carbon emission trading market will play an essential role in incentivizing cooperation projects. Promoting comprehensive regional renewable energy cooperation in the three countries will not be easy; the system is large and complex but can significantly promote renewable energy and rapidly develop the economy in East Asian regions.

References

BP (2019) Statistical review of world energy. https://www.bp.com/en/global/corporate/energy-economics/statistical-review-of-world-energy.html. Accessed 14 April 2020

Calder K, Ye M (2008) The making of Northeast Asia. Stanford University Press, Stanford

China Electricity Council (2019) Data and statistics. https://english.cec.org.cn/menu/index.html?251. Accessed 24 May 2020

China-Korea Joint Ocean Research Center (CKJORC) (2020) The status quo of R&D of South Korea's marine technology and the necessity of the transformation of Sino-Korea cooperation (in Chinese). http://www.ckjorc.org/cn/admin/news/edit/UploadFile/kuaixun/201894159542.pdf. Accessed 14 June 2020

Economic and Social Commission for Asia and the Pacific (2017) Regional cooperation for sustainable energy in Asia and the Pacific. United Nations Publication, New York

Falkner R (2008) Business power and conflict in international environmental politics. Springer, pp 191–210

Freeman C (2015) Building an energy cooperation regime in Northeast Asia. Energy security cooperation in Northeast Asia, pp 158–191

IEA (2019) China power system transformation. https://www.iea.org/reports/china-power-system-transformation. Accessed 14 May 2020

IRENA (2019) Renewable capacity statistics 2019. International Renewable Energy Agency (IRENA), Abu Dhabi

JEMA (2019) Major laws related to new energy (in Japanese). https://www.jema-net.or.jp/Japanese/res/outline/kanren.html. Accessed 14 June 2019

KEMCO (2014) Overview of new and renewable energy in Korea 2013. Korea Energy Management Corporation, Yongin

Kim J, Park J, Choi H (2014) Towards a framework for energy cooperation in Northeast Asia: challenges and opportunities. The Asian Institute for Policy Studies, Seoul

Meng X, Serafettin Y (2020) Renewable energy cooperation in Northeast Asia: incentives, mechanisms, and challenges. Energ Strat Rev 29(1–7)

National Renewable Energy Center (2018). http://boostre.cnrec.org.cn/wp-content/uploads/2018/11/CREO-2018-Summary-CN.pdf (in Chinese). Accessed 14 June 2019

NEDO (2012) Marine energy: renewable energy technology white paper. https://www.nedo.go.jp/content/100544821.pdf. Accessed 14 June 2020

Qian X, Zhou W, Nakagami K (2015) International environmental cooperation between Japan and China toward east Asian sustainable development. J Policy Sci 9:81–95

The Ministry of Technology (2008) China-Japan-Korea renewable energy and new energy technology cooperation forum held in Beijing (in Chinese). http://www.gov.cn/gzdt/2008-08/06/content_1065352.htm. Accessed 14 June 2020

Trilateral Cooperation Secretariat (2019) Trilateral cooperation. https://tcs-asia.org/en/cooperation/overview.php?topics=15. Accessed 14 June 2019

UN Environmental, Frankfurt School-UNEP Center, BloombergNEF (2020) Global trends in renewable energy investment 2019. https://wedocs.unep.org/bitstream/handle/20.500.11822/29752/GTR2019.pdf?sequence=1&isAllowed=y. Accessed 14 June 2020

Zhang G (2018) Building an energy cooperation community between China and Southeast Asian neighboring countries (in Chinese). Energ Rev 01:46–47

Zhou W (2019) Global sustainability and creation of policy engineering. Speech, 1st forum by Research Center for Sustainability Science, OIC Campus, Ritsumeikan University

Chapter 20
Potential for Technical Cooperation Between Japan and China in a Third-Country Market

Yishu Ling, Xuepeng Qian, and Weisheng Zhou

Abstract In order to realize a low-carbon society over a wide area, it is crucial to transfer appropriate existing technologies internationally as a technological strategy in addition to developing innovative technologies. As analyzed in Chap. 1, Japan already has the world's highest level of energy-saving and high-efficiency technology, and it is costly to reduce CO_2 further domestically. In Japan, it is impossible to realize a dramatic reduction of CO_2 emission now. In other words, it has a market that contributes to the realization of a broader regional low-carbon society by transferring Japan's advanced low-carbon technology overseas. Bilateral cooperation still has excellent potential like Japan and China, but the potential for Japan and China to cooperate in low-carbon projects in third-country markets other than Japan and China remains excellent potential. In this study, we conducted a quantitative analysis to clarify the potential of four countries (Japan, China, India, Russia) via third-country market cooperation in the field of energy and environment under the Belt and Road Initiative proposed by China as a regional economic zone concept. Due to the potential for energy and environment reduction, we take India as a case study to find out the measures of technical cooperation between Japan and China on a third market. When Japan, China, and India cooperated to build a 660 MW supercritical coal-fired power plant in India, the economic profit of Japanese and Chinese companies is 237 million dollars per unit in total. Also, Japanese and Chinese companies will receive a total of 215 million dollars in environmental benefits annually. India has a total annual economic benefit of 633 million dollars and the environmental benefit of 13.391 million dollars. It was shown from the cooperation case with India in this study that the cooperation between Japan and

Y. Ling
Graduate School of Policy Science, Ritsumeikan University, Ibaraki, Osaka, Japan
e-mail: ps0290ih@ed.ritsumei.ac.jp

X. Qian
College of Asia Pacific Studies, Ritsumeikan Asia Pacific University, Beppu, Oita, Japan
e-mail: qianxp@apu.ac.jp

W. Zhou (✉)
College of Policy Science, Ritsumeikan University, Ibaraki, Osaka, Japan
e-mail: zhou@sps.ritsumei.ac.jp

W. Zhou et al. (eds.), *East Asian Low-Carbon Community*,
https://doi.org/10.1007/978-981-33-4339-9_20

China in the third-country market in the field of energy and environment has excellent potential in constructing a regional low-carbon society across borders. Therefore, we propose a "market principle + sharing of economic and environmental benefits + risk-sharing" method as a trilateral market cooperation method. The characteristics of this method are also consistent with the community spirit of sharing benefits and risks.

20.1 The Belt and Road Initiative and Third-Country Market Cooperation

The Belt and Road Initiative (BRI) is a global development strategy adopted by China in 2013. It promotes the orderly free flow of economic factors, efficient allocation of resources, and deep market integration, facilitates coordination of economic policies within the BRI, and creates joint openness through encouraging deeper regional cooperation. The concept aims to create a framework for regional economic cooperation that is comprehensive, balanced, and reciprocal and protects free trade and an open global economy. The goal of the BRI is to support the achievement of sustainable global development; the theoretical basis supporting the BRI is sustainability science. The Sustainable Development Goals (SDGs) formulated by the United Nations and the BRI may provide opportunities for growth of the global economy, which will require active input and support from China. BRI construction is an extraordinary approach to achieve the SDGs.

At present, there are 131 countries of the Belt and Road and 187 signed cooperation agreements under the BRI. Projects have started in several countries, such as Turkey, Saudi Arabia, and Bangladesh. Belt and Road countries cover Southeast Asia, South Asia, Central Asia, the Middle East, and Central and Eastern Europe across which there are large economic, technical, and policy disparities. According to the per capita annual income standard of the World Bank, 8% (5 countries) are low-income countries (less than 995 USD), 28% (18 countries) are low-middle-income countries (996–3895 USD), 35% (23 countries) are high-middle-income countries (3896–12,055 USD), and 29% (19 countries) are high-income countries (more than 12,055 USD). Belt and Road countries face economic development problems (i.e., poverty), regional environmental problems (i.e., air, water, and soil pollution, and waste treatment), and global environmental problems (i.e., climate change, loss of biodiversity, desertification, ozone depletion, and cross-border pollution). As countries of the Belt and Road are simultaneously facing domestic and international and short-term and long-term sustainable development challenges, there is great potential for bilateral and multilateral cooperation (third-country market cooperation), despite economic, technical, and social disparities among countries. The BRI is necessary for large-scale reform and the opening up of China and also provides an important platform for countries of the Belt and Road to solve various domestic issues and global sustainable development issues through international cooperation.

Table 20.1 International trends in third-country market cooperation

Date	Summary
September 2013	China advocates the Belt and Road Initiative
June 2015	Announcement of *Joint Statement on China-France Third-Country Market Cooperation*
December 2017	The Japanese government compiles the *Japan-China Private Economic Cooperation Guidelines in Third Countries* to promote cooperation of private companies in Japan and China in energy conservation, the environment, industrial development, and logistics and announces provision of support such as financing
May 2018	Prime Minister Lee Keqiang visits Japan, and the Chinese and Japanese governments conclude a memorandum of understanding (MOU) on third-party economic cooperation between Japan and China and agree to establish a Belt and Road Public-Private Council
October 2018	Prime Minister Abe holds the first Japan-China Third-Country Market Cooperation Forum in Beijing, signing 52 MOUs for cooperation projects
April 2019	Workshop on Japan-China Business Cooperation in Thailand is held, and Japan-China third-country market cooperation for Thailand begins
November 2020	Second Japan-China Third-Country Market Cooperation Forum due to be held in Tokyo

Source: https://www.yidaiyilu.gov.cn and https://www.meti.go.jp

"Third-country market cooperation" is a Chinese concept whereby Chinese and foreign companies cooperate economically (i.e., financing, technology transfer, capacity building) in a third-country market to realize industrial and infrastructure development and public welfare of the third country. It is a new system of international cooperation that enables cooperation among three countries to achieve an effect greater than the sum of its parts by improving people's livelihoods. At present, China has signed a memorandum of understanding regarding third-country market cooperation with 14 countries including Japan, Australia, Austria, Belgium, Canada, France, Italy, the Netherlands, Portugal, South Korea, Singapore, Spain, Switzerland, and the United Kingdom.

Table 20.1 shows international trends in third-country market cooperation. Since 2017, Japan has become more interested in third-country market cooperation, and the Japanese government has compiled the *Guidelines for Japan-China Private Economic Cooperation in Third Countries*, to guide cooperation with China in energy, the environment, industrial development, and logistics. Japan has announced that it will actively promote cooperation between Japanese and Chinese private companies.

In this chapter, we calculate the carbon dioxide and energy reduction potential of each country of the Belt and Road. Then, we estimate the economic and energy environment benefits to two countries with the highest potential, India and Russia. By analyzing a case study of cooperation between Japanese and Chinese high-efficiency coal-fired power generation businesses in India, we calculate the economic and energy environmental benefits obtained by the three countries involved and calculate the CO_2 reduction effect and economics of third-country market cooperation.

20.2 Potential Energy and Environment Reductions in Belt and Road Countries

To calculate the reduction potential of each country of the Belt and Road in the field of energy and environment, we divided the countries into six regions: Southeast Asia, South Asia, Central Asia, the Middle East, the former Soviet Union, and Central and Eastern Europe. Then, we calculated the CO_2 reduction and energy-saving potential when the six regions reach carbon dioxide emissions per gross domestic product (GDP) and energy consumption per GDP equivalent to those in Japan in the same year.

Factors that influence CO_2 emissions per GDP (which is a measure of CO_2 emission intensity) are energy consumption structure, industrial structure, and energy use efficiency. Factors that influence energy consumption per GDP (which is a measure of energy consumption intensity) are industrial structure and energy efficiency. Therefore, a third country that could cooperate with Japan and China in this field would need an industrial structure similar to Japan. In this study, industrial structure is judged mainly by the ratio of increased industry value to GDP. Since 2000, the ratio of the increased value of the Japanese manufacturing industry to GDP has increased by approximately 19–22%. Japan is the third largest economy globally; it is impossible to choose a third country with the same ratio of increased value in the manufacturing industry to GDP as Japan. Therefore, it is reasonable to classify a country as having a similar industrial structure to Japan if its ratio of manufacturing industry value to GDP has been 15% or more since 2000 (Table 20.2). Table 20.3 shows the reduction in energy consumption and CO_2 emissions of these countries in 2014 if the intensity of their CO_2 emissions and energy consumption was equal to that of Japan.

Table 20.2 Target countries with potential to reduce carbon dioxide emissions and energy consumption through third-country market cooperation

Regions	Countries
Southeast Asia	Indonesia, the Philippines, Brunei, Vietnam, Singapore, Thailand, Cambodia Malaysia
South Asia	India, Pakistan, Bangladesh, Sri Lanka
Central Asia	Kazakhstan, Turkmenistan, Kyrgyzstan
Middle East	Israel, Qatar, Egypt, Turkey
Central and Eastern Europe	Poland, Lithuania, Estonia, Czech Republic, Slovak Republic, Hungary, Slovenia, Serbia, Romania
The former Soviet Union	Russia, Ukraine, Belarus, Armenia

Source: summarized by the author

Table 20.3 Reduction in CO_2 emissions and energy consumption when national CO_2 emission intensity (CO_2 emissions per GDP) and energy consumption intensity (energy consumption per GDP) equal those of Japan in 2014

Regions	CO_2 reduction (Mt)	Carbon dioxide reduction rate (%)	Energy consumption reduction (oil-equivalent Mt)	Energy consumption rate (%)
Southeast Asia	801	58	329	61
South Asia	1862	75	699	75
Central Asia	258	79	82	77
Middle East	838	60	305	60
Central and Eastern Europe	265	44	114	48
Former Soviet Union	1429	71	639	75

According to the World Bank database, there is no data of Vietnam. Therefore, we did not calculate Vietnam in Southeast Asia

The three countries in Central Asia have the highest carbon dioxide and energy reduction rates but low energy efficiency (Table 20.3). Therefore, the potential for reduction in energy consumption is low. Countries in Central and Eastern Europe also have low potential because of small energy consumption reductions.

Conversely, many developing countries in South Asia consume a large amount of energy, and their energy efficiency remains low. Notably, India is the biggest emitter in South Asia and depends on coal-fired power generation, which is a large source of CO_2 emissions, ranking India as the third highest emitter of CO_2 globally in 2014. As a result, South Asia has the highest reduction potential when compared to Japan. Countries of the former Soviet Union inherited industrial systems of the Soviet era. Many problems exist in the former Soviet Union due to significant delays in development of energy technology and high energy consumption coupled with low energy efficiency. Therefore, the amount of carbon dioxide reduction and energy consumption reduction of the former Soviet Union countries is smaller than that of South Asia but is much higher than in other regions.

South Asia has the highest reduction potential when reaching Japan's level of CO_2 emission intensity and energy consumption intensity, with India accounting for 92% and 91%, respectively, of these totals. Former Soviet Union countries have the second-highest reduction potential. Russia, as the largest emitter and consumer of the former Soviet Union group of countries, accounts for 82% of reductions in energy consumption and 83% of CO_2 reductions. Therefore, India and Russia have the highest potential for reductions in CO_2 emission intensity and energy consumption intensity of the Belt and Road countries. Accordingly, the remainder of this chapter analyzes the potential of energy and environmental cooperation in a third-country market in India and Russia through Japan-China cooperation.

20.3 Potential Outcomes in India and Russia of Cooperation in Energy and the Environment with China and Japan

20.3.1 Economic and Energy Profiles of Japan, China, India, and Russia

Tables 20.4 and 20.5 summarize the basic economic and energy statistics of Japan, China, India, and Russia in 2014. Japan has the highest energy efficiency and lowest CO_2 emissions. Therefore, it is reasonable to assume that India and Russia both have significant potential to reduce both CO_2 emissions and energy consumption if able to reach the same levels as Japan. In contrast, China has lower energy intensity than India and Russia, which means a relatively significant potential if India and Russia reach the levels of China. However, development of both China and India is dependent on fossil fuel, which results in a higher carbon intensity than Russia, where development depends on liquefied natural gas (LNG). Therefore, when taking the emissions of China as a reference, only India shows potential reductions.

Table 20.4 Basic economic and energy statistics of Japan, China, India, and Russia in 2014

	Nominal GDP (US$ billion)	Real GDP (US$ billion)	Population (million)	Primary energy consumption (oil-equivalent Mt)	CO_2 emissions (Mt)
Japan	48,504	59,163	127	440	1214
China	104,824	83,172	1364	3051	10,292
India	20,391	21,250	1293	825	2238
Russia	20,637	16,907	144	710	1705

Table 20.5 Energy consumption per GDP and CO_2 emissions per GDP in Japan, China, India, and Russia in 2014

	Energy consumption per nominal GDP (oil-equivalent kg/US$)	Energy consumption per real GDP (oil-equivalent kg/US $)	CO_2 emission per nominal GDP (kg/US$)	CO_2 emission per real GDP (kg/US$)
Japan	0.09	0.07	0.25	0.21
China	0.29	0.37	0.98	1.24
India	0.4	0.38	1.10	1.05
Russia	0.34	0.42	0.83	0.99

20.3.2 *Calculating Overall Reduction Potentials for India and Russia*

As described in Sect. 20.3.1, India has the potential to reduce both energy consumption per GDP and CO_2 emissions per GDP if they match the levels of these in Japan and China. Russia could reduce energy consumption per GDP by matching that in Japan and China. These potential reductions are shown in Fig. 20.1.

If India and Russia had matched the CO_2 emission intensity of Japan from 1992 to 2014, the total amount of CO_2 emissions reductions would be 57,718 Mt, or 85% of the cumulative CO_2 emissions in India and Russia from 1992 to 2014. In contrast, if India and Russia had matched the CO_2 emission intensity of China over the same period, there would have been a reduction only from 1998 to 2003.

If India and Russia had matched the energy consumption intensity of Japan from 1992 to 2014, the total amount of energy consumption reduction would have been 23,402 oil-equivalent Mt. In contrast, as seen with CO_2 emission reduction, if India and Russia had matched the energy consumption intensity of China, there would have been a reduction only from 1997 to 2004.

There are reductions from 1997 to 2003 or 2004 for India and Russia when matching China's carbon intensity and energy consumption. Economic crisis in Russia caused a sharp increase in energy consumption per GDP, which creates a potential for reduction if Russia reached the energy consumption intensity of China during this period. Since 2012, energy consumption per GDP has continued to decrease due to increases in tertiary industry and improvement of energy utilization technology in China. As a result, there is also reduction potential for energy consumption in both India and Russia after 2012 (Fig. 20.2).

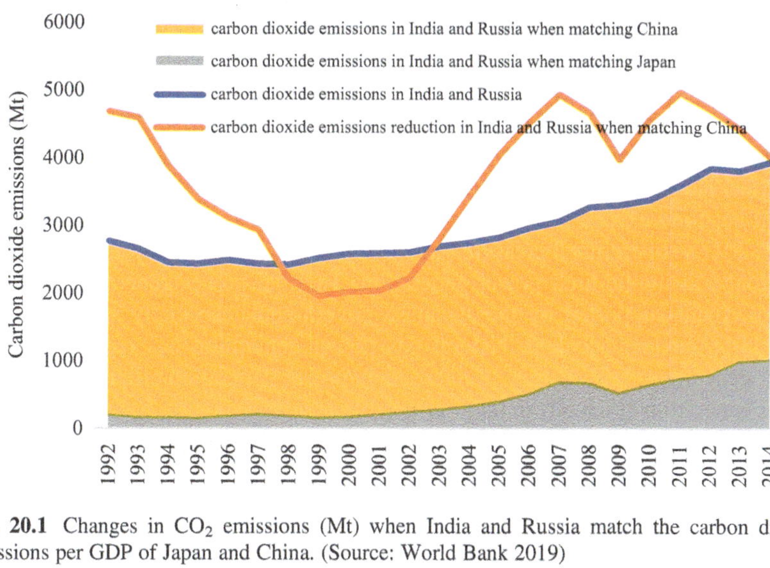

Fig. 20.1 Changes in CO_2 emissions (Mt) when India and Russia match the carbon dioxide emissions per GDP of Japan and China. (Source: World Bank 2019)

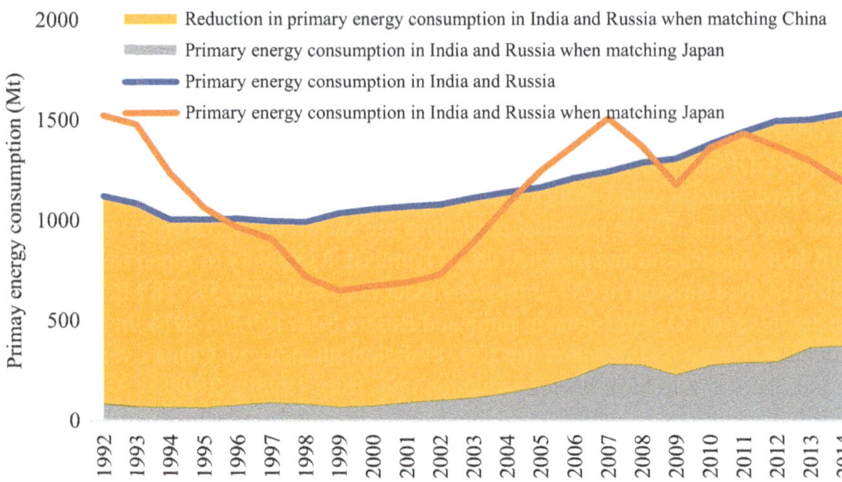

Fig. 20.2 Changes in energy consumption (Mt) when India and Russia match the energy consumption per GDP of Japan and China. (Source: World Bank 2019)

20.3.3 Reduction Potential of Thermal Power Generation Projects in India and Russia

20.3.3.1 Overview of Thermal Power Generation in India and Russia

The CO_2 emission and energy consumption intensities of India and Russia are significantly higher than those of Japan and are influenced by energy utilization efficiency, industrial structure, and energy structure. Here, we focus on energy efficiency, of which power generation efficiency is an essential factor. Both India and Russia have a high proportion of thermal power generation. Therefore, we calculated the economic, environmental, and energy benefits to India and Russia if they were to reach the power generation efficiency of Japan (Fig. 20.3).

Prior to 2011, Japan has the highest thermal power generation among the three countries. However, following rapid development in India, its thermal power generation began to exceed that in Japan from 2011. In India's Twelfth Five-Year Plan (2012–2017), supercritical pressure coal-fired power generation accounts for 39% of new coal and lignite-fired power generation. Thermal power generation has been promoted and supported by the Japanese government, especially since the 2011 Fukushima nuclear accident. Natural gas power generation has replaced nuclear power generation, leading to a 3% increase in power generation efficiency in Japan. Power generation efficiency in India, which has received support, has also improved. In contrast, due to the impact of the Lehman Shock in 2008, a shale gas revolution in 2011, and economic sanctions from the United States in 2014, domestic power generation in Russia has remained unchanged, and power generation efficiency has not improved.

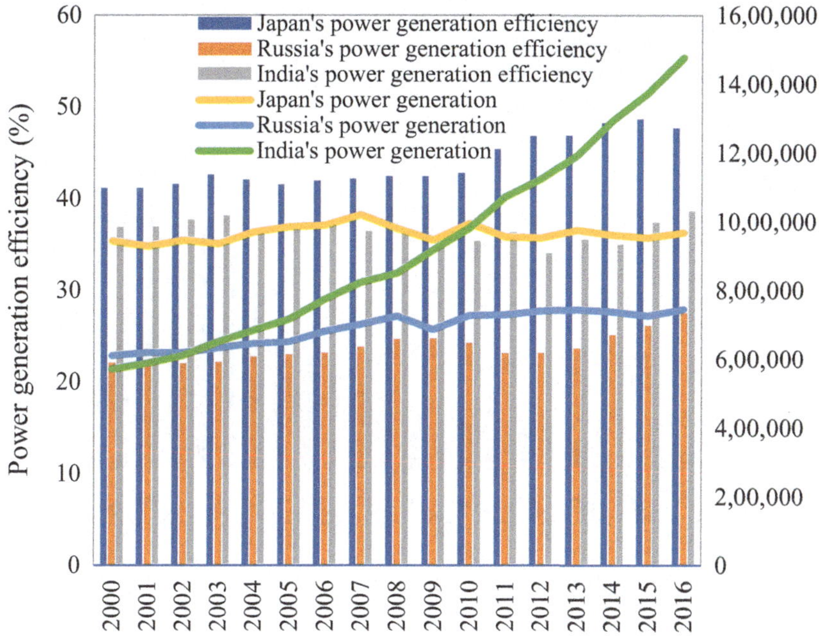

Fig. 20.3 Total thermal power generation and efficiency in Japan, Russia, and India. (Source: IEA 2019)

20.3.3.2 Potential Reductions If India and Russia Match Power Generation Efficiency of Japan

Since India depends mainly on coal-fired power generation and Russia on natural gas-fired power generation, we estimate the potential for economic and energy environment benefits if India and Russia can achieve the same level of coal-fired power generation efficiency and LNG power generation efficiency, respectively, as Japan.

From 1992 to 2016, if the coal-fired power generation efficiency of India improved to a level equivalent to that of Japan, 743 oil-equivalent Mt of energy consumption and 556 Mt of carbon dioxide emissions could have been reduced (Fig. 20.4). By 2006, the amount of coal reduction would have exceeded the volume of coal imported. After 2006, coal imports could have been reduced by more than 55%. In India's Twelfth Five-Year Plan (2012–2017), Japan provided support for coal-fired power generation. If Japan and China cooperate in a third-country market in India, the low cost of Chinese technology for coal-fired power generation could further improve the situation in India.

Russia could have reduced 3950 oil-equivalent Mt energy consumption and 1367 Mt carbon dioxide emissions from 1992 to 2016 if its LNG power generation efficiency had matched that of Japan (Fig. 20.5). If all of the excess LNG was exported, the total economic benefit would exceed 727.3 billion USD.

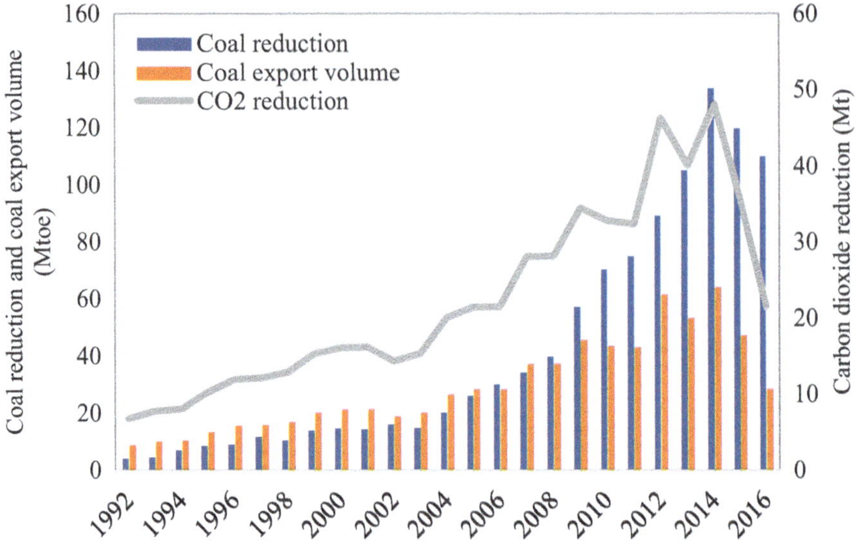

Fig. 20.4 Potential for improvements in the energy environment when India reaches the coal-fired power generation efficiency of Japan. (Source: World Bank 2019)

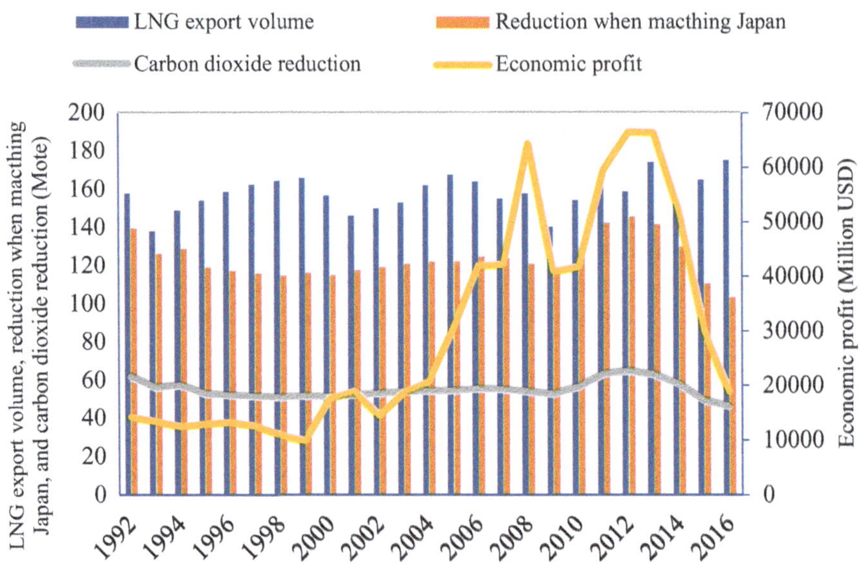

Fig. 20.5 Improvements in economic and energy environment when Russia matches the LNG power generation efficiency of Japan. (Source: World Bank 2019)

An improvement in economics and the energy environment is observed when Russia reaches the same LNG power generation efficiency as Japan. The oil shock of 2003–2008 increased the price of natural gas, which would have significantly increased Russia's economic profits. However, the 2009 financial crisis caused the price of natural gas to fall and Russia's economic profits to decline. From 2010 to 2013, the price of natural gas and potential profits for Russia rose again with economic recovery. However, the shale revolution in 2013 will have a considerable impact on natural gas, significantly lowering the price of LNG and reducing the potential economic profits for Russia again.

20.4 Cooperation of Japan and China in a Third-Country Market

20.4.1 Types of Japan-China Cooperation in a Third-Country Market

Based on our analysis, there are significant reduction potentials if India and Russia could reach levels of energy consumption intensity and CO_2 emission intensity in Japan. These reduction potentials could create a vast environmental market. However, when Japan cooperates with India and Russia alone, there are many engineering, procurement, and construction (EPC) issues such as barriers to plant cost (economic efficiency) and transfer of appropriate technology that considers localization, such as financing. Therefore, for Japanese companies to access new markets in the global economy, a new form of business is needed that surpasses the limits of self-sufficiency and overcomes challenges by cooperating with, and complementing the strengths of, foreign companies. It is desirable to promote third-country cooperation to improve competitiveness regardless of infrastructure (Ministry of Economy, Trade, and Industry 2019).

For China to promote development of its domestic economy and realize the BRI of a wide-reaching economic zone, it should utilize the advanced technologies of developed countries. Then, the development needs of Belt and Road countries are effectively linked, and individualized benefits can be coordinated in such a way as to achieve a win-win result with reciprocal complementation. Detailed quantitative analysis of a case study is provided later in this chapter.

Examples of Japan-China cooperation in third-country markets in the fields of energy and the environment are summarized in Table 20.6 and can take the following forms.

Type 1 Japan supervises the investment, and China manages the project.
Type 2 Japan and China oversee investment and project management together.
Type 3 Japan provides the technology, and China provides EPC.

Table 20.6 Examples of Japan-China cooperation in third-country markets in the fields of energy and the environment

Sort	Field	Overview
Type 1	Resource mining	Japan financed a resource mining operation in Egypt with approximately $50 million. China manages the project generally
Type 2	Natural gas mining	The Dutch Royal Dutch Shell, China Petroleum and Natural Gas Corporation (CNPC) Mitsubishi Trading Co., Korea Gas Corporation (KOGAS), and Canada jointly carry out a natural gas mining business. The result was 12 million tons of LNG generated per year and obtained a 25-year export license (no export destination restrictions)
Type 3	Hydropower	China Hydroelectric Process Advisory Group Co., Ltd., and Toshiba Hydroelectric Equipment (Hangzhou) Co., Ltd., have provided power generation facilities with a total output of 260 MW to the Chun Son Hydroelectric Power Plant in Vietnam
Type 3	Hydropower	Zhejiang Guomao Dongfang Electric Machinery Co., Ltd., and Toshiba Hydroelectric Equipment (Hangzhou) Co., Ltd., will provide a total output of 288 MW to the Upper Yeywa Hydroelectric Power Plant in Myanmar

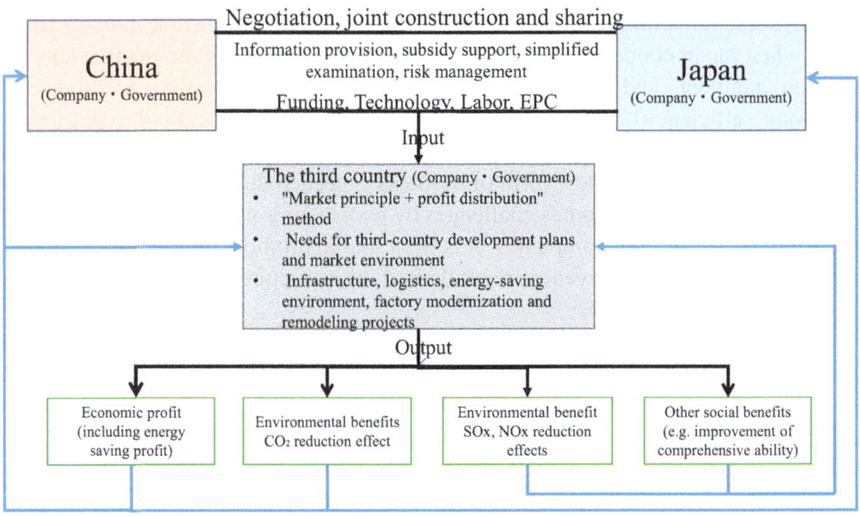

Fig. 20.6 Schematic diagram of Japan-China cooperation in a third-country market. (Source: Zhou 2019)

20.4.2 *Process of Japan-China Cooperation in a Third-Country Market*

The process of cooperation between Japan and China in a third-country market is shown in Fig. 20.6. Japanese and Chinese companies provide funds, technology, and EPC following the basic principles of mutual commerce, mutual development, and mutual enjoyment and support or undertake projects in fields such as infrastructure, logistics, and energy conservation that meet the needs of third-country development plans in civilian environments. The Chinese and Japanese governments provide policy support, such as tax incentives, and simplification of inspections. Furthermore, Japan, China, and third countries share the economic benefits generated by third-country market cooperation. Specifically, Japan and China share the environmental benefits of carbon dioxide reduction, and the third countries enjoy the environmental benefits of SO_x and NO_x reduction and other benefits.

20.5 Quantitative Analysis of Japan-China Cooperation in a Third-Country Market

20.5.1 Introduction of Case Study in India

As a case study, we analyzed high-efficiency coal-fired power generation businesses developed by Japan in India. In 2012, Toshiba of Japan provided two 660 MW supercritical thermal power generation facilities to a primary coal-fired power plant of the Indian Thermal Power Corporation (hereafter NTPC) for an order price of 3.15 billion USD. Commercial operation began in 2016. The main form of power generation in India is subcritical coal-fired power generation, which has a thermal efficiency of 32% and a carbon dioxide emission coefficient of 1.32 kg/kWh. In comparison, the thermal efficiency of the supercritical pressure power generation facilities provided by Japan is 38%, and the carbon dioxide emission coefficient is 1.06 kg/kWh, rendering it better for the environment.

We analyzed the profits generated by a high-efficiency coal-fired power generation business when Japan and China cooperate and the benefits of remodeling India's coal-fired power plants. Although it is not currently in the framework of the Clean Development Mechanism (CDM) of the Kyoto Protocol, it is possible to save carbon dioxide by introducing supercritical coal-fired power generation equipment rather than replacing the original subcritical equipment. The environmental benefits calculated under the CDM framework allow trading of CO_2 reduction as credits (Ren et al. 2011).

Expected profit in the case of cooperation between Japan and China in India's market is estimated based on the schematic diagram shown in Fig. 20.7. In this example of third-country market cooperation, a 660 MW supercritical pressure coal-fired power generation facility is installed in an Indian power plant, replacing the original 660 MW subcritical pressure coal-fired power generation facility. Both Japan and China will enter into India's market, and the equipment will be locally produced. While China will produce the EPC and raise part of the funds, Japan will provide technology and materials for the equipment and raise another part of the funds. India will locally produce the equipment. Quantitative analysis determines the total provision of land, personnel, electricity, and funds by India. Then, the economic profit of selling electricity and the environmental profit of carbon dioxide

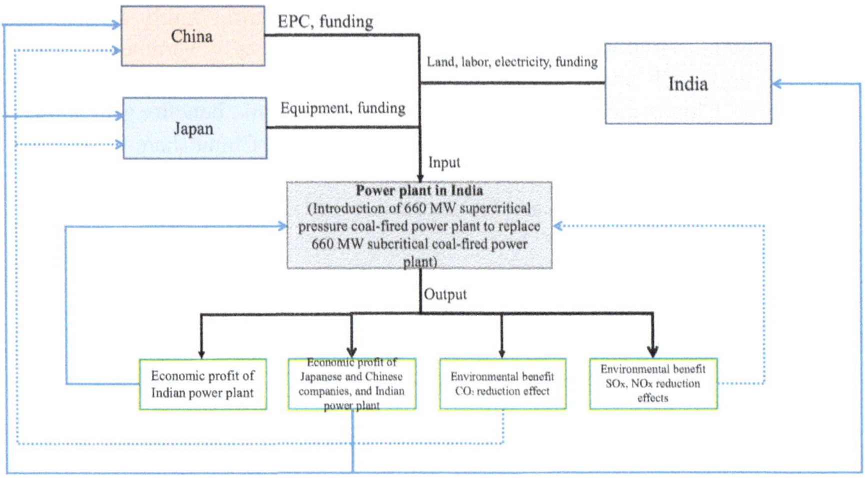

Fig. 20.7 Schematic diagram of Japan-China cooperation in a third-country market for high-efficiency coal-fired power generation in India. (Source: Zhou 2019)

reduction comprise the total profit of the project. Profit distribution among the three countries is based on the contribution of each country to the project. Japan and China distribute the environmental benefits of reducing carbon dioxide, and India benefits from desulfurization, denitrification, and dedusting through the environmental benefits of SO_x, NO_x, and dust reduction.

20.5.2 *Methodology*

We estimated the economic and environmental benefits to Japanese, Chinese, and Indian companies when Japan and China (1) provide new power generation facilities for coal-fired power generation in India and (2) modify existing power plants in India.

20.5.2.1 Economic Benefits to Japanese, Chinese, and Indian Companies

Economic profit refers to the profit that Japanese, Chinese, and Indian companies receive from the project and is calculated using the following equations:

$$C_{\mathrm{m}} = C_{\mathrm{mj}} + \left(C_{\mathrm{in}} - C_{\mathrm{j}}\right) \times S_{\mathrm{q}} + \left(S_{\mathrm{ain}} - S_{\mathrm{aj}}\right) \times \mathrm{POP}_{\mathrm{m}} + \left(P_{\mathrm{elin}} - P_{\mathrm{elj}}\right) \times \mathrm{El} \quad (20.1)$$

$$C_{\text{mat}} = C_{\text{mj}} - C_{\text{j}} \times S_{\text{q}} - S_{\text{aj}} \times \text{POP}_{\text{m}} - P_{\text{elj}} \times \text{El} \tag{20.2}$$

$$B_{\text{c}} = P - (C_{\text{m}} + C_{\text{EPC}}) \times (1 + R)^{n} \tag{20.3}$$

where, B_c is the corporate profit; P is the order amount; C_{mj} and C_m are the cost of manufacturing equipment in Japan and India, respectively; S_q is the area of the plant for manufacturing equipment; C_j and C_{in} are the prices of land in Japan and India, respectively; S_{aj} and S_{ain} are per capita salaries for manufacturing in Japan and India, respectively; POP_m is the number of people required for manufacturing; El is the electricity required for manufacturing equipment; P_{elj} and P_{elin} are the price of electricity in Japan and India, respectively; C_{EPC} is the cost of EPC; R is the interest rate; and n is the number of years.

Profits are allocated to Japanese, Chinese, and Indian companies based on their investment ratio, which in this case study is equal to the ratio of equipment cost and EPC business cost. The economic profits of each company in Japan, China, and India are calculated using the following equations:

$$B_{\text{cj}} = B_{\text{c}} \times \frac{C_{\text{mat}} \times (1 + R)^{n}}{(C_{\text{m}} + C_{\text{EPC}}) \times (1 + R)^{n}} \tag{20.4}$$

$$B_{\text{cc}} = B_{\text{c}} \times \frac{C_{\text{EPC}} \times (1 + R)^{n}}{(C_{\text{m}} + C_{\text{EPC}}) \times (1 + R)^{n}} \tag{20.5}$$

$$B_{\text{cin}} = B_{\text{c}} \times \frac{(C_{\text{in}} \times s_{\text{q}} + S_{\text{ain}} \times \text{POP} + p_{\text{elin}} \times \text{El}) \times (1 + R)^{n}}{(C_{\text{m}} + C_{\text{EPC}}) \times (1 + R)^{n}} \tag{20.6}$$

where, B_{cj} is the economic profit for Japanese companies, B_{cc} is the economic profit for Chinese companies, B_{cin} is the economic profit for Indian companies, and C_{mat} is the cost of material.

20.5.2.2 Environmental Benefits to Japanese and Chinese Companies

The total amount of power generated in a third country, in this case India, is given in the following equation:

$$Q_{\text{tot}} = F_{\text{cap}} \times R_{\text{q}} \times Y \times T \tag{20.7}$$

where, Q_{tot} indicates the amount of power generated, F_{cap} is the installed capacity, R_q is the operating rate, Y is the number of years of operation, and T indicates the annual operating hours.

The environmental benefits to Japanese and Chinese companies are the carbon dioxide reduction effects of power plants after remodeling, calculated using the following equations:

$$B_{\mathrm{jenv1}} = (Q_{\mathrm{tot}} \times \mathrm{CO}_{2\mathrm{sub}} - Q_{\mathrm{tot}} \times \mathrm{CO}_{2\mathrm{sc}}) \times \left(C_{\mathrm{jCO_2}} - C_{\mathrm{inCO_2}}\right) \tag{20.8}$$

$$B_{\mathrm{jenv2}} = (Q_{\mathrm{tot}} \times \mathrm{CO}_{2\mathrm{sub}} - Q_{\mathrm{tot}} \times \mathrm{CO}_{2\mathrm{sc}}) \times \frac{C_{\mathrm{mat}} \times (1+R)^n}{(C_{\mathrm{m}} + C_{\mathrm{EPC}}) \times (1+R)^n} \times \left(C_{\mathrm{jCO_2}} - C_{\mathrm{inCO_2}}\right) \tag{20.9}$$

$$B_{\mathrm{cenv}} = (Q_{\mathrm{tot}} \times \mathrm{CO}_{2\mathrm{sub}} - Q_{\mathrm{tot}} \times \mathrm{CO}_{2\mathrm{sc}}) \times \frac{C_{\mathrm{EPC}} \times (1+R)^n}{(C_{\mathrm{m}} + C_{\mathrm{EPC}}) \times (1+R)^n} \times \left(C_{\mathrm{cCO_2}} - C_{\mathrm{inCO_2}}\right) \tag{20.10}$$

where, B_{ienv1} is the environmental benefit to Japan when operating alone; B_{jenv2} and B_{cenv} are the environmental benefit to Japan and China, respectively, when the two countries cooperate; $\mathrm{CO}_{2\mathrm{sub}}$ and $\mathrm{CO}_{2\mathrm{sc}}$ are the carbon dioxide coefficient before and after remodeling, respectively; and $C_{\mathrm{jCO_2}}$, $C_{\mathrm{cCO_2}}$, and $C_{\mathrm{inCO_2}}$ are the CO_2 reduction costs in Japan, China, and India, respectively.

20.5.2.3 Environmental Benefits to Indian Power Plants

The environmental benefits to Indian power plants are mainly the reduction of sulfur oxides, nitrogen oxides, and dust. Since SO_x and NO_x differ depending on the type of coal, we used the emission factor of standard coal. The calculation method for sulfur oxides, nitrogen oxides, and dust equals the energy consumption saved (tons of coal equivalent) multiplied by the emission factors for sulfur oxides, nitrogen oxides, and dust, respectively. The benefit of sulfur oxide reduction is calculated using the following equation:

$$B_{\mathrm{tenvsox}} = \mathrm{Re}_{\mathrm{en}} \times \mathrm{Fa}_{\mathrm{so}_x} \times C_{\mathrm{so}_x} \times Y \tag{20.11}$$

where, B_{tenvsox} shows the benefits of sulfur oxide reduction in the third country (India), $\mathrm{Re}_{\mathrm{en}}$ shows the energy consumption reduction, $\mathrm{Fa}_{\mathrm{so}_x}$ shows the sulfur oxides emission factor for standard coal, and C_{so_x} shows the cost of sulfur oxide treatment.

Benefits of NO_x reduction can be calculated using the following equation:

$$B_{\mathrm{tenvnox}} = \mathrm{Re}_{\mathrm{en}} \times \mathrm{Fa}_{\mathrm{no}_x} \times C_{\mathrm{no}_x} \times Y \tag{20.12}$$

where, B_{tenvnox} is the benefit of reduction of nitrogen oxides, $\mathrm{Fa}_{\mathrm{no}_x}$ shows the nitrogen oxides emission factor for standard coal, and C_{no_x} is the treatment cost of nitrogen oxides.

The dust reduction effect is calculated according to the following equation:

$$B_{\mathrm{tenvdust}} = \mathrm{Re}_{\mathrm{en}} \times \mathrm{Fa}_{\mathrm{dust}} \times C_{\mathrm{dust}} \times Y \tag{20.13}$$

where, $B_{tenvdust}$ is the benefit of dust reduction, Fa_{dust} shows the dust emission factor for standard coal, and C_{dust} is the dust disposal cost.

The environmental benefit to the third country, India, is the sum of the reduction effects of sulfur oxides and nitrogen oxides and can be calculated using the following equation:

$$B_{tenv} = B_{tenvsox} + B_{tenvnox} + B_{tenvdust} \tag{20.14}$$

where, B_{tenv} is the total environmental benefit to the third country (in this case India).

20.5.2.4 Economic Profits of Indian Power Plants

Fuel cost equals energy consumption saved multiplied by coal price. Sales of power are obtained by multiplying the amount of power generated by one coal-fired power plant (Eq. 20.7) by the power sale price and power transmission efficiency. It is assumed that model type does not affect the operating cost, which is calculated by adding the annual operating cost in India to the number of operators multiplied by operator salary. The annual profit of the power plant (including desulfurization, denitrification, and dust removal) equals the annual sales of the power plant minus the difference of annual operating cost and costs of desulfurization, denitrification, and dust removal. The final economic profit for India equals profit from CO_2 sales plus profit from the Indian power plants. The economic profit of the Indian power plant is estimated here mainly as the profit of the company and can be calculated using the following equation:

$$B_t = Q_{tot} \times P_e \times R_{tra} - (P + \mathrm{POP} \times S_a \times Y + \mathrm{EN}_{sc} \times P_c \times Y + M \times Y) - C_a \tag{20.15}$$

$$B_{tenvco2} = Q_{tot} \times \mathrm{Fa}_{CO_2} \times P_{CDM} \tag{20.16}$$

$$B_{te} = B_t - B_{tenv} + B_{tenvco2} \tag{20.17}$$

where, B_t is the economic profit of the third country (India); P_e is the selling price of electricity; POP is the number of employees at the power plant; S_a is the employee salary; EN_{sc} is the annual coal consumption of the power plant after remodeling; P_c is the price of coal; M is the annual operating cost excluding personnel costs; C_a is the original power generation facility disposal cost; Fa_{CO_2} shows the carbon dioxide emission factor for standard coal; P_{CDM} is the price of per CDM credit; and B_{te} shows the annual profit of the power plant (excluding desulfurization, denitrification, and dust removal).

20.5.3 Data Sources

The thermal efficiency, carbon dioxide emission coefficient, total facility cost, annual operating cost, and operating years of the supercritical and subcritical coal-fired power generation facilities are taken from the *International Projects for Increasing the Efficient Use of Energy, Project Formation Research on High-efficiency Clean Coal Technology*, and the *Follow-up Study Concerning Development of Project Utilizing Super Critical Coal-fired Thermal Power Plant and CCS Technologies in Bulgaria* (NEDO 2014a, b). EPC costs are taken from the *project formation study for ultra-supercritical power generation project by co-firing of imported coal in Vietnam* (NEDO 2014a, b), and the selling price of electricity in India is taken from the *Survey on Indian Electric Power Business Environment* (NEDO 2014a, b). The cost of disposing of coal-fired power generation equipment refers to "a list of specifications of each power source" (METI 2019). The annual income of the operator is India's per capita income, and the transmission efficiency of India is estimated by IEA. The price of coal is taken from the BP yearbook. We also take Toshiba's projects in India as refereneces (Toshiba Hydroelectric Equipment (Hangzhou) Co., Ltd. 2018, Toshiba Systems Corporation 2018). The calculation assumes that Japan has provided 38 power generation facilities to India so far, and that one-third of these will be supercritical coal-fired. Production of all 13 sets of supercritical coal-fired power generation equipment is assumed to take place in India. Table 20.7 lists the variables used in the calculation.

20.5.4 Benefits to Japan, China, and India from the Introduction of Supercritical Coal-Fired Power Generation

20.5.4.1 Benefits to India

This study assumes that the annual operating costs (excluding fuel costs) and number of operators are constant with the introduction of 660 MW units of supercritical coal-fired power generation equipment. Economic and environmental profits for India resulting from installation are given in Table 20.8.

Our calculations show that the installation of a 660 MW supercritical coal-fired power plant can reduce 1.43 Mt of carbon dioxide at a cost of 10.75 USD/ton. Assuming that SO_x, NO_x, and dust reduction costs are 100 USD/ton, annual treatment costs for sulfur oxides, nitrogen oxides, and dust in India would be 523,000, 505,000, and 311,000 USD, respectively. Accounting for these treatment costs, the carbon dioxide reduction cost of introducing a supercritical coal-fired power generation facility is then 9.82 USD/ton. Introducing supercritical pressure coal-fired power generation equipment can reduce 0.32 coal-equivalent Mt of coal per unit per year. If the import price of coal is 75 USD/ton, then annual fuel cost can be

Table 20.7 Variables related to high-efficiency coal-fired power generation

No.	Explanation	Hypothetical data
1	Order for supercritical coal-fired power generation equipment (in case of modification to power plant)	15.75 billion dollars/unit
2	The total cost of supercritical coal-fired power generation facilities manufactured in Japan	8.54 billion dollars/unit
3	Years of operation	40 years
4	Annual operating costs (excluding personnel costs)	1.3 billion dollars
5	The thermal efficiency of subcritical coal-fired power plant	32%
6	The thermal efficiency of supercritical coal-fired power plant	38%
7	The CO_2 emission factor of subcritical coal-fired power plant	1.32 kg/kWh
8	The CO_2 emission factor of supercritical coal-fired power plant	1.06 kg/kWh
9	EPC cost in China	700 dollars/kWh
10	EPC cost in Japan	1000 dollars/kWh
11	Required area for equipment manufacturing	7000 m^2
12	Number of people required for manufacturing	100 people
13	Electric power required for manufacturing	10,000 kWh/month
14	Electricity price in India	0.09 dollars/kWh
15	Electricity price in Japan	0.13 dollars/kWh
16	Land price in India	400 dollars/m^2
17	Land price in Japan	1746 dollars/m^2
18	Annual labor cost to manufacture in India	1995 dollars/year
19	Annual labor cost to manufacture in Japan	40,779 dollars/year
20	The annual income of the power plant operator	1995 dollars/year
21	CO_2 reduction costs in China	70 dollars/ton
22	CO_2 reduction costs in Japan	200 dollars/ton
23	Interest rate	1%
24	Number of operators	100 people
25	Transmission efficiency in India	75.2%
26	Coal price	75 dollars/ton
27	Disposal cost	0.23 billion/unit
28	Machine manufacturing time	1 year
29	Project operation years (including machine manufacturing time)	2 years
30	Units manufactured in India	13 units
31	Units manufactured in Japan	38 units
32	Capacity utilization rate	95%
33	Dust emission coefficient of standard coal	0.0096 tons/ton of coal equivalent (tce)
34	Sulfur oxide emission factor of standard coal	0.0165 tons/tce
35	Nitrogen oxide emission factor of standard coal	0.0156 tons/tce

Table 20.8 Economic and environmental benefits to India of introduction of one 660 MW supercritical coal-fired power plant

Category	Before Introduction	After Introduction	Difference between before and after Introduction
Carbon dioxide emission (Mt/year)	7.25	5.82	−1.43
Sulfur oxide emissions (kt/year)	33.75	28.52	−5.23
Nitrogen oxide emissions (kt/year)	32.02	26.97	−5.05
Dust emissions (kt/year)	19.71	16.60	−3.11
Energy consumption (Mt tce/year)	2.05	1.73	−0.32
Equipment cost (billion dollars/year)	0	0.39	0.39
Fuel cost (billion dollars/year)	1.54	1.30	−0.24
Electricity sales (billion dollars/year)	3.71	3.71	0.00
operating expenses (billion USD/year)	1.32	1.32	0.00
The annual profit of power plant (excluding the cost of desulfurization, denitrification, and dedusting) (billion USD/year)	0.85	0.7	−0.15
The annual profit of power plant (including the cost of desulfurization, denitrification, and dedusting) (billion USD/year)	0.76	0.63	−0.13
Annual economic benefits of power plants under the CDM framework (including desulfurization, denitrification, dedusting) (billion USD/year)	0.76	0.67	−0.09

reduced by 240 million USD. When trading the CO_2 reductions under the CDM framework as carbon credits, the power plants could generate an annual environmental benefit of 67 million USD.

20.5.4.2 Changes in Profits with Japan-China Cooperation and with Support from Japan Only

Tables 20.9 and 20.10 show the changes in profits with Japan-China cooperation and if only Japan is involved. Although the machine cost itself is fixed, it is affected by differences in land prices between India and Japan. The machine cost with only Japanese involvement is assumed as in Table 20.7. As shown in Eq. (20.1), the cost of local production in India equals EPC cost per kilowatt multiplied by facility capacity, and interest repayment equals interest rate multiplied by the sum of machine cost and EPC cost. In the case of Japan's single entrance to the Indian market, economic profit for Japan equals the price of the order minus the sum of the cost of manufacturing in Japan and the interest rate. In the case of Japan-China cooperation, Eqs. (20.4), (20.5), and (20.6) are used. Carbon dioxide credits for Japanese and Chinese companies are calculated using Eqs. (20.8), (20.9), and (20.10). As shown in Table 20.9, when installing one supercritical coal-fired power plant, the economic benefit of Japan-China cooperation is approximately

Table 20.9 Changes in required investments for introduction of a supercritical coal-fired power generation unit if Japan works alone or cooperates with China in the Indian market

Category	Case 1 Japan entries into the Indian market alone	Case 2 Japan-China Cooperation on Indian Market	Difference between case 1 and 2
Personnel expenses (10,000USD)	407.79	19.95	−387.84
Land price for one unit (10,000 USD)	32.16	21.54	−10.62
Electricity expenses (USD)	15,600	10,800	−4800
The material cost of the machine (billion USD)	8.495	8.495	0
EPC expenses (billion USD)	6.6	4.62	−1.98
Interest repayment (billion USD)	0.3	0.26	0.04
Total (billion USD)	15.44	13.38	−2.06

Table 20.10 Changes in profit from introduction of a supercritical coal-fired power generation unit if Japan works alone or cooperates with China in the Indian market

	Case 1 Japan entries into Indian market alone (billion USD)	Case 2 Japan-China Cooperation on Indian Market (billion USD)	Difference between case 1 and 2 (billion USD)
Japan (billion USD)	0.31	1.53	1.22
China (billion USD)	–	0.83	0.83
India (10,000 USD)	–	23.7	23.7
Total (billion USD)	0.31	2.37	2.06

206 million USD higher than that of Japan operating alone. Japanese companies increase their profit by 122 million USD through cooperation, and Chinese companies receive 83 million USD profit.

The environmental benefits to Japan and China if this project is conducted under the CDM framework are shown in Table 20.11. Due to differences in profit distribution, the environmental profits of cooperation are 280 million USD/year, which is 980 million USD/year higher than if Japan entered alone. China receives an environmental benefit of 33 million USD/year. Since the cost of reducing carbon dioxide is higher in Japan than in China, when distributed under the CDM framework, the

Table 20.11 Environmental profit distribution between Japan and China under the CDM framework

	Case 1 Japan entries into the Indian market alone (billion USD)	Case 2 Japan-China Cooperation on Indian Market (billion USD)	Difference between case 1 and 2(billion USD)
Japan's annual CO_2 credit	2.80	1.82	−0.98
China's annual CO_2 credit	0.00	0.33	0.33
Total annual CO_2 credit	2.80	2.15	−0.65

environmental benefit from Japan-China cooperation is about 605 million USD/year lower than that of Japan operating alone.

20.5.5 *Profit Distribution Through Japan-China Cooperation on the Indian Market*

20.5.5.1 Cooperation Methods and Their Characteristics

Three cooperation cases are assumed in the analysis, as follows:

Cooperation method 1 (India alone) India uses the original 660 MW subcritical coal-fired power generation unit, with a thermal efficiency of 32%.

Cooperation method 2 (Japan alone) Japan provides India with a 660 MW supercritical coal-fired power plant with a thermal efficiency of 38%. India switches to this from subcritical pressure coal-fired power generation equipment with a thermal efficiency of 32%. Economic profits accrue to Japanese companies that provide supercritical coal-fired power generation facilities.

Cooperation method 3 (Japan-China cooperation in Indian market) Japan, China, and India cooperate to build a 660 MW supercritical coal-fired power plant in India. Japan cooperates with a Sino-Indian company to provide India with supercritical pressure coal-fired power generation equipment with a thermal efficiency of 38%, switching from subcritical pressure coal-fired power generation equipment with an original thermal efficiency of 32%. Chinese and Japanese companies providing supercritical coal-fired power generation equipment generate economic profits. Also, when considering the CDM, environmental profits accrue to Japanese and Chinese companies.

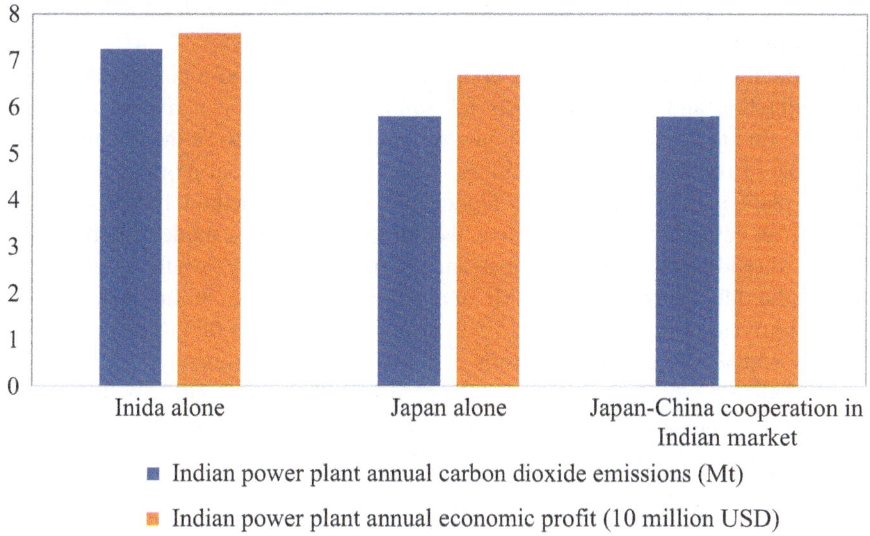

Fig. 20.8 Annual CO_2 emissions and economic profit of an Indian power plant under three methods of cooperation

20.5.5.2 Profit Distribution and Evaluation of Cooperation Methods

Economic and Environmental Benefits to Indian Power Plants

Figure 20.8 shows the economic and environmental impacts of the three cooperation methods on Indian power plants. In method 1, since India does not introduce supercritical coal-fired power generation equipment, there is no reduction in carbon dioxide, sulfur oxides, nitrogen oxides, and dust. Therefore, there are no carbon dioxide credits and sulfur oxides, nitrogen oxides, and dust avoidance costs. Since they continue to use the original power plant, there is no need to purchase a machine and no fuel cost reduction. The annual carbon dioxide emissions of the Indian power plant are 7.25 Mt Annual sulfur oxides, nitrogen oxides, and dust emissions are 33.75, 32.02, and 19.71 kt, respectively. Economically, the annual profit of the Indian power plant is 76 million USD.

As the same type of machine is introduced in cooperation methods 2 and 3, the economic and environmental benefits to India's power plants are the same for both. The Indian power plant has improved its carbon dioxide emission coefficient and the thermal efficiency of the power generation facility by introducing a new model, resulting in reductions in carbon dioxide, sulfur oxides, nitrogen oxides, and dust. As a result, there are carbon dioxide credits, and sulfur oxides, nitrogen oxides, and dust avoidance costs generated by the project. However, introducing a new model to the power plant involves a purchasing cost, although the resultant improvement in thermal efficiency also results in a reduction of fuel costs. The annual carbon dioxide emissions of the Indian power plant are 5.82 Mt. Annual sulfur oxides, nitrogen oxides, and dust emissions are 28.52, 26.97, and 16.6 kt, respectively. The annual economic profit of the Indian power plant, including carbon dioxide credits, is 67 million USD.

From an economic perspective, cooperation methods 2 and 3 both result in fuel cost reduction and avoidance costs for sulfur oxides, nitrogen oxides, and dust disposal due to the introduction of new equipment models. Nevertheless, there is a high cost of purchasing these machines. As a result, the annual profit of the Indian power plant is 0.99 million USD higher in cooperation method 1 than in 2 and 3. From an environmental perspective, the introduction of new models improves the carbon dioxide emission coefficient and thermal efficiency, resulting in annual carbon dioxide reductions of 1.43 Mt, or approximately 19.7%. Sulfur oxides reduce by 5.23 kt/year or approximately 15.5%, nitrogen oxides by 5.05 kt/year or approximately 15.8%, and dust emissions by 3.11 kt/year or approximately 15.8%.

For Indian power plants, cooperation method 1 has the highest annual profit but also the most substantial environmental burden. In cooperation methods 2 and 3, annual profit in India decreases by 12.5%, but the burden on the environment is also significantly reduced (carbon dioxide by 19.7%, sulfur oxide by 15.5%, nitrogen oxide by 15.8%, and dust emissions by 15.8%). Therefore, considering the joint perspectives of the environment and economy, cooperation method 2 or 3 should be selected.

Economic and Environmental Benefits to Japanese and Chinese Companies

Figure 20.9 shows the economic and environmental benefits to Japan, China, and India through the three methods of cooperation.

Cooperation method 1 uses the original equipment of the Indian power plant and does not introduce new models, so there are no economic benefits to Sino-Japanese

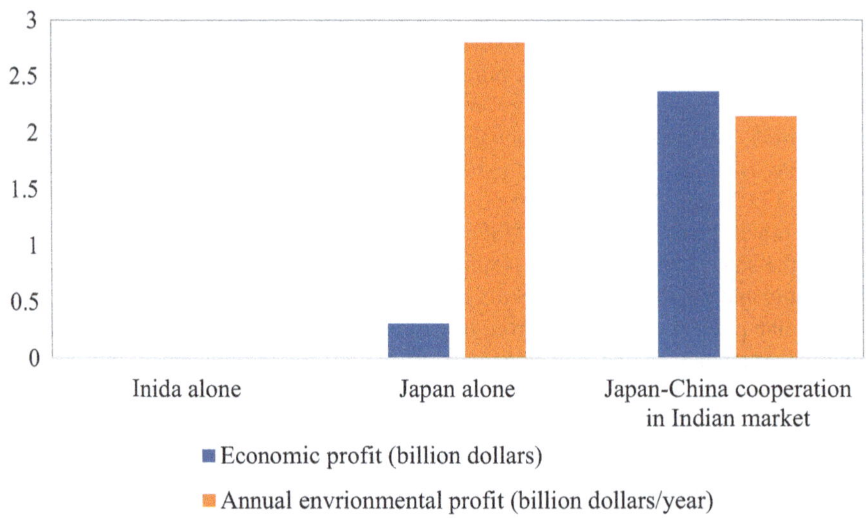

Fig. 20.9 Total economic and environmental profit of three methods of cooperation

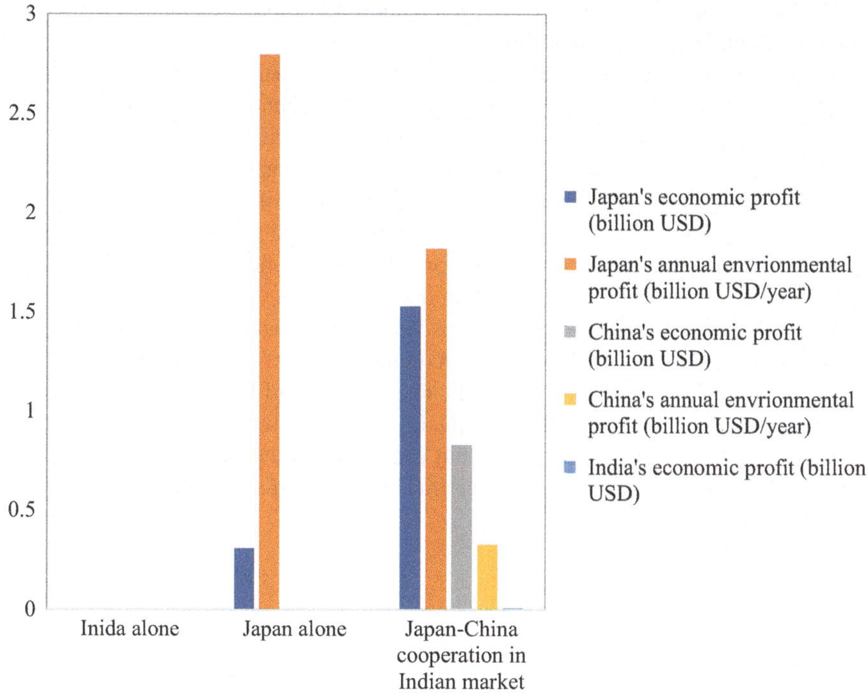

Fig. 20.10 Economic and environmental profit of Chinese, Japanese, and Indian companies under three methods of cooperation

companies. In cooperation method 2, a Japanese company provides supercritical coal-fired power generation facilities to a power plant in India. The Japanese company receives an economic profit of 31 million USD. Furthermore, the Japanese company generates an additional 280 million USD/year in environmental benefits under the CDM.

In cooperation method 3, Japanese, Chinese, and Indian companies cooperate in providing supercritical coal-fired power generation facilities to power plants in India. The total economic profit of Japanese, Chinese, and Indian companies is 237 million USD. Under the framework of the CDM, the total environmental profit of Japanese, Chinese, and Indian companies is 215 million USD/year. Compared to cooperation method 2, the Japanese company provides technology and materials, manufacture occurs locally in India, and China oversees the EPC, so costs are significantly reduced. As a result, the economic benefits of cooperation method 3 are approximately 7.6 times higher than those of method 2, generating an extra 206 million USD. However, because the cost of reducing carbon dioxide in China is lower than that in Japan, the annual environmental benefit of cooperation method 3 is approximately 605 million USD lower than that of method 2.

The profits of Japanese, Chinese, and Indian companies are shown in Fig. 20.10. In cooperation method 2, Japan manages everything from machine manufacturing to

EPC. The economic profit of Japanese companies is 31 million USD, and the annual environmental benefit is 280 million USD.

In cooperation method 3, a Chinese company cooperates with Japanese and Indian companies. The Japanese company provides technology and materials, the Indian company undertakes local manufacture, and China manages EPC. The economic profit of the Japanese and Chinese companies is approximately 153 million USD and 83 million USD, respectively. The economic profit of Indian companies is 237,000 USD. Under the CDM framework, Japanese and Chinese companies have annual environmental profits of 182 million USD and 33 million USD, respectively. Indian companies do not generate carbon dioxide credits.

Overall, the environmental benefits of cooperation method 3 are approximately 23% lower than those of method 2, but economic benefits are 7.6 times greater.

Focusing on the distribution of profit to Japanese and Chinese companies, the environmental profits of Japanese companies in cooperation method 3 are approximately 35% lower than in method 2, but economic profits are approximately four times higher. Therefore, from both economic and environmental perspectives, cooperation method 3 is appropriate.

Selecting the Optimum Cooperation Method

In view of the economic and environmental impacts on Indian power plants, cooperation methods 2 and 3 are more appropriate than method 1. From the viewpoint of economic and environmental benefits received by companies, cooperation method 3 is more suitable than methods 1 and 2. Therefore, cooperation method 3 is the most appropriate.

Benefits to Japan, China, and India via Cooperation Method 3

In the case of cooperation method 3, the input and profit distribution of the Indian power plant is shown in Fig. 20.11. If China and Japan cooperate in providing India with one unit of a supercritical coal-fired power plant, China will invest 462 million USD and obtain an economic profit of 83 million USD and an annual environmental profit of 33 million USD. Japan will invest 845 million USD, which will bring an economic benefit of 153 million USD and an environmental benefit of 182 million USD per year. The Indian company will invest 424,300 USD to generate an economic benefit of 237,000 USD. The refurbished Indian power plant will generate an annual economic benefit of 67 million USD and an annual environmental benefit of 13.39 million USD.

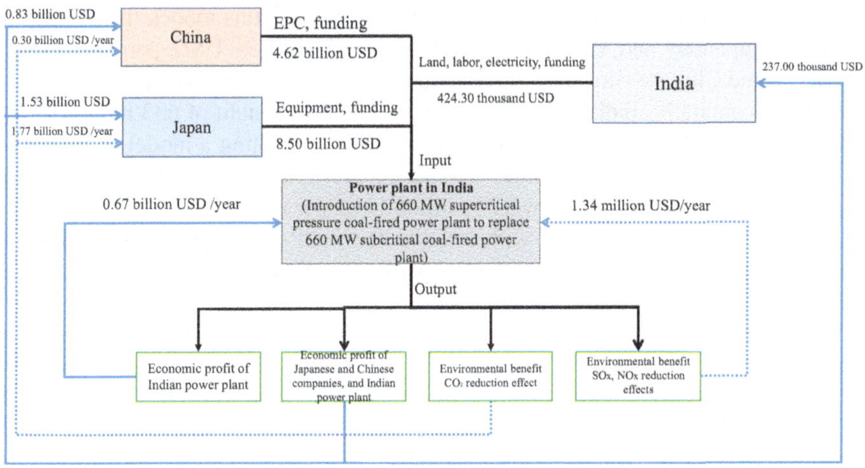

Fig. 20.11 Schematic diagram of Japan-China cooperation in a third-country market for power plant replacement project in India. (Source: Zhou 2019)

20.6 Conclusion

This chapter shows that India and Russia have enormous potential for reductions in energy use and environmental impact when using Japanese technology in a third-country market. If India had matched the energy consumption per GDP and carbon dioxide emissions per GDP of Japan from 1992 to 2016, there would have been reductions of 57,718 Mt CO_2 emissions and 23,402 oil-equivalent Mt energy consumption. If India had achieved the same level of power generation efficiency as Japan in the same period, a total of 743 oil-equivalent Mt energy consumption and 556 Mt CO_2 emissions could have been eliminated. Similarly, Russia had the potential to reduce 3950 petroleum-equivalent Mt energy consumption and 1367 Mtoe CO_2 over the same period and would have generated more than 727.3 billion USD in economic benefits every year after the shale gas revolution.

The following model of Japan-China cooperation in a third-country market is proposed. Japanese and Chinese companies provide funds, technology, and EPC following the basic principles of co-commerce, co-building, and coexistence. Participating in fields such as infrastructure, logistics, and energy-saving is based on the needs outlined in third-country development plans and the local environmental context. The Japanese and Chinese governments provide support for policy, tax incentives, and simplification of inspections. Finally, Japan, China, and the third countries share the economic benefits generated in the third-country market. Specifically, Japan and China share the environmental benefits of CO_2 reduction, and the third countries enjoy the environmental benefits of SO_x and NO_x reduction. Third countries also receive other benefits.

We used a case study to show that Japan-China cooperation in a third-country market is the most appropriate cooperation method when comprehensively considering environmental and economic perspectives. In this model, the economic profit of Japanese and Chinese companies is 237 million USD per unit. Also, Japanese and Chinese companies receive a total of 215 million USD in environmental benefits annually. India has a total annual economic benefit of 633 million USD and environmental benefits of 13.391 million USD. Adopting a model of cooperation, Chinese and Indian companies can make a profit, and Japanese companies can increase their economic profit by 122 million USD per unit.

The cooperation case study with India in the field of energy and environment shows excellent potential for constructing an international, regional low-carbon society, an Asian low-carbon community.

In Southeast Asian, South Asian, and Central Asian countries, annual energy consumption and carbon dioxide emissions could be reduced by 60–70% if countries could match the energy consumption per GDP and carbon dioxide emission per GDP of Japan. India has low energy efficiency and has the most significant reduction potential. In this case, if Japan and China make full use of each other's advantages to promote trilateral market cooperation, all three countries benefit. Cooperation can contribute to achieving the SDGs. We propose a "market principle + sharing of economic and environmental benefits + risk-sharing" method as a trilateral market cooperation method. The characteristics of this method are consistent with the community spirit of sharing benefits and risks.

References

IEA Database (2019). https://www.iea.org/statistics

Ministry of Economy, Trade and Industry (2019) About Japan-China third country market cooperation forum. https://jja.ne.jp/industry/giinfo/gi20191115.pdf

NEDO (2014a) FY2013 results report: International energy consumption efficiency technology/system demonstration project Basic project High-efficiency coal utilization system project formation survey project Supercritical coal-fired power generation and project formation rationalization survey in CCS project in Bulgaria. https://www.nedo.go.jp/library/seika/shosai_201408/20140000000229.html

NEDO (2014b) Project formation study of ultra-supercritical power generation project by co-firing imported coal and domestic anthracite in Vietnam. https://app5.infoc.nedo.go.jp/disclosure/SearchResultDetail

Ren H, Zhou W, Gao W, Wu Q (2011) Promotion of energy conservation in developing countries through the combination of ESCO and CDM: a case study of introducing distributed energy resources into Chinese urban areas. Energy Policy 39(12):8125–8136

Toshiba Hydroelectric Equipment (Hangzhou) Co., Ltd. (2018). http://www.toshiba-thpc.com

Toshiba Systems Corporation (2018). https://www.toshiba-energy.com

World Bank open database (2019). https://data.worldbank.org/

Zhou W (2019) First RCS forum materials

Chapter 21
Opportunities for Renewable Energy Introduction Through Feed-in Tariff (FIT) Scheme

Weisheng Zhou, Xuepeng Qian, and Ken'ichi Nakagami

Abstract In order to realize the East Asian low-carbon community, significant uptake of renewable energy is essential. However, uncertainties remain around how to promote the introduction of renewable energy. The fixed price purchase system (feed-in tariff, FIT) is a policy that stipulates the purchase price of electricity from renewable energy sources in order to promote the spread of renewable energy. Globally, it is the most widely used policy to promote renewable energy. In Japan, the FIT system was approved by the National Diet on August 26, 2011, and officially implemented from July 1, 2012. Under the FIT, the introduction of renewable energy, especially solar power, has progressed at an unprecedented speed. However, the unification problem (overreliance on a single power source) and mismatches in supply and demand, which are biased toward photovoltaic power generation, have become apparent. This chapter examines the genealogy of renewable energy introduction under the FIT system and quantitatively analyzes the regional characteristics of Japan, the endowment of natural resources, policy issues, and the global pioneer of the FIT system. These findings can contribute to future policy development through learning from the experiences of Germany.

21.1 Introduction of Renewable Energy in Japan

21.1.1 Single Renewable Energy Power Source

Significant introduction of renewable energy is indispensable for realizing the East Asian low-carbon community. However, how to promote the introduction of renewable energy is a big issue. Feed-in tariff (FIT) is the most widely used policy to

W. Zhou (✉) · K. Nakagami
College of Policy Science, Ritsumeikan University, Ibaraki, Osaka, Japan
e-mail: zhou@sps.ritsumei.ac.jp; nakagami@sps.ritsumei.ac.jp

X. Qian
College of Asia Pacific Studies, Ritsumeikan Asia Pacific University, Beppu, Oita, Japan
e-mail: qianxp@apu.ac.jp

W. Zhou et al. (eds.), *East Asian Low-Carbon Community*,
https://doi.org/10.1007/978-981-33-4339-9_21

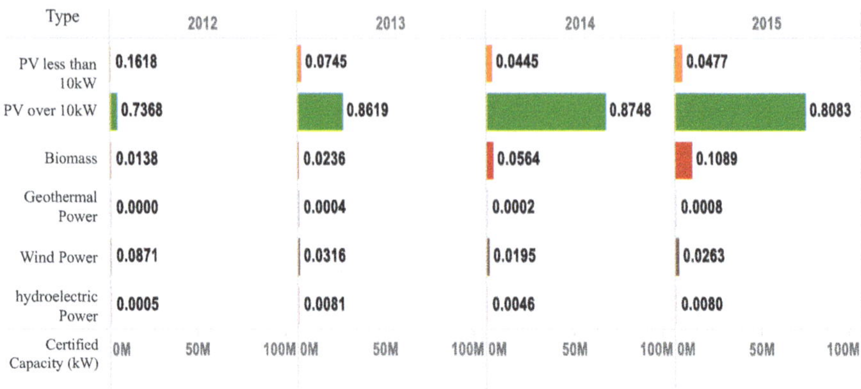

Fig. 21.1 Certified capacity of renewable energy in Japan in 2014 and 2015. (Source: Agency for Natural Resources and Energy)

promote renewable energy (Gross et al. 2012, REN21 2015). In Japan, the FIT system was approved by the National Diet on August 26, 2011, and officially implemented from July 1, 2012. Under the FIT, the introduction of renewable energy, especially solar power, has progressed at an unprecedented speed. As shown in Fig. 21.1, in Japan, the certified capacity of solar power generation (hereafter PV) is much higher than that of other renewable energy power generation facilities. In 2014, PV facilities of 10 kW or more accounted for approximately 90% of total installed renewable energy. By the end of 2015, it still accounted for more than 80%.

In Japan, seasonal changes in the output of photovoltaic power generation do not match seasonal changes in power demand. In summer and winter, when power consumption is high, PV output is low, so a sufficient alternative power supply must be prepared. Regardless of the amount of solar power generation installed, it is not possible to reduce other power generation facilities, such as thermal power generation. In contrast, in seasons with high output of solar power generation and low power consumption (spring and autumn), it is necessary to put other power generation equipment into standby. In addition, since solar power output may exceed demand, countermeasures such as suppressing its output may be needed. In both situations, it is not efficient to introduce large amounts of solar power alone. Instead, a certain amount of renewable energy power generation equipment—such as biomass, which has little seasonal fluctuation in output and can be easily controlled, or wind power, whose seasonal fluctuation in output differs from that of solar power generation—should be installed at a fixed ratio. This is a more efficient way to stabilize the power system.

Although solar power generation has merit in terms of resource utilization when installed in large quantities in areas with high solar radiation, such as Kyushu, it can also be installed in areas with relatively poor solar radiation. One challenge is that it has a much higher potential capacity than other renewable energies.

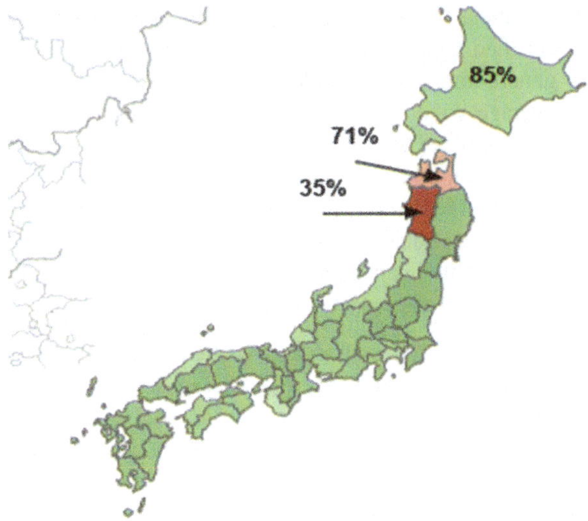

Fig. 21.2 Photovoltaic power generation certification status in each region of Japan in December 2015. Certified solar power generation is 75% or greater in all areas where numbers are not displayed. (Source: Agency for Natural Resources and Energy)

As shown in Fig. 21.2, the amount of certified solar power generation is incredibly high, at 75% or more in most parts of Japan. In particular, even in regions such as Hokkaido and Tohoku that have abundant wind power resources and large potential for wind power generation, the overall amount of certified solar power generation is large (Ministry of the Environment, "2010 Renewable Energy Introduction Potential Survey Report").

In order to investigate the relationship between the amount of certified renewable energy (solar and wind power generation) and natural resource endowment (solar radiation and wind speed) (NEDO solar radiation database 2017, NEDO local wind velocity map 2017) in each region, the following method was used.

First, 15 regions were randomly extracted from each of Hokkaido, Aomori prefecture, and Akita prefecture. The comparative advantage of wind resources over solar resources in these 45 municipalities is quantified as the scaled wind speed minus the scaled solar radiation. The correlation between this comparative advantage of wind power resources and the certified capacity of solar and wind power generation was analyzed. The results are as follows:

1. Overall, the certified capacity of wind power generation of 20 kW or more is less than that of solar power generation of 10 kW or more.
2. As the comparative advantage of wind speed increases, the certified capacity of wind power generation of 20 kW or more does not significantly increase; there is no correlation between wind speed and the certified capacity of wind power generation.
3. Many cities, towns, and villages have a larger certified capacity for solar power generation than for wind power generation, even when the comparative advantage of wind speed is positive and the average annual wind speed is 6 m/s or more. Examples include Noshiro City and Kazuno City in Akita Prefecture and Hirauchi Town, Higashitsugaru District, Aomori Prefecture.

4. In cities, towns, and villages with good wind speed and solar radiation, the certified capacity of solar power generation is higher than that of wind power generation. For example, in Hachinohe City, Aomori Prefecture, the annual average wind speed is 6 m/s or more, but the certified capacity for wind power generation of 20 kW or more is zero, while the certified capacity of solar power generation of 10 kW or more is 207 MW.
5. Among 18 municipalities where wind speed is 6 m/s or more, the total certified capacity of solar power generation of 10 kW or more is about 419 MW, while the total certified capacity of wind power generation of 20 kW or more is approximately half that, at 204 MW.

This analysis shows that the certified capacity of photovoltaic power generation is large in general, even in areas where solar radiation conditions are not good and wind power generation would be suitable. However, the capacity of Japan to adopt renewable energy is limited because of the poor regional grid. Therefore, when solar power is used alone in large quantities, there is insufficient room on the grid to accept energy from other renewable sources such as wind power. It will be difficult to introduce biomass or wind power generation in order to maintain the stability of the entire electric power system unless the connection capacity is increased.

21.1.2 Mismatch Between Certified Renewable Energy Power Generation and Power Demand

There is a mismatch between the amount of certified renewable energy generation equipment and power demand in all regions of Japan. For example, in the area under the jurisdiction of Kyushu Electric Power, power demand is small, and the certified capacity of solar power generation facilities of 10 kW or more is the largest in the country. In contrast, certified capacity in areas covered by Tokyo Electric Power Company and Kansai Electric Power Company, which have high power demands, is small.

Such a mismatch would not be a problem if electric power could be freely exchanged between regions. However, the lack of interregional transmission lines limits the potential for large-scale power interchange. Therefore, most electricity generated by renewable energy facilities must be consumed locally. In areas where power consumption is low, the ability to adjust frequencies is usually small, so intensive introduction of renewable energy generation in these areas may increase the risk of exceeding frequency adjustment capacity. In order to reduce the burden of frequency adjustment, priority connection rules for power from renewable energy must be changed.[1] In some cases, it is also necessary to suppress the output of

[1]On January 22, 2015, the Ministry of Economy, Trade and Industry promulgated the Renewable Energy Special Measures Law, *Ministerial Ordinance to revise a part of the enforcement rules*, changing the conventional priority connection rules for renewable energy and giving priority to

renewable energy power generation equipment. Eventually, this will lead to loss of power generation opportunities for some renewable energy facilities.

We analyzed the amount of renewable energy introduced in regions with low power demand, considering factors that influence the introduction of solar power generation. The amount of solar radiation is directly related to the output of solar power generation. In addition, since a considerable amount of land is required for solar power generation, land rent is a substantial cost. In order to clarify relationships between the amount of certified solar power generation of 10 kW or more and solar radiation and land rent in each region of Japan, multiple regression analysis was performed.

The location of a meteorological station that observes sunshine hours (AMeDAS) and a total of 843 points (municipalities) were used as the survey population. After assigning a number to each point in this population, 80 points were randomly sampled. The capacity of certified solar power generation facilities of 10 kW or more in each municipality in December 2015 is the response variable. Ordinance-designated cities, such as Kyoto and Sakai, were also sampled, but the size of these areas is clearly larger than other municipalities, and the range in land prices is great. Data from government-designated cities such as Sakai City were not used.

The explanatory variable of solar radiation was obtained from the New Energy and Industrial Technology Development Organization (NEDO) solar radiation database, which records month, azimuth angle, and tilt angle. For convenience, an azimuth angle of 0°, tilt angle of 30°, and annual average total solar radiation are used. Although there is some regional variation, maximum output of solar power generation equipment can be achieved when installed with an azimuth of 0° and an inclination of 30°.

There are no specific data available for the second explanatory variable, land rent, so average land price in each municipality was used instead (land price can reflect land rent rates). Average land price data were taken from SUMAITI, a real estate information site. Factors such as topography and the ratio of vacant land may also affect the amount of certified solar power generation facilities of 10 kW or more, but are not considered here since data on these variables are difficult to obtain.

Figures 21.3 and 21.4 show the relationship between the total capacity of certified solar power facilities of 10 kW or more and the amount of solar radiation and average land price. Transformations were applied to the explanatory variables in order to maintain linearity of the Ordinary Least Squares (OLS) regression model, which accounts for the possibility of alternating terms between variables.

The number of variables input into the multiple regression model was large (Table 21.1), so the theoretically possible number of models was very large.[2] The

---

nuclear power. In addition, power companies are allowed to refuse connection to renewable energy power lines.

[2]Since there are 19 variables, there is a possibility of 2^{19} state species in theory.

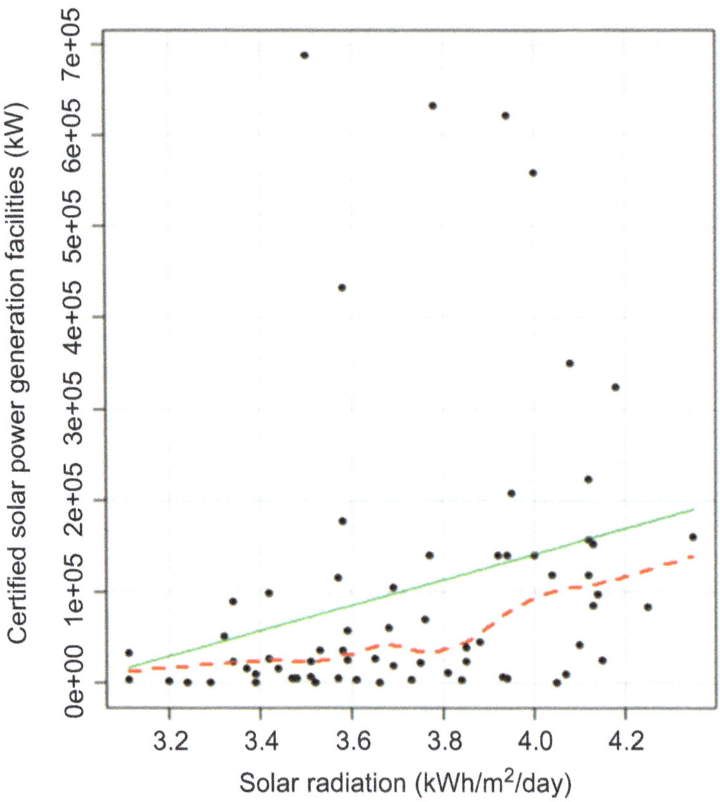

Fig. 21.3 Relationship between solar radiation and certified solar power generation facilities using local regression smoothing (loess) (horizontal axis: $kWh/m^2/day$, vertical axis: kW)

optimal model was selected using the p value (with a threshold of 0.15) and Akaike's information criterion (AIC) value using the stepwise method.[3]

Using studentized residuals for the selected models,[4] we found that Tono City in Iwate Prefecture and Kikugawa City in Shizuoka Prefecture are outliers. When abnormal values are measured, Tono City, Iwate Prefecture and Edogawa Ward, Tokyo, show strong influence. When these abnormal values and strong influence values were removed and regression analysis performed, an improved model was obtained:

[3]Using the step AIC function in the car package in R.

[4]Outliers and strong influence points were analyzed using the outlier test function and influence plot function, respectively, in the car package in R.

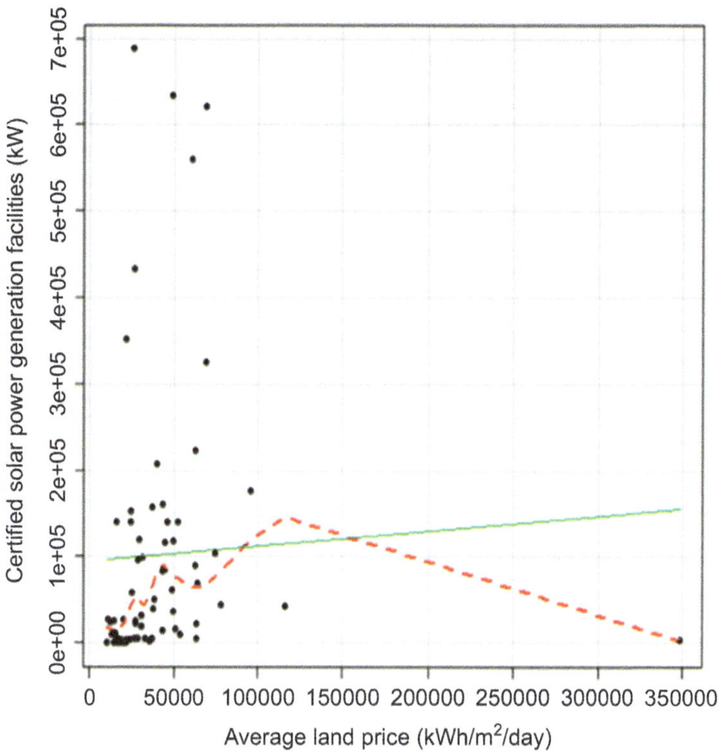

Fig. 21.4 Relationship between average land price and certified solar power generation facilities using local regression smoothing (loess) (horizontal axis, JPY/m^2; vertical axis, kW)

Table 21.1 Variables entered into regression analysis model by stepwise method

Insolation (x)	Land price (z)	Cross terms
x	z	$x \times z$
x^2	z^2	$x \times z^2$
$\exp(x)$	z^3	$x \times z^3$
	$\log(z)$	$x \times \log(z)$
		$x^2 \times z$
		$x^2 \times z^2$
		$x^2 \times z^3$
		$x^2 \times \log(z)$
		$\exp(x) \times z$
		$\exp(x) \times z^2$
		$\exp(x) \times z^3$
		$\exp(x) \times \log(z)$

Table 21.2 Summary statistics of model variables

	Coefficient	Standard deviation	*t* value	*p* value
Solar radiation	1.388e+05	5.699e+04	2.436	0.0178*
Average land price	−1.032e+00	2.450e+00	−0.421	0.6750
log(Average land price)	1.013e+05	9.331e+04	1.086	0.2820
Intercept	−1.434e+06	8.845e+05	−1.621	0.1103

Multiple R^2: 0.1942, Adjusted R^2: 0.1539
F-statistic: 4.821 on 3 and 60 DF, p value: 0.004505
*p value less than 0.05 was considered statistically significant

Table 21.3 Power consumption in 15 Japanese prefectural office areas with the highest national insolation

Prefecture	Solar radiation (kWh/m^2/day)	Consumption amount rank
Yamanashi	4.44	44
Kochi	4.4	45
Miyazaki	4.36	19
Shizuoka	4.3	11
Aichi	4.21	2
Kagoshima	4.21	31
Gifu	4.21	25
Ehime	4.18	26
Tokushima	4.18	37
Gunma	4.17	21
Mie	4.17	16
Hiroshima	4.17	12
Wakayama	4.15	41
Kagawa	4.12	5
Kumamoto	4.11	29

Source: Insolation data from NDEO, electricity consumption from METI 2012 electricity consumption data (METI 2017)

$$y = -1,434,000 + 138,800x - 1.032z + 101,300 \times \log(z) \tag{21.1}$$

where, x, z, and y are the amount of solar radiation, land price, and certified solar power of 10 kW or more.

As shown in Table 21.2, the p value of the coefficient of solar radiation is less than 0.05, so solar radiation is related to the amount of certified solar power generation of 10 kW or more installed in municipalities. Since the coefficient is positive, the higher the solar radiation, the greater the total capacity of certified solar power generation of 10 kW or more. In contrast, the p value of the coefficient of average land price is large, so it is not possible to determine the relationship of land price with installed solar power.

This analysis confirms that solar power generation of 10 kW or more tends to be concentrated in places with high solar radiation. However, in Japan, power consumption is generally low in these areas, so large amounts of solar power generation equipment have been introduced to areas with low power demand (Table 21.3).

The mismatch between the amount of certified renewable energy power generation facilities and demand for electric power is related to the endowment of natural resources in Japan.

21.1.3 Designing a Fixed Price System for Renewable Energy

21.1.3.1 Learning Effect of PV System Technology

The nationwide bias in the introduction and certification of renewable energy toward solar power is closely related to the learning effect of introducing solar power generation facilities and the method of determining the purchase price in Japan. When the FIT system was introduced, the Procurement Price Calculation Committee set the purchase price for each type of renewable energy by adding an internal rate of return (IIR) of roughly the same level to the cost base. If the investment effect is the same, differences in the amount of certified renewable energy of different types should be small, but the standard cost is that at the time of price setting. However, the purchase price did not reflect that the cost of PV systems continues to reduce at a significantly faster rate than that of other renewable energy power generation facilities, due to the learning effect.

The average cost of a solar power generation system of between 10 and 50 kW has decreased by 31% (procurement price, etc.) in 2 years (from July–September 2012 to September–December 2014; Price Procurement Committee 2015). Initially, the purchase price of renewable energy was determined by the price at the time at which the power generation company submitted its application. This led some companies to wait for declines in power generation system prices before submitting their applications. A typical example of this was seen in March 2014, when there was a last-minute rush of renewable energy introduction before a new purchase price was applied. Since 2014, the system has been gradually revised, the certification of some inappropriate projects has been revoked, and expiration dates have been introduced.

In addition to cost reductions resulting from the learning effect, other benefits of solar power generation were not reflected in the purchase price. For example, it is easier to pass the environmental assessment for solar power generation than wind power generation, and the construction period is very short. Energy companies naturally chose to invest in solar power, as the purchase price was calculated at roughly the same rate of return, but virtually any investment in solar power would yield higher returns than other renewable energy sources.

21.1.3.2 Cost-Benefit Evaluation

The Procurement Price Calculation Committee (hereafter the "committee"), based on the provisions of the Renewable Energy Act, reviews energy consumption based on "costs that would normally be required for efficient supply" and sets the purchase price for energy (cost basis). Renewable energy generation costs consist of equipment costs, operation and maintenance costs, land preparation costs, land rents, etc. In calculating the cost of each part, "the relevant cost data reflecting the actual cost" and "data reflecting the new market conditions" are required (Price Procurement Committee 2015).

As the committee targeted renewable energy introduction nationwide based on the overall average value of actual costs and using a more competitive method, some expensive cases were excluded from the cost calculations. The calculated cost was actually less than the overall average cost. For example, when the committee calculated the purchase price of PV of 10 kW or more in 2015, the national mean actual land rent was 219 JPY/m^2/year (median was 152 JPY/m^2/year). The committee ruled out expensive cases and used a cost of 150 JPY/m^2/year, which is lower than the median. When calculating the fixed price, the average value (or lower) of costs of each part were combined and used as the standard, instead of using the total cost of introduced cases. As a result, the cost estimated by the committee is clearly lower than the average cost of a whole installation. This method of deciding the fixed purchase price, which emphasizes efficiency and competition, is thought to reduce introduction costs on a national scale. However, while pursuing a reduction in overall installation costs, weak interregional transmission lines (difficulty in supplying power over wide areas), essential power demand, and endowment of natural resources were not considered. The fixed price does not reflect the fact that electricity demand is low in regions with good natural resource endowment.

When a FIT system was introduced in China in 2013, electricity demand and the endowment of natural resources were reflected in the system design. Instead of adopting a national uniform purchase price like Germany, the Chinese government set three solar power generation purchase prices. The lowest purchase price is in the west at 0.9 CNY/kWh, the purchase price in the central part of the country is 0.95 CNY/kWh, and that in the east is 1 CNY/kWh. In China, power demand increases from west to east, but the amount of solar radiation decreases. Thus, their scheme aims to prevent a mismatch between the amount of installed solar power generation and power demand. If Japan were to similarly set different purchase prices according to electricity demand and natural resource endowment in each region, the mismatch problem would become less of an issue.

21.2 Introducing Renewable Energy in Germany

21.2.1 Retaining a Diversity of Renewable Energy Power Sources

In Germany in the 1990s, the ratio of hydroelectric and onshore wind power generation was the highest among renewable energy power sources. In the twentieth century, the proportion of photovoltaic and biomass power generation gradually increased, while the proportion of hydroelectric power generation decreased. Renewable energy power generation in Germany in 2014 was mainly composed of onshore wind (34%), biomass (26.8%), solar (21.8%), and hydroelectric (12.8%). Unlike in Japan, where solar power generation dominates, renewable energy in Germany has maintained a degree of diversity.

Prior to 2000, renewable energy in Germany was biased toward onshore wind. From 2008, the speed of introduction of solar power generation facilities has gradually outpaced that of other renewable energy facilities. Since then, the German government has unsuccessfully tried to reduce the amount of solar power generation installed, but by 2012, solar accounted for 46.7% of renewable energy. The share of photovoltaic power generation in renewable energy has since gradually decreased, and the balance of renewable energy power has improved. Germany has succeeded in maintaining the current diversity of power sources through the PDCA (plan-do-check-act cycle) process. In particular, the history of revisions to their Renewable Energy Law (EEG Law) is of great significance to Japan (Takeuchi 2015).

The Electricity Supply Act, which is thought of as the origin of FIT, obliged electric power companies to purchase power generated by renewable energy sources in their supply area at a fixed ratio (90%) of the retail price of electric power. Since all renewable energy purchase prices were set at the same level, only the cheapest, onshore wind power, was introduced in large volumes.

The EEG of 2000 was created against this background. The law established the main framework for FIT today. Unlike the Electricity Supply Act, the purchase price of renewable energy power under the EEG Law can be adjusted on a cost basis for each type of renewable energy, rather than being set at a fixed ratio. In this way, biomass and solar power generation were gradually introduced. For solar power generation, which was expensive at the time, an upper limit of 350 MW was set in order to reduce total introduction cost (in July 2002, the upper limit was raised to 1000 MW).

A revision of the EEG Law in 2004 removed the upper limit of installed capacity for solar power generation facilities. The purchase price of photovoltaics, especially small-scale photovoltaics, was also significantly increased and that of geothermal power was raised. Purchase prices were segmented and subdivided according to the output and installation form of the equipment used. Finally, bonuses were introduced to encourage innovations (landfill gas, biogas, etc.). This facilitated the gradual introduction of various types of large-scale power generation facilities.

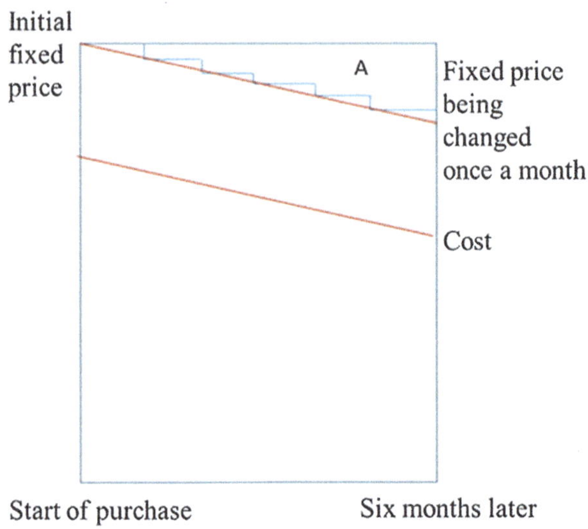

Fig. 21.5 Fixed price purchase system that is revised monthly

Eliminating the upper limit on solar power generation facilities has certainly increased the number of solar power installations. As of 2008, the cumulative installed capacity of German solar power generation facilities became the highest in the world (REN21 2009). However, there was a rapid increase in levies required due to mass introduction of expensive solar power generation. In neighboring Spain, vocal opposition to the government resulted in an effective reduction in the purchase price by imposing an environmental tax on renewable power generation businesses. Since 2009 in Germany, solar power levies have accounted for more than half of all levies. Under these circumstances, EEG 2009, which was implemented in January 2009, applied a large reduction rate to the purchase price for photovoltaic power generation, especially for facilities of 1 MW or more. Although the purchase price reduction for solar power generation slowed the rate of increase in solar power generation, the ratio of solar power to total renewable energy power generation facilities continued to expand. As Germany, like Japan, had not reflected the cost reduction of solar power generation systems in purchase prices at that time, investing in solar power generation returned a higher rate of profit than other forms of renewable energy. For this reason, EEG 2012 changed the frequency of purchase price revision from once every 6 months to monthly. In addition, a reduction rate of −0.5 to 2.8% was applied to the purchase price according to newly installed capacity in the most recent year.

As Fig. 21.5 shows, the excessive profit (A) of photovoltaic power generation has decreased due to continuous cost reduction. Figure 21.6 shows the actual effect of this system, with which the amount of new certification for solar power generation has gradually decreased.

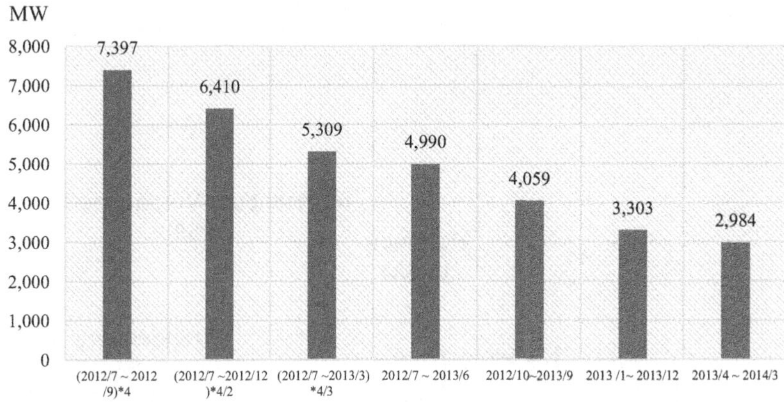

Fig. 21.6 Transition in introduction of solar power generation during the reduction rate reference period in Germany (EEG 2012). (Source: Federal Network Agency)

21.2.2 Resolving the Mismatch Problem in Germany

Figure 21.7 shows electricity consumption in Germany. North Rhine-Westphalia consumes the most electricity, followed by Bayern and Baden-Württemberg in the south and Lower Saxony (Niedersachsen) in the northwest.

In Japan, solar power generation facilities of 10 kW or more are the most commonly certified renewable energy and tend to be concentrated in areas with low power consumption. In contrast, in Germany, the introduction of renewable energy is largest in regions with high power consumption. Only wind power is installed foremost in areas with low power demand. The power generated by other renewable energies roughly matches regional power demands.

The fact that Germany, unlike Japan, was not exposed to a serious mismatch in supply and demand is related to its natural resource endowment. For example, in Germany, as in Japan, the introduction of photovoltaic power generation was mainly concentrated in areas with high solar radiation. However, southern Germany, where solar radiation is high, is a densely populated industrial area with high electricity demand. Thus, the introduction of large amounts of solar power in the southern region is rather appropriate.

Perhaps more important than the natural conditions is the stance of the German government in strongly promoting renewable energy. Japan did not announce a specific target for renewable energy introduction until 2015, whereas this was legislated in Germany in 2009. Following several revisions, the target announced in EEG 2014 is an extremely high target of 40–45% renewable energy power by 2025 and 55–60% by 2035. To achieve these goals, the German government has put a lot of effort into strengthening the transmission system. In 2009, the Transmission Line Expansion Plan (EnLAG) detailed 24 transmission line projects (covering a total of 1876 km; in the latest amendment, the final project was replaced by another plan).

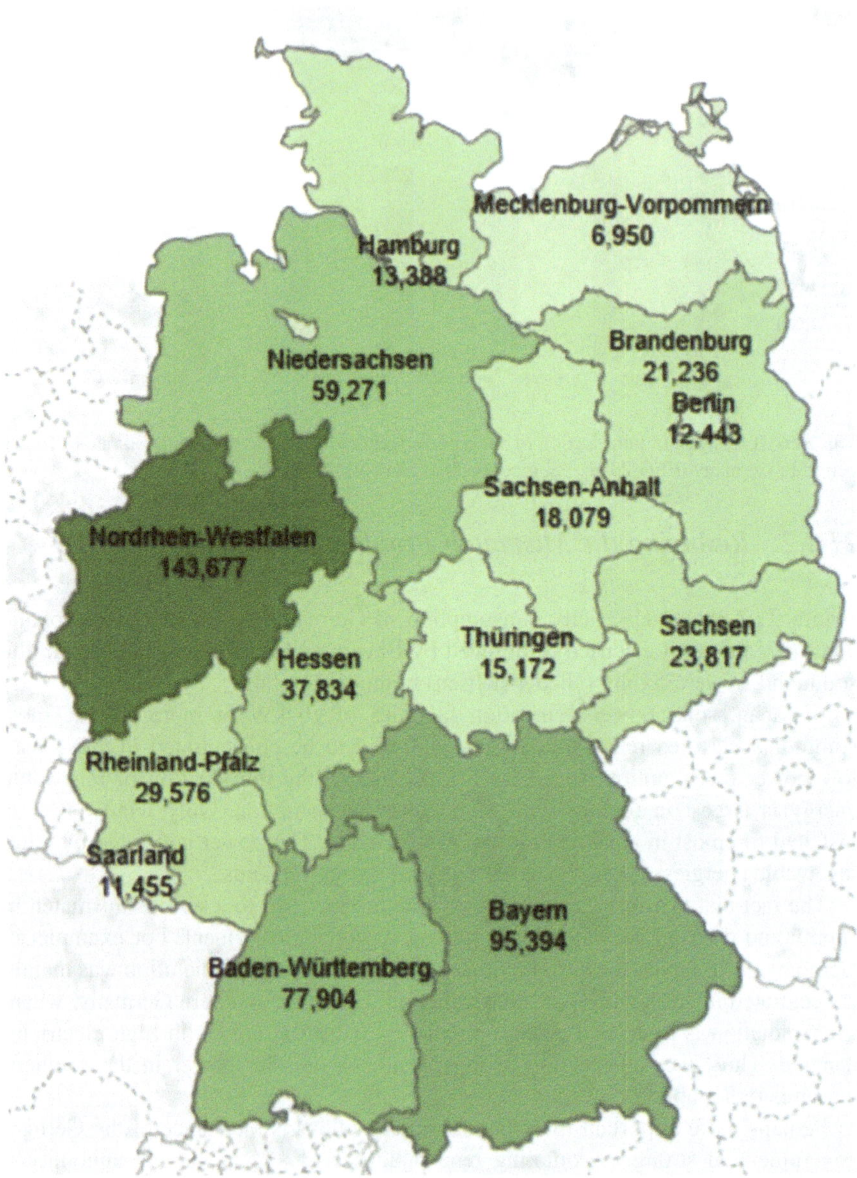

Fig. 21.7 Regional electricity consumption in Germany (2013, unit: million kWh). (Source: German New Energy Agency 2017)

The transmission line was constructed from north to south in order to balance power supply. A large amount of wind power is produced in the north where power demand is low and is sent to the south where power demand is high. North Rhine-Westphalia has the highest electricity consumption, but since the amount of

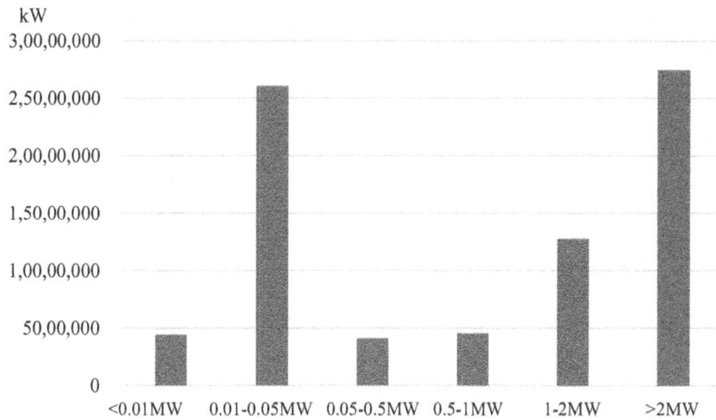

Fig. 21.8 Certified photovoltaic power generation by installation scale (December 2015). (Source: Agency for Natural Resources and Energy)

renewable energy introduced is low, construction plans for the power grid in North Rhine-Westphalia are also detailed.

According to the German Federal Network Agency's 2015 Monitoring Report, by the third quarter of 2015, 30% of the Transmission Line Expansion Plan was completed. In order to adapt to the ever-changing state of renewable energy and power transmission technology, the German transmission line reinforcement plan will be updated annually to meet the needs of the following 10–20 years.

Construction of a large-scale transmission and distribution system similar to that in Germany leads to huge costs. In 2014, the four German transmission systems (TSOs) invested a total of 1769 million EUR. The 808 Power Distribution System (DSO) invested 6193 million EUR (Federal Network Agency 2015). Japan is also reinforcing its transmission system by increasing the capacity of the Kitamoto interconnection line from 600,000 to 900,000 kW, but this is on a much smaller scale than efforts in Germany. The existing transmission system in Japan requires maximization of renewable energy introduction, whereas in Germany, in order to maximize the introduction of renewable energy, a transmission system suitable to meet a high target is constructed and reinforced. The biggest difference between Japan and Germany is in the promotion of renewable energy. However, improving the supply and demand mismatch of renewable energy does not always necessitate huge costs. One possible solution is to introduce renewable energy mainly in regions with high power demand, shifting to local production for local consumption.

Figure 21.8 shows that most certified large-scale solar power generation in Japan is by 2 MW facilities, followed by equipment from 10 to 50 kW. Solar power generation for residential use of 10 kW or less and equipment installed from 50 kW to 1 MW are both very small. Clearly, introduction to date has not followed local production for local consumption; instead, companies and other businesses have taken the lead in selling electricity.

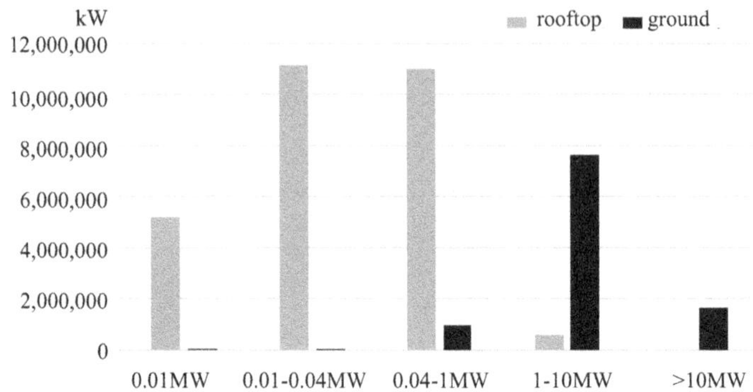

Fig. 21.9 Amount of certified solar power generation in Germany in 2014. (Source: German New Energy Agency 2017)

The high number of large-scale solar power generation installations is related to the setting of fixed purchase prices in Japan. The purchase price of solar power generation is high and differentiated at two scales: less than 10 kW and more than 10 kW. However, as the scale of installation increases, the average cost of solar power generation tends to decrease, yet the purchase price is set at the same level. Thus, introduction of large-scale equipment leads to higher profits than for small-scale facilities, so large-scale solar power generation becomes more popular.

When a facility reaches 50 kW, an enormous amount of additional cost is involved, so the number of installations drops after 50 kW and increases again for much larger facilities. In order to promote local production for local consumption, rooftop space should be better utilized (especially in large cities with high power demand)[5] than under the current system.

In Germany, small- to medium-sized (>0.01 and <0.04 MW) rooftop installations have the highest total capacity (Fig. 21.9). Overall, equipment with a capacity of 1 MW or less accounts for 74% of the total. By installation type, rooftop solar power generation facilities account for 73%. Purchase price is subdivided according to the five scales and installation configurations (rooftop and ground) shown in Fig. 21.9. The smaller the scale, and the higher the purchase price for rooftop installation, the larger the amount of small-scale, rooftop solar power generation installed.

There are merits to introducing mainly large-scale power generation facilities. Large-scale facilities have lower selling prices than small-scale installations. For example, the original purchase price for solar power generation of less than 10 kW was 42 JPY/kWh, whereas the purchase price for solar power generation of 10 kW or more was 40 JPY/kWh. With the generation of tens of billions of kWh of solar power annually and a purchase period of 20 years, it is possible to avoid enormous costs by introducing large-scale solar power generation. However, if the amount of

[5]When installing rooftop photovoltaic panels, an area of 5–10 m^2 is required per 1 kW.

renewable energy introduced exceeds the connectable amount in the region, measures such as suppressing the output of solar power generation and reinforcing the power transmission system become necessary. Therefore, the introduction cost of a scenario that focuses on large-scale power generation facilities is expected to rise sharply. In the case of local production for local consumption, the initial introduction cost is high, but since the output power is consumed locally, the connectable amount should be greater than in the former scenario. When the amount of installation is fixed, the cost of introducing mainly large-scale power generation facilities may exceed the cost of introducing local production for local consumption.

Since autumn 2014, there have been frequent problems with renewable energy power generation in Japan, and the amount of renewable energy seems to have exceeded the connectable amount in many areas. When first introducing a FIT system, it is reasonable to set an introduction target according to national needs, assume various introduction scenarios according to that goal, and select a scenario with low cost. However, in the *Long-term Energy Supply and Demand Outlook* announced in July 2015, it was decided that the target of solar power generation for 2030 would be 7% of the power supply mix.

21.3 Conclusion

This chapter deals with the current situation of renewable energy introduction under the feed-in tariff (FIT) system in Japan, with particular reference to the unification problem (overreliance on a single renewable energy source) and mismatch between the amount of certified renewable energy and electricity demand. The results of this analysis are summarized below.

1. Solar power is installed nationwide and is increasing in regions that are more suitable for the introduction of other renewable energy (wind power). Given the lack of interregional power grids in Japan, the amount of renewable energy that can be connected is limited. The system would be more efficient if it reflected regional suitability for different renewable energy sources.
2. In calculating the fixed price of solar power generation, the committee introduced a competitive method, setting a uniform purchase price nationwide to encourage effective use of natural resources. In areas where solar radiation is high, there is little electricity demand, and as a result, a large amount of solar power generation was introduced in areas with low electricity demand. In order to solve this mismatch, purchase prices should be set regionally, reflecting natural resource endowment and electricity demand, as in China.
3. Centrally introducing a large-scale power generation facility (centralized installation) saves installation costs initially, but if the amount of installed power exceeds a certain level, the output of renewable energy is suppressed. As such, there is a possibility that the overall introduction cost will rise sharply, so it is

important to define an introduction target and assume an introduction scenario with low cost.

4. Germany, the pioneer of the FIT system, avoided serious problems by increasing the frequency of purchase price revisions, segmenting the purchase price, and encouraging small-scale, rooftop solar power generation. This is closely related to natural resource endowment and the developed power grid in Germany, as well as appropriate system revisions such as reducing high purchase prices.
5. Since the environments in which renewable energy is introduced are different, it is very important for Japan to learn from Germany's experience and design a system in line with environmental conditions in Japan.

References

German New Energy Agency renewable energy electricity introduction status. https://www.foederal-erneuerbar.de/uebersicht/bundeslaende. Accessed 20 Oct 2017

Gross R, Stern J, Charles C, Nicholls J, Candelise C, Heptonstall P, Greenacre P (2012) On picking winners: the need for targeted support for renewable energy. Center For Energy Policy and Technology, Imperal College, London

METI power survey statistics. http://www.enecho.meti.go.jp/statistics/. Accessed 20 Oct 2017

NEDO local wind velocity map. http://app8.infoc.nedo.go.jp/nedo/. Accessed 20 Oct 2017

NEDO solar radiation database. http://www.nedo.go.jp/library/nissharyou.html?from=b. Accessed 20 Oct 2017

Price Procurement Committee (2015) Opinion on procurement price and procurement period of 2015. https://www.meti.go.jp/shingikai/santeii/pdf/report_004_01_00.pdf. Accessed 20 Oct 2017

REN21 (2009) Renewables 2009 global status report

REN21 (2015) Renewables 2015 global status report

Renewable Energy Source Act (EEG2004). https://www.clearingstelle-eeg.de/files/node/8/EEG_2004_Englische_Version.pdf

Renewable Energy Source Act (EEG2009). https://www.clearingstelle-eeg.de/files/node/8/EEG_2009_Englische_Version.pdf

Renewable Energy Source Act (EEG2012). https://www.clearingstelle-eeg.de/files/node/8/EEG_2012_Englische_Version.pdf

Renewable Energy Source Act (EEG2014). https://www.clearingstelle-eeg.de/files/node/8/EEG_2014_Englische_Version.pdf

Takeuchi J (2015) Chapter 5: Ideals and realities of German energy policy. In: Yamauchi H, Sawa A (eds) Evaluation of electric power system reform. Hakuto Shobo, Tokyo, pp 113–138

Chapter 22
Integration of Local and Global Perspectives

Weisheng Zhou

Abstract Global warming cannot be solved without multidimensional cooperation such as international cooperation; however, it is also a typical interdisciplinary problem, and as such, an interdisciplinary approach is an essential condition for the realization of a sustainable decarbonized society. This chapter provides an overall summary of the previous chapters and raises several key academic queries which are core to a regional, low-carbon sustainable society. It explores prospects from the transdisciplinary perspective through the integration of different fields, from technology to social systems. This chapter is the conclusion for this whole book and emphasizes the importance of integrating local and global environmental measures with several initial schemes and design ideas.

22.1 Integrated Strategy for Countermeasures Against Global Warming and International Environmental Cooperation

East Asia has been attempting to address several issues: economic development (overcoming poverty), local environmental issues (overcoming pollution), and global environmental issues (CO_2 emissions reduction), alongside air and water pollution, soil contamination, ecosystem destruction, DSS, desertification, transboundary pollution, and the long-term security of nuclear power and energy resources. The implementation of multidimensional and composite measures and wide-based international cooperation are becoming increasingly important. The realization of environmental cooperation between Japan, China, Korea, and the broader East Asian region is expected to lead to the formation of a sustainable and vibrant international society. In such a society, economic development, pollution reduction, and social harmony are balanced, in addition to measures to mitigate

W. Zhou (✉)
College of Policy Science, Ritsumeikan University, Ibaraki, Osaka, Japan
e-mail: zhou@sps.ritsumei.ac.jp

W. Zhou et al. (eds.), *East Asian Low-Carbon Community*,
https://doi.org/10.1007/978-981-33-4339-9_22

global warming. Three basic ideas are presented for the top design of international reciprocal complementary environmental cooperation that integrates the economy and environment.

22.1.1 Integration of Local and Global Environmental Measures

Developed countries have experienced economic development, local environmental problems (such as local pollution), and global environmental problems (such as global warming) in this order. In contrast, developing countries not only have these three priorities to meet; they are also confronting all three issues simultaneously. Global acidification and global warming are caused by the burning of fossil fuels, deeply rooted in modern civilization. As such, it is important to consider measures that integrate local and global environmental measures with a view to solving both problems simultaneously. It is important to understand that environmental problems are both global issues and issues that need to be addressed in our local communities. In particular, to encourage developing countries to participate in the realization of a low-carbon society, supporting local measures is imperative.

22.1.2 Utilization of Market Mechanisms for Technology Transfer

In a globalized economic system, international competitiveness is the basic condition for the survival of a company. As technology is owned by the private sector, technology transfer can lead to a decline in international competitiveness on the part of the transferring countries, hollowing out industry and technology. As such, there has not been great progress in this area as there are many factors that inhibit technology transfer. These factors include the lack of adequate protection of intellectual property rights in developing countries, lack of institutions and financial mechanisms to facilitate technology transfer, inadequate systems to facilitate technology transfer and digestion, and lack of efficiency in technology transfer. It may be postulated that the market economy mechanism should be used to promote technology transfer and make efficient use of the transferred technology.

22.1.3 Design of Reciprocal Cooperation Mechanisms Between Developed and Developing Countries

In reciprocal cooperation, projects are selected based on market needs and economic benefits for both parties. This is in addition to the reduction of environmental pollutants such as CO_2 and SO_x and the economic and environmental benefits garnered from projects that are shared by both parties. Under this approach, developed countries experience a risk reduction associated with technology transfer, while the financial burden on developing countries for technology transfer is greatly diminished. This enables more efficient retrofitting of old and obsolete thermal power plants and other energy-intensive sectors. In turn, this provides a greater number of business opportunities for developed countries. To promote this type of technology transfer, the governments of both countries need to secure human resources and provide specific policy incentives including financial incentives (e.g., subsidies and preferential policies in terms of tax), to promote this type of technology transfer.

22.2 China, Japan and the USA Should Establish a Community to Tackle Climate Change

This book describes the concept of an "East Asia Low-Carbon Community" centered around the cooperation among the three CJK countries and the research regarding the realization of this concept. In the climate change framework, one country that must not leave the table in terms of responsibility and capacity is the USA. Although the USA is the world's largest cumulative and per capita emitter of CO_2, it has withdrawn from the Kyoto Protocol and the Paris Agreement; these are the two most important frameworks on climate change. The Kyoto Protocol and the Paris Agreement were signed by democratic administrations and withdrawn by republican administrations, showing that domestic party politics greatly influence international cooperation and global interests.

While the pros and cons of the US withdrawal will not be discussed in this book, we would like to raise the possibility of cooperation among the three countries in light of their responsibilities and capabilities. While we are currently in the midst of a confrontation between the USA and China, the USA is a representative of Western civilization, and China is a representative of Eastern civilization. Japan is unique in that it is a civilization that fuses the three countries to a certain extent.

The CO_2 emissions from China, the USA, and Japan rank as the global highest, the second highest, and the fifth highest, respectively. Their combined emissions account for 45% of the total global emissions. These countries are also are the top three economies in the world, with a combined GDP that is 40% of the world total. Japan has a much higher energy-saving efficiency than the USA and China. If Chinese and US energy efficiency and CO_2 discharges were to be the same as

Japan, their combined energy consumption and CO_2 emission are estimated to decrease by over 50%. Therefore, given the capital and technological capabilities and emissions reduction responsibilities, China, the USA, and Japan should collaborate to tackle climate change.

Most of the previous actions aim at creating a low-carbon, sustainable, and vigorous society, instead of merely tackling climate change at the expense of the quality of life for human beings. As such, society aims to strike a balance between economy, environment, and the people that comprise this society. If the three countries were to cooperate in areas such as energy technology, energy mixes, and industrial structures, two problems may be addressed with one strategic and collaborative approach.

From a global perspective, the tripartite cooperation could significantly reduce CO_2 emissions. At the country scale, it will improve the energy mix, energy sustainability, and industrial structure, to enhance energy efficiency, reduce hazardous events such as acid rain, and improve the livelihood of communities. Regardless of global warming, this is the ideal situation that human society should be pursuing.

An important point to note is that although the USA has withdrawn from the Kyoto Protocol and the Paris Agreement, we may identify the country's two-sidedness in this regard. Up until 2010, the USA was the world's largest per capita emitter of CO_2; as such, the country has a greater responsibility as well as a greater capacity than any other nation attempting to address climate change.

Addressing climate change involves two possible approaches: reducing and stabilizing levels of heat-trapping GHGs in the atmosphere (mitigation) and/or adapting to the climate change that is already occurring (adaptation). Although the USA acts passively on mitigation measures such as reducing CO_2 emissions, it attaches great significance to the research and development of adaptation-related technologies, with enormous capital and talent input.

This duality is also manifested by the different approaches taken by the US federal government and state governments on mitigation policies. While the federal government has disregarded the Kyoto Protocol and the Paris Agreement, some states, such as California and New Mexico, are proactively implementing emissions-reduction measures and making commitments in this regard. Therefore, to return the USA back to the negotiating table is of utmost significance to promote the coordinated development of mitigation measures and economic growth. This is also a very important task for China and Japan when dealing with climate change.

In a nutshell, China, Japan, and the USA should establish a community in tackling climate change via bilateral, multilateral, and third-party cooperation on green financing, technology transfer, capacity improvement, and emissions trading. Although climate change has different impacts on different countries, no country can walk away from this battle unscathed. Despite political, military, economic, and trade conflicts, China, the USA and Japan should forge a community with a shared future with aims to cope with the climate change, particularly given the shared benefits, responsibilities, and risks. The USA was the driving force underpinning postwar globalization and the architect of international order; it has also been the greatest beneficiary of these actions. The USA will not be less likely to abandon this international order such as the climate framework convention.

22.3 Policy Science Approach to Realize a Low-Carbon Society Across a Wide Area

The realization of a low-carbon society will combat global warming and create a sustainable and vibrant society in which the economy, the environment, and society are in harmony. Technological measures to achieve a low-carbon society include promoting the conversion to renewable energy and carbon fixation, while reducing energy consumption. Novel institutional designs such as carbon neutrality, carbon offsetting, and carbon footprints are also becoming established as social measures.

This book examines the potentials for a regional, low-carbon community based on urban-rural cooperation, and the East Asian low-carbon community based on cross-border international cooperation in order to create a wide-area, low-carbon society (Zhou and Nakagami 2009). Policy science provides a comprehensive approach to addressing these issues and is a flexible discipline with the ability to cross extant disciplines. It aims develop a deeper understanding of society and use such an understanding to instigate change while maintaining deep connections with the real world.

22.3.1 Policy Science Issues for Realizing a Wide-Area, Low-Carbon Society

The policy science approach is characterized by three major features: (1) a comprehensive response to the situation of the times, (2) an interdisciplinary research area, and (3) a problem-solving orientation. This section describes the characteristics considered necessary to realize a regional low-carbon society based on the three features:

1. A shared understanding of the concept of a wide-area, low-carbon society: it is necessary to foster a shared understanding of the concept and goals of the term "low-carbon society" as well as increase awareness regarding the significance of this terms among the community.
2. International significance and responsibility to achieve a wide-area, low-carbon society: it is necessary to clarify the benefits and obligations of countries involved in achieving a wide-area, low-carbon society, as well as to share awareness of the international significance and responsibilities of common efforts.
3. Designing international and social institutions for realization of common goals: it is necessary to establish treaties and national legal systems as the basis for institutional design.

4. Technical and economic design for realization: there is a need to establish a database of information required for technical and economic design and a quantitative study of financial and human resources.
5. Clarification of issues: a low-carbon society involves all socioeconomic bases, as such, it is necessary to identify common problems that need to be addressed.
6. Promoting common methods of problem-solving and means of implementing solutions: technical standards, decision-making rules, and means of implementing solutions need to be clarified. Many international problems, particularly transboundary environmental problems among the CJK countries, such as acid rain and yellow sand, have been previously discussed. Consistent implementation of processes from the institutional design stage to the implementation and human resource development stage is the way forward, while the reality of implementation is challenging. To date, the issues have been concrete and explicit environmental problems; however, the low-carbon society is a summary of the solution aiming to address implicit environmental problems.

It is necessary to elaborate on these characteristics and create a roadmap for the realization of these issues. The willingness and social support to work together is also required.

22.3.2 Toward the Creation of "Policy Engineering"

Sustainability, especially the construction of a wide-area, low-carbon society, is a contemporary social issue that is complex involving politics, government, economics, management, the short and long terms, and the local and global scales. Policies to solve these problems are pluralistic and complex, and we need to shift from policies that rely on experience and intuition to evidence-based policies based on the scientific method. We need to find a means transfer the knowledge and techniques obtained from various sciences into policy development and feedback issues into the scientific method. For this purpose, it is important to establish original research methods such as experimental policy studies and experimental policy theory.

Policy engineering is a discipline that analyzes, plans, implements, and evaluates (verifies) the efficiency, effectiveness, and fairness of the policy lifecycle on a spatial, temporal, and countermeasure axes. Simply put, policy engineering is the study of optimizing policies using engineering methods to maximize their effectiveness and minimize their cost (risk). Policy engineering is an approach to derive optimal solutions to social problems by analyzing objective data and models across a wide range of disciplines, from the natural to social sciences. As opposed to discussing solutions from a broad and shallow perspective, this approach extracts the common structure underlying each problem, positions the problems systematically based on that common structure, and highlights policy development scenarios for systematic solutions (Zhou et al. 2014).

Policy engineering refers to comprehensiveness (multiple policies on a single issue or one policy addressing multiple issues) and quantitative approaches; as society becomes more complex, it is becoming essential to understand the quantitative cost-effectiveness of those policies. The discipline is also characterized by optimality (maximizing social utility and minimizing cost (risk)) and connectivity between theory and practice. In other words, it is a discipline that integrates various approaches to seek optimal solutions (policies) to solve complex social problems in the most appropriate way.

Policy engineering is not a discipline for the sake of discipline, it is a practice. Further, it is the application, implementation, and realization of discipline in the real world, with an aim to improve the real world. For example, the IPCC, in its series of assessment reports, cited a number of analyses of integrated assessment models, which combine natural and social scientific knowledge to test the effectiveness of socioeconomic policies. Although the IPCC assessment reports did not discuss the nature of policy-making, they did provide a number of examples of policy engineering in the real-world context. Future policy recommendations must be strategic, systematic, quantitative, and concrete, based on quantitative evaluation simulation analysis, to withstand social implementation.

In the future, a cross-disciplinary academic field will be created alongside a method to modify the policy-making process into an objectively verifiable, evidence-based one for policy-makers in society at large. I am currently offering "Introduction to Policy Engineering" as a foundational course at Ritsumeikan University. I hope to see the creation and development of this nascent discipline in the future.

References

Zhou W, Nakagami K (2009) Modification of energy and resource systems and policy empirical research for realizing low carbon society. Report of the Japan Global Environment Research Fund (JGPSSI): Hc-084 EcoDesign of low carbon society based on regional partnership between urban and rural areas

Zhou W, Ren H, Su X, Qian X Yamazaki M, Ibano K, Sun F, Haga H (2014) Toward the realization of East Asian low carbon community and the establishment of 'Policy Engineering': summary and prospect of R-GIRO Project 'Fundamental Technology Development and Strategic Innovation for Low Carbon Society'. Policy Sci 21(3):213–230

The manufacturer's authorised representative in the EU is Springer Nature Customer Service Centre GmbH, Europaplatz 3, 69115 Heidelberg, Germany. If you have any concerns regarding our products, please contact ProductSafety@springernature.com

Printed and bound by CPI Group (UK) Ltd, Croydon, CR0 4YY
30/09/2026
02225293-0014